Trends in Mathematics

Trends in Mathematics is a book series devoted to focused collections of articles arising from conferences, workshops or series of lectures.

Topics in a volume may concentrate on a particular area of mathematics, or may encompass a broad range of related subject matter. The purpose of this series is both progressive and archival, a context in which to make current developments available rapidly to the community as well as to embed them in a recognizable and accessible way.

Volumes of TIMS must be of high scientific quality. Articles without proofs, or which do not contain significantly new results, are not appropriate. High quality survey papers, however, are welcome. Contributions must be submitted to peer review in a process that emulates the best journal procedures, and must be edited for correct use of language. As a rule, the language will be English, but selective exceptions may be made. Articles should conform to the highest standards of bibliographic reference and attribution.

The organizers or editors of each volume are expected to deliver manuscripts in a form that is essentially "ready for reproduction." It is preferable that papers be submitted in one of the various forms of TEX in order to achieve a uniform and readable appearance. Ideally, volumes should not exceed 350-400 pages in length.

Proposals to the Publisher are welcomed at either:
Birkhäuser Boston, 675 Massachusetts Avenue, Cambridge, MA 02139, U.S.A.
math@birkhauser.com
or
Birkhäuser Verlag AG, PO Box 133, CH-4010 Basel, Switzerland
math@birkhauser.ch

Analysis and Geometry in Several Complex Variables

Proceedings of the 40th Taniguchi Symposium

Gen Komatsu
Masatake Kuranishi
Editors

Birkhäuser
Boston • Basel • Berlin

Gen Komatsu
Department of Mathematics
Osaka University
Toyonaka, Osaka 560
Japan

Masatake Kuranishi
Department of Mathematics
Columbia University
New York, NY 10027
U.S.A.

Library of Congress Cataloging-in-Publication Data
Analysis and geometry in several complex variables / Gen Komatsu, Masatake Kuranishi, editors.
p. cm. -- (Trends in Mathematics)
Includes bibliographical references.
ISBN-13: 978-1-4612-7441-4 e-ISBN-13: 978-1-4612-2166-1
DOI: 10.1007/978-1-4612-2166-1
1. Mathematical analysis Congresses. 2. Geometry, Differential Congresses. 3. Functions of several complex variables Congresses. I. Komatsu, Gen, 1949 - II. Kuranishi, Masatake, 1924- III. Taniguchi Symposium "Analysis and Geometry in Several Complex Variables" (40th : 1997 : Katata, Japan) IV. Series.
QA299.6 .A54 1999 99-31211
515'.94--dc21 CIP

AMS Subject Classifications: 32-06, 35-06, 53-06

Printed on acid-free paper.

Softcover reprint of the hardcover 1st edition 1999

Birkhäuser

Reformatted from electronic files in LaTeX by TeXniques, Cambridge, MA

9 8 7 6 5 4 3 2 1

Contents

Preface

This volume consists of a collection of articles for the proceedings of the 40th Taniguchi Symposium *Analysis and Geometry in Several Complex Variables* held in Katata, Japan, on June 23–28, 1997.

Since the inhomogeneous Cauchy-Riemann equation was introduced in the study of Complex Analysis of Several Variables, there has been strong interaction between Complex Analysis and Real Analysis, in particular, the theory of Partial Differential Equations. Problems in Complex Analysis stimulate the development of the PDE theory which subsequently can be applied to Complex Analysis. This interaction involves Differential Geometry, for instance, via the CR structure modeled on the induced structure on the boundary of a complex manifold. Such structures are naturally related to the PDE theory. Differential Geometric formalisms are efficiently used in settling problems in Complex Analysis and the results enrich the theory of Differential Geometry.

This volume focuses on the most recent developments in this interaction, including links with other fields such as Algebraic Geometry and Theoretical Physics. Written by participants in the Symposium, this volume treats various aspects of CR geometry and the Bergman kernel/ projection, together with other major subjects in modern Complex Analysis. We hope that this volume will serve as a resource for all who are interested in the new trends in this area.

We would like to express our gratitude to the Taniguchi Foundation for generous financial support and hospitality. We would also like to thank Professor Kiyosi Ito who coordinated the organization of the symposium. Finally, we greatly appreciate all the efforts of the referees.

Gen Komatsu
Masatake Kuranishi
Editors

Symposium

The 40th Taniguchi Symposium
Analysis and Geometry in Several Complex Variables
June 23–28, 1997
at Kyuzeso Seminar House, Katata, Japan

Organizers

Gen Komatsu
Masatake Kuranishi

Participants

Takao Akahori
Department of Mathematics, Himeji Institute of Technology

David W. Catlin
Department of Mathematics, Purdue University

Michael G. Eastwood
Department of Pure Mathematics, University of Adelaide

Charles L. Fefferman
Department of Mathematics, Princeton University

Kengo Hirachi
Department of Mathematics, Osaka University

Joseph J. Kohn
Department of Mathematics, Princeton University.

Gen Komatsu
Department of Mathematics, Osaka University

Masatake Kuranishi
Department of Mathematics, Columbia University

Joachim Michel
Université du Littoral

Kimio Miyajima
Department of Mathematics, Kagoshima University

Shin Nayatani
Mathematical Institute, Tohoku University

Takeo Ohsawa
Graduate School of Mathematics, Nagoya University

Tsuji, Hajime
Department of Mathematics, Tokyo Institute of Technology

Sidney M. Webster
Department of Mathematics, University of Chicago

Keizo Yamaguchi
Department of Mathematics, Hokkaido University

Analysis and Geometry in Several Complex Variables

CHAPTER I

The Bergman Kernel and a Theorem of Tian

David Catlin

Introduction

Given a domain Ω in $\mathbb{C}^n$, the Bergman kernel is the kernel of the projection operator from $L^2(\Omega)$ to the Hardy space $\mathcal{A}^2(\Omega)$. When the boundary of Ω is strictly pseudoconvex and smooth, Fefferman [2] gave a complete description of the asymptotic behavior of $K(z,z)$ as z approaches the boundary. This work was then extended by Boutet de Monvel and Sjöstrand [1] who showed that, for the same domains, a similar asymptotic expansion for $K(z,w)$ holds off the diagonal. Moreover, they showed that the Bergman kernel is a Fourier integral operator with a complex phase function. The first goal of this paper is to prove the following theorem:

Theorem 1. *Suppose E is a holomorphic vector bundle defined over a smoothly bounded strictly pseudoconvex manifold $\Omega = \{z; R(z) < 1\}$, and suppose that the L^2-norm is defined in terms of both a smooth Hermitian metric on E and a smooth metric g on the base manifold Ω. Then the Bergman kernel $K(z,w)$ of the projection onto $\mathcal{A}^2(\Omega, E)$ is a Fourier integral operator and can be represented by*

$$K(z,w) = \frac{F(z,w)}{(1 - R(z,w))^{n+1}} + G(z,w)\log(1 - R(z,w)). \tag{0.1}$$

As in [1] and [2], the function $R(z,w)$ is almost analytic along the boundary diagonal. The coefficients F and G are smooth sections of the vector bundle whose fiber at (z,w) is $\mathrm{Hom}(E_w, E_z)$.

Theorem 1 is hardly a surprising result. It seems certain that the proof of Boutet de Monvel and Sjöstrand would carry over to the situation of Theorem 1 with few changes. The proof given here assumes the theorem of Boutet de Monvel-Sjöstrand and also makes use of a few simple facts about Fourier integral operators.

Secondly, we use the result of Theorem 1 to study the asymptotic behavior of a family of finite-dimensional Bergman kernels on circular domains. Let E and L be holomorphic vector bundles of rank p and 1, respectively, over a complex manifold M, and let R be a smooth Hermitian metric on L. We assume that R has been extended onto a smooth function $L \times L$ that is almost-analytic along the diagonal, linear in the first entry and anti-linear in the second.

We let $\overline{\Omega} = \{\xi \in L;\ R(\xi, \xi) \leq 1\}$, and then, using $\pi : \overline{\Omega} \to M$, we define $\widetilde{E} = \pi^* E$ and also a metric $\widetilde{G} = \pi^* G$ on $\widetilde{E}$. Thus we obtain an L^2-norm by setting $\|\Phi\|^2 = \sum\limits_{\Omega} |\Phi|^2 \mathrm{vol}_{\tilde{g}}$, where $\tilde{g}$ is a suitably chosen metric on $T\Omega$.

Let $\mathcal{A}_d(\Omega, \widetilde{E})$ denote the space of holomorphic section of $\widetilde{E}$ on Ω that are homogeneous of order d on each fiber L_z, and let $K_d(\xi, \theta)$, having values in Hom $(\widetilde{E}_\theta, \widetilde{E}_\xi)$, denote the kernel of the projection

$$(P_d f)(\xi) = \sum_{\Omega} K_d(\xi, \theta) f(\theta) \mathrm{vol}_{\tilde{g}}(\theta) \tag{0.2}$$

of $L^2(\Omega, \widetilde{E})$ onto $\mathcal{A}_d(\Omega, \widetilde{E})$.

Theorem 2. *Suppose that the curvature of R is negative on M. Then for all $\ell = 0, 1, \ldots$, there exist smooth sections $a_\ell(\xi, \theta)$ having values in Hom$(\widetilde{E}_\theta, \widetilde{E}_\xi)$ and constant along fibers of L such that*

$$K_d \sim R^d \sum_{\ell=0}^{\infty} d^{n+1-\ell} a_\ell, \tag{0.3}$$

where (0.3) means that for any integers q, $N \geq 0$,

$$\left\| K_d - R^d \sum_{\ell=0}^{N} d^{n+1-\ell} a_\ell \right\|_{C^q} \leq M_{N,q} d^{n+q-N}. \tag{0.4}$$

Moreover, at any point $\xi \in b\Omega$, a_0 satisfies

$$a_0(\xi, \xi) = \frac{1}{2\pi} |\lambda_1(z) \ldots \lambda_n(z)|\ Id, \tag{0.5}$$

where $\lambda_1(z), \ldots, \lambda_n(z)$ are the eigenvalues of the curvature form of R at $z = \pi(\xi)$.

We note that the negativity of the curvature of R means that Ω is strictly pseudoconvex. The fact that Ω is invariant under the map $\xi \to e^{i\phi}\xi$ means that $\mathcal{A}^2(\Omega, \widetilde{E})$ is the orthogonal sum of the finite-dimensional spaces

$\mathcal{A}_d(\Omega, \widetilde{E})$. Using the kernel formula from Theorem 1 for the projection onto $\mathcal{A}^2(\Omega, \widetilde{E})$, we show that when K is written as a Taylor series in the fiber variable ζ, the only terms that act on $\mathcal{A}_d(\Omega, \widetilde{E})$ are the terms of order ζ^d. This leads to (0.3).

It is well-known that each section $\varphi \in L^2(M, E \otimes L^{*d})$ can be identified with a section $I_d(\varphi) \in L^2_d(\Omega, \widetilde{E})$, which is defined to be the set of sections in $L^2(\Omega, \widetilde{E})$ that are homogeneous of order d on each fiber L_z. Given G, R, and g, there is a naturally defined L^2-norm $\| \quad \|_{E\otimes L^{*d}}$ on $L^2(M, \; E \otimes L^{*d})$. If $\tilde{\varphi} \in L^2_d(\Omega, \widetilde{E})$, we obtain a norm $\|\tilde{\varphi}\|^2 = \|I_d^{-1}(\tilde{\varphi})\|_{E\otimes L^{*d}}$ which turns out to be a slight perturbation of the usual $\widetilde{E}$-norm. We let $K_{M,d}$ denote the kernel of the projection $P_{M,d}$ of $L^2_d(\Omega, \widetilde{E})$ onto $\mathcal{A}_d(\Omega, \widetilde{E})$ with respect to this new norm. Thus $P_{M,d}$ is just the projection onto $H^0(M, \; E \otimes L^{*d})$, transferred over to Ω.

Theorem 3. *Under the assumptions of Theorem 2, the kernel $K_{M,d}$ of $P_{M,d}$ satisfies*

$$K_{M,d} \sim \frac{2\pi}{d+2} K_d. \tag{0.6}$$

Hence $K_{M,d}$ has an asymptotic expansion of the form

$$K_{M,d} \sim R^d \sum_{\ell=0}^{\infty} A_\ell d^{n-\ell}, \tag{0.7}$$

where A_ℓ is constant along fibers and where

$$A_0(z, z) = |\lambda_1(z) \dots \lambda_n(z)| \; Id. \tag{0.8}$$

Corollary. *Let $\varphi_1 \dots, \varphi_N$ be an orthonormal basis of $H^0(M, \; E \otimes L^{*d})$ and define $B(z) = \sum_{k=0}^{N} |\varphi_k(z)|^2_{E\otimes L^{*d}}$. Then*

$$B(z) \sim \sum_{\ell=0}^{\infty} tr A_\ell(z) d^{n-\ell}. \tag{0.9}$$

In the case when $E = L^*$, the above result and its corollary and also Theorem 4 which follow were obtained independently by Zelditch [5] and the author. The asymptotic description of $K_{M,d}$ in [5] is based on the study of the Szegö kernel of the disk bundle Ω.

For the final result of this paper, we use Theorem 3 to describe the asymptotic behavior of a sequence of metrics g_d introduced by Tian [4]. When the curvature of R is negative, then for large d, a basis $\Phi_1, \ldots, \Phi_{N_d}$ of $H^0(M, \ E^* \otimes L^{*d})$ leads to an embedding ϕ_d of M into the Grassmanian G_{p,N_d}. When $\Phi_1, \ldots, \Phi_{N_d}$ is an orthonormal basis, the map ϕ_d should have nice regularity properties. In particular, the pullback $g_d = \frac{1}{d}\phi_d^* g_{Gr}$ of the standard metric g_{Gr} on G_{p,N_d} can be computed. (The factor of $\frac{1}{d}$ is a normalization.)

Theorem 4. *If the Ricci curvature $Ric(R)$ is negative on M, then there are smooth $(1,1)$-forms $m_\ell, \ell = 1, 2, \ldots,$ on M such that*

$$g_d = -p \ Ric(R) + \sum_{\ell=1}^{\infty} d^{-\ell} m_\ell. \tag{0.10}$$

It follows that g_d approaches $-p \ Ric(R)$ in the C^∞ topology.

When $E = L$, this result was obtained by Tian [4] in the C^2-topology, and as noted above by Zelditch in the C^∞-topology. In [6], the Bergman projection on $L^2(\Omega, \widetilde{E})$ was used to prove an isometric embedding theorem for holomorphic vector bundles.

I would like to express my gratitude to the Taniguchi Foundation for having invited me to attend the conference in Japan last summer. I would like to thank Larry Tong for some very helpful discussions and also Betty Gick and Judy Snider for patiently typing several versions of this paper.

1. The Bergman projection

Let E be a holomorphic vector bundle over a complex manifold Ω and let $(\ , \ )$ be a Hermitian metric on E. Given a metric g on the tangent bundle, we obtain a volume form vol_g, so we can define a norm on sections of E on Ω by

$$\|F\|^2 = \int_\Omega (F, F) \ \mathrm{vol}_g. \tag{1.1}$$

If Φ_ν, $\nu = 1, 2, \ldots$ is an orthonormal basis of $\mathcal{A}^2(\Omega, E)$, the set of holomorphic sections of E, then the Bergman projection $P \colon L^2(\Omega, E) \to \mathcal{A}^2(\Omega, E)$ can be written as $(PF)(z) = \sum_{\nu=1}^{\infty} \langle F, \Phi_\nu \rangle \Phi_\nu(z)$.

In order to describe the kernel of P, let $e_1', \ldots, e_p'$ and $e_1'', \ldots, e_p''$ be frames for E in neighborhoods U and V, respectively. We let F_U, $(PF)_V$, etc., denote the column vector of coefficients of F and PF with respect to $e_1', \ldots, e_p'$ and $e_1'', \ldots, e_p''$. If we define a $p \times p$ matrix $A_U(w)$ by $[A_{jk}(w)] =$

$(e_k(w), e_j(w))$, and if F is supported in U, then we obtain

$$(PF)_V(z) = \int_\Omega \sum_\nu \Phi_{\nu,V}(z)\Phi^*_{\nu,U}(w)A_U(w)F_U(w) \text{ vol}_g(w).$$

Thus the kernel $K(z,w)$ defined by

$$(PF)(z) = \int_\Omega K(z,w)F(w) \text{ vol}_g$$

takes values in Hom (E_w, E_z), and the local representation of K in $V \times U$ is

$$K_{V,U}(z,w) = \sum_\nu \Phi_{\nu,V}(z)\Phi^*_{\nu,U}(w)A_U(w).$$

Moreover, it follows immediately that the usual property that K is holomorphic in z and anti-holomorphic in w becomes $K_{V,U}(z,w)A_U^{-1}(w)$ is holomorphic in z and anti-holomorphic in w.

Closely related to the Bergman kernel is the quantity $B(z) = \sum_\nu |\Phi_\nu(z)|^2$, which, relative to the frame $e_1, \dots, e_p$ in U, equals $\sum_\nu \Phi^*_{\nu,U}(z)A_U(z)\Phi_{\nu,U}(z)$. By taking the trace, we see that

$$\text{tr } (\Phi_{\nu,U}(z)\Phi^*_{\nu,U}(z)A_U(z)) = \Phi^*_{\nu,U}A_U(z)\Phi_{\nu,U}(z),$$

so that by summing over ν, we obtain

$$\text{tr } K(z,z) = B(z). \tag{1.2}$$

Our goal is to show that $K(z,w)$ can be written as a Fourier integral operator, just as in the well known case when E is the trivial bundle and $\langle\ ,\ \rangle$ and g are the standard metrics, as proved by Boutet de Monvel and Sjöstrand in [1].

We first consider the local problem and assume that D is a strictly pseudoconvex domain in $\mathbb{C}^n$ and suppose that smooth sections $e_1, \dots, e_p$ of a vector bundle E are defined on $\overline{D}$. We assume that $(\ ,\ )$ is a metric defined on E and we define a matrix A' by $A'_{jk}(z) = (e_k(z),\ e_j(z))$. If in addition there is a volume element $\text{vol}(z) = b(z)\text{ vol}_0$, where vol_0 is the Euclidean volume element in $\mathbb{C}^n$, then a global norm on sections of E is given by $\|F\|^2 = \int_D F^*AF \text{ vol}_0$, where $A = A'b$, and where in the integral F denotes the column vector given by the coefficients $F_1, \dots, F_p$ of $F = \sum_{k=1}^n F_k e_k$.

We now show that P can be written in terms of a multiplication operator and the usual Bergman projection operator P_0, corresponding to the case when the matrix A in the above norm equals I_p (so that P_0 acts component-wise on $F_1, \dots, F_p$).

Let $\overline{\partial}^*$ be the usual adjoint of $\overline{\partial}$ corresponding to $A = I$. It is well known that the adjoint $\overline{\partial}^*_A$ with respect to the norm in (1.1) satisfies $\overline{\partial}^*_A = A^{-1}\overline{\partial}^* A$. It follows that

$$(P_0 A^{-1} P_0 A)(\overline{\partial}^*_A) = P_0 A^{-1} P_0 A A^{-1} \overline{\partial}^* A = P_0 A^{-1} P_0 \overline{\partial}^* A = 0,$$

since $P_0 \overline{\partial}^* = 0$. Thus $P_0 A^{-1} P_0 A$ annihilates the orthogonal complement of $A^2(D, E)$. On the other hand,

$$P_0 A^{-1} P_0 A = P_0 + T, \tag{1.3}$$

where $T = P_0[A^{-1}, P_0]A$. We now show that in the operator norm, T is small if D is sufficiently small. If A_0 is a $p \times p$ matrix of constants, then by thinking of A_0 as the linear map $f \to A_0 f$, it follows that $P_0 A_0 = A_0 P_0$. For some fixed point $z_0 \in D$, set $A_0 = A(z_0)$ and let $B = A_0^{-1} A$. Then

$$\begin{aligned} T &= P_0 A^{-1} P_0 A - P_0 = P_0 A^{-1} A_0 P_0 A_0^{-1} A - P_0 \\ &= P_0 B^{-1} P_0 B - P_0 = P_0 (B^{-1} - I) P_0 + P_0 B^{-1} P_0 (B - I). \end{aligned}$$

If the diameter of the domain equals σ, then $\|B - I\| + \|B^{-1} - I\| \leq C\sigma$. It follows that if σ is small, then $\|T\| < C_1 \sigma$. Using this fact, it follows that

$$P = (I - T + T^2 - \dots)(P_0 + T). \tag{1.4}$$

In fact, as noted above $P_0 + T$ annihilates $\mathcal{A}^2(D, E)^\perp$. If $f \in \mathcal{A}^2(D, E)$, then

$$(I - T + T^2 \dots)(P_0 + T)f = (I - T + T^2 + \dots +)(I + T)f = f.$$

Thus the expression in (1.4) is the Bergman projection operator.

Suppose that $r(z)$ is a smooth defining function for $D = \{z; r(z) < 0\}$. As noted by Boutet de Monvel and Sjöstrand, one can define an *almost analytic extension* $\psi(z, w)$, which is a smooth function satisfying

(i) $\psi(z, z) = \frac{1}{i} r(z)$, $z \in \mathbb{C}^n$

(ii) $\psi(z, w) = -\overline{\psi(w, z)}$, $z, w \in \mathbb{C}^n$, and

(iii) $\overline{\partial}_z \psi(z, w)$ and $\partial_w \psi(z, w)$ both vanish to infinite order when $z = w$.

In addition, since D is strictly pseudoconvex, one can choose ψ so that there is a positive constant c so that

(iv) $\operatorname{Im}\psi(z,w) \geq c(|r(z)| + |r(w)| + |z-w|^2),\ z,w \in \overline{D}$.

The result of Boutet de Monvel and Sjöstrand is that there exists a classical symbol $b(z,w,t) \in S^n(\overline{D} \times \overline{D} \times \mathbb{R}^+)$ with expansion $b(z,w,t) \sim \sum_{k=0}^{\infty} t^{n-k} b_k(z,w)$ such that the Bergman kernel $B(z,w)$ (of P_0) differs from $\int_0^{\infty} e^{it\psi(z,w)} b(z,w,t)dt$ by a smooth function on $\overline{D} \times \overline{D}$. By integrating in t, it follows easily that there exist smooth almost analytic functions $F(z,w)$ and $G(z,w)$ such that

$$B(z,w) = F(z,w)(-i\psi(z,w))^{-n-1} + G(z,w)\log(-i\psi(z,w)).$$

We now show that the same results hold for the Bergman projection P of $L^2(D,E)$ onto $\mathcal{A}^2(D,E)$.

Note that the operator $S = [A^{-1}, P_0]$ is a (matrix-valued) Fourier integral operator, since its symbol is $b(z,w,t)(A^{-1}(z) - A^{-1}(w))$. By the method of stationary phase described in Theorem 2.3 of [3] (Melin–Sjöstrand) the above symbol which arises as a commutator is in fact asymptotically equivalent to a symbol $S(z,w,t) \in S^{n-1}(\overline{D} \times \overline{D} \times \mathbb{R}^+)$. Thus the operator T in (1.3) is the composition of two Fourier integral operators with the same phase function $\psi(z,w)$ and with symbols $b(z,w,t)$ and $S(z,w,t)A(w)$. In Proposition 4.8 of [1], it is shown that the composition of two operators with this same phase function $\psi(z,w)$ is also a Fourier integral operator with phase function $\psi(z,w)$. According to p. 209 of [3], the symbol of the composition of operators with classical symbols of orders m_1' and m_2' will be of order $m_1' + m_2' - n$. In particular, the symbol $q(z,w,t)$ of T is of order $n-1$.

From (1.4), it follows that we can write

$$P = P_0 + \sum_{k=1}^{\infty}(-1)^k(T^kP_0 - T^k) = P_0 + \left(\sum_{k=1}^{\infty}(-1)^k T^k\right)(P_0 - I). \quad (1.5)$$

Following the same reasoning as above, it follows that modulo C^∞ functions on $\overline{D} \times \overline{D}$, each operator $Q_k = T^k$ is a Fourier integral operator with phase function ψ and with a symbol $q_k(z,w,t)$ in $S^{n-k}(\overline{D} \times \overline{D} \times \mathbb{R}^+)$.

Since the degree of each symbol q_k is $n-k$, which as a function of k approaches $-\infty$, we can choose a sequence d_k, $k = 1,2,\ldots$, growing sufficiently rapidly so that if $\chi \in C^\infty(\mathbb{R})$ with $\chi(t) = 2$ for $t \geq 1$ and $\chi(t) = 0$ for $t \leq \frac{1}{2}$, then the series

$$\tilde{q}(z,w,t) = \sum_{k=1}^{\infty}(-1)^k \chi(t/d_k) q_k(z,w,t)$$

converges to a symbol in S^n.

There is the technical difficulty however that each operator T^k differs from the corresponding operator with symbol $\tilde{q}_k(z,w,t) = \chi(t/d_k)q_k(z,w,t)$ by an operator H_k. We still need to show that the sum $\sum_{k=1}^{\infty} H_k$ converges to an operator H with a smooth kernel. If this can be done, then $P = P_0+(\widetilde{Q}+H)(P_0-I)$, where $\widetilde{Q}$ is the operator with symbol $\tilde{q}$. This would show that P is a Fourier integral operator with symbol in S^n.

To estimate H, we first note that if s_1 and s_2 are nonnegative integers, then there is an integer k_0 (dependent on s_1 and s_2) so that $\|T^{k_0}f\|_{s_1} \leq C_{s_1,s_2}\|f\|_{s_2}$. Since the same result also holds for the adjoint T^*, it follows from duality that this inequality extends to the case of arbitrary integers s_1 and s_2.

Let m_1 and m_2 be positive integers. By choosing k_1 to be the value of k_0 that works for $s_1 = -m_1$ and $s_2 = 0$, and k_2 to be the value that works for $s_1 = 0,\ s_2 = m_2$, it follows that if $k \geq k_1 + k_2$,

$$\begin{aligned}\|T^k f\|_{-m_1} &= \|T^{k_1}(T^{k-k_1}f)\|_{-m_1} \leq C\|T^{k-k_1}f\|_0 \\ &\leq C2^{-(k-k_1-k_2)}\|T^{k_2}f\|_0 < C'2^{-(k-k_1-k_2)}\|f\|_{m_2},\end{aligned}$$

where in the second inequality, we have used the fact that $\|T\| < \frac{1}{2}$. Let $\widetilde{Q}_k$ be the operator with symbol $\tilde{q}_k$. The order of $\widetilde{Q}_k$ approaches $-\infty$, so if we choose d_k to be sufficiently large, then for large k, $\|\widetilde{Q}_k f\|_{-m_1} \leq C'2^{-k}\|f\|_{m_2}$. Hence $H_k = (-1)^k(T^k - \widetilde{Q}_k)$ satisfies $\|H_k f\|_{-m_1} \leq C''2^{-k}\|f\|_{m_2}$, where C'' depends on m_1, m_2.

The above estimate clearly implies that $\sum H_k$ converges in the C^∞-topology to H. Thus we have proved that the Bergman projection operator P associated with the vector bundle E on D with metric A is a Fourier integral operator.

We conclude that there exist smooth matrix–valued functions F and G on $\overline{D} \times \overline{D}$ so that the kernel $K_D(z,w)$ satisfies

$$K_D(z,w) = \frac{F(z,w)}{(-r(z,w))^{n+1}} + G(z,w)\log(-r(z,w)). \tag{1.6}$$

We note that the coefficient of t^n in the symbol of P is exactly the same as the symbol for P_0, since the only operator of order n in (1.5) is just P_0 acting componentwise. For the operator P_0, it is shown in (4.2) of [1] that

$$F(z,z) = \frac{n!}{2^{n+1}\pi^n}\,\lambda(z)|dr(z)|^2,\ z \in bD,$$

where $\lambda(z)$ is the determinant of the Levi form $\partial\bar{\partial}\rho$, restricted to an orthonormal basis of $S_z = \mathbb{C}T_z^{1,0} \cap \mathbb{C}T(bD)$. Using the metric g, one can show that $\lambda(z)|dr(z)|^2 \mathrm{vol}_0 = \lambda_g(z)|dr(z)|_g^2 \mathrm{vol}_g(z)$, where λ_g is now the determinant of $\partial\bar{\partial}\rho$ restricted to a g–orthonormal basis of S_z. Hence, if we write the kernel K_0 of P_0 using the volume form vol_g as in (1.1), and use the fact that P and P_0 have the same top terms, then the coefficient matrix $F(z,w)$ in (1.6) satisfies

$$F(z,z) = \frac{n!}{2^{n+1}\pi^n} \lambda_g(z)|dr(z)|_g^2 I_p, \quad z \in bD. \tag{1.7}$$

In order to extend (1.6) to the case of a bundle E over a strongly pseudoconvex manifold $\Omega = \{z; R(z) < 1\}$, we first note that R can be extended to be almost analytic along the boundary diagonal and so that $|R(z,w)| < 1$ off the boundary diagonal.

Fefferman proved in Lemma 1 of [2] that if D is a subdomain of Ω such that the boundaries coincide inside an open subset W of $\overline{D}$, then the associated Bergman kernels satisfy

$$K_D(z,w) - K_\Omega(z,w) = e(z,w), \ (z,w) \in W \times W,$$

where $e(z,w)$ is smooth in $W \times W$. If we choose a finite set of small strictly pseudoconvex neighborhoods D_i so that bD_i and $b\Omega$ coincide in W_i, where W_i, $i = 1, \dots, N$ cover $b\Omega$, then we can assume that in each set $W_i \times W_i$, $R(z,w)$ equals $r_i(z,w)$, the defining function for D_i. Consequently, in each set $W_i \times W_i$, we can write

$$K_\Omega(z,w) = \frac{F_i(z,w)}{(1-R(z,w))^{n+1}} + G_i(z,w)\log(1-R(z,w)) + e_i(z,w).$$

By using a partition of unity defined near $b\Omega$, we conclude that there are smooth sections $F(z,w)$, $G(z,w)$ and $e(z,w)$ (having values in $\mathrm{Hom}(E_w, E_z)$) defined in a neighborhood diagonal of $\overline{\Omega} \times \overline{\Omega}$ such that

$$K(z,w) = \frac{F(z,w)}{(1-R(z,w))^{n+1}} + G(z,w)\log(1-R(z,w)) + e(z,w). \tag{1.8}$$

For (z,w) not lying on the boundary diagonal, Kerzman's theorem implies that $K(z,w)$ is smooth, so that (1.8) holds on all of $\overline{\Omega} \times \overline{\Omega}$.

Thus one immediately obtains Theorem 1 as stated in the introduction. We note also that if $A(w)$ is defined in terms of a local frame by $A_{jk}(w) = (e_k(w), e_j(w))$, then $F(z,w)A(w)^{-1}$ and $G(z,w)A(w)^{-1}$ are almost-analytic along the boundary diagonal.

2. Bergman projections onto finite-dimensional subspaces

We now consider a special case of the above results. Let E and L be vector bundles of rank p and 1, respectively, over a compact manifold M, with corresponding smooth Hermitian metrics G and R. If $\pi\colon L \to M$ is the projection map for L, then we let $\Omega = \{\xi \in L;\ R(\xi,\xi) < 1\}$ and we define $\widetilde{E} = \pi^* E$. We can extend R to a smooth function on $\overline{\Omega} \times \overline{\Omega}$ that is almost-analytic on the diagonal, linear in the first entry and anti-linear in the second, and also so that $|R| < 1$ off the boundary diagonal. In order to define a smooth Hermitian metric on $T^{1,0}\Omega$, let $\chi(t)$ be a smooth nonnegative function so that $\chi(t) = 1$ if $t < \frac{1}{4}$ and $\chi(t) = 0$ if $t \geq \frac{1}{2}$. Let g_1 be any smooth Hermitian metric on $T^{1,0}\Omega$ and define the metric

$$\tilde{g} = \pi^* g + \partial R \wedge \overline{\partial} R + \chi(R) g_1. \tag{2.1}$$

Letting $\tilde{G} = \pi^* G$ we obtain a metric on $\widetilde{E}$ and thus also an inner product $\langle F', F'' \rangle = \int_\Omega \tilde{G}(F', F'')\mathrm{vol}_\Omega$ for sections F', F'' of $\widetilde{E}$ on Ω.

Given a local frame e in $W \subset M$, we obtain the map $(z,\zeta) \to \zeta e(z)$. Thus, Ω locally corresponds to $R_1(z)|\zeta|^2 < 1$, where $R_1(z) = R(e(z),\ e(z))$.

If $e_1, \ldots, e_p$ is a local frame for E, we define sections $\tilde{e}_1, \ldots, \tilde{e}_p$ to be the pullbacks of $e_1, \ldots, e_p$. Since $\tilde{G} = \pi^* G$, the matrix $A(w,\omega)$ defined by

$$A_{jk}(w,\omega) = \tilde{G}(\tilde{e}_k(w,\omega),\ e_j(w,\omega)) = G(e_k(w),\ e_j(w))$$

is constant along the fiber, so we denote it by $A(w)$.

We now introduce some notation. For $(z,w) \in \overline{\Omega} \times \overline{\Omega}$, define $\pi_1(z,w) = w$ and $\pi_2(z,w) = z$. If $(z,w) \in M \times M$, define $p_1(z,w) = w$ and $p_2(z,w) = z$. Now let $E_1 = p_1^* E,\ E_2 = p_2^* E,\ \widetilde{E}_1 = \pi_1^* E$, and $\widetilde{E}_2 = \pi_2^* E$. We let $\mathcal{A}_\Delta(\overline{\Omega}^2,\ \mathrm{Hom})$ be the set of sections F in $C^\infty(\overline{\Omega} \times \overline{\Omega},\ \mathrm{Hom}(\widetilde{E}_1, \widetilde{E}_2))$ such that in the local frame, $F((z,\zeta),\ (w,\omega))A(w)^{-1}$ is almost-analytic along the boundary diagonal of $\overline{\Omega} \times \overline{\Omega}$. Let $\mathcal{A}_\Delta(M^2,\ \mathrm{Hom})$ denote the set of sections $F \in C^\infty(M \times M,\ \mathrm{Hom}(E_1, E_2))$ such that $F(z,w)A(w)^{-1}$ is almost analytic along the diagonal. Finally let $\mathcal{A}^*_\Delta(\overline{\Omega}^2,\ \mathrm{Hom})$ denote the set of sections $\pi^* F$, where $F \in \mathcal{A}_\Delta(M^2,\ \mathrm{Hom})$ and where $\pi((z,\zeta),\ (w,\omega)) = (z,w)$. Thus the elements of $\mathcal{A}^*_\Delta(M^2,\ \mathrm{Hom})$ are those sections of $\mathcal{A}_\Delta(\overline{\Omega}^2,\ \mathrm{Hom})$ that are constant in the fiber variables ζ and ω.

We shall study the asymptotic description of a family of Bergman kernels, i.e., projections onto a family of finite-dimensional subspaces. Let $\mathcal{A}_d(\Omega, \widetilde{E})$ be the set of sections $F \in \mathcal{A}(\Omega, \widetilde{E})$ (the holomorphic sections of $\widetilde{E}$ on Ω) that are homogeneous of degree d, i.e., in local coordinates of Ω, $F(z, t\zeta) = t^d F(z,\zeta)$ for all $t \in \mathbb{C}$ with $|t| \leq 1$. Since $\mathcal{A}_d(\Omega, \widetilde{E})$

can be identified with $\mathcal{A}(M, E \otimes L^{*d})$, it clearly is finite-dimensional. By using the Taylor expansion in ζ about $\zeta = 0$, it is easy to see that if $F \in \mathcal{A}(\Omega, \widetilde{E})$, then we can write $F = \sum_{d=0}^{\infty} F_d$, where $F_d \in \mathcal{A}_d(\Omega, \widetilde{E})$. Note also that since ζ^j and ζ^k are orthogonal with respect to the standard inner product on the disk $|\zeta| < r$ in $\mathbb{C}$, it follows that the subspaces $\mathcal{A}_j(\Omega, \widetilde{E})$ and $\mathcal{A}_k(\Omega, \widetilde{E})$ are orthogonal when $j \neq k$. From this one can easily show that if $F \in \mathcal{A}^2(\Omega, \widetilde{E})$, then the series $\sum_{d=0}^{\infty} F_d$ converges to F in the L^2-norm.

If Φ_ν^d, $\nu = 1, \ldots, N_d$, is an orthonormal basis of $\mathcal{A}_d(\Omega, \widetilde{E})$, one sees that the Bergman projection P and the projection P_d onto $\mathcal{A}_d(\Omega, \widetilde{E})$ satisfy

$$PF = \sum_{d=0}^{\infty} \sum_{\nu=1}^{N_d} \langle F, \ \Phi_\nu^d \rangle \Phi_\nu^d, \qquad P_d F = \sum_{\nu=1}^{N_d} \langle F, \ \Phi_\nu^d \rangle \Phi_\nu^d.$$

Letting K_d denote the associated kernel, it follows easily that if (z, ζ) and (w, ω) denote points in Ω, then $K_d(z, \zeta, w, \omega)$ (For convenience, we shall write $K_d(z, \zeta, w, \omega)$ instead of $K_d((z, \zeta), (w, \omega))$, and similarly, write $K(z, \zeta, w, \omega)$, $R(z, \zeta, w, \omega)$ and $F(z, \zeta, w, \omega)$.) is homogeneous of degree d in both ζ and $\overline{\omega}$. Thus, given $K(z, \zeta, w, \omega)$, in order to compute K_d, one need only write down those terms that are homogeneous of degree d in ζ and $\overline{\omega}$.

Theorem 1 implies that there exist sections F_1 and F_2 in $\mathcal{A}_\Delta(\overline{\Omega}^2, \text{ Hom})$ such that $K = F_1(1-R)^{-n-2} + F_2 \log(1-R)$. Because Ω has the automorphisms $(z, \zeta) \to (z, e^{i\theta}\zeta)$, it follows that $K(z, e^{i\theta}\zeta, w, e^{i\theta}\omega) = K(z, \zeta, w, \omega)$. Letting $S_\theta(z, \zeta, w, \omega) = (z, e^{i\theta}\zeta, \omega, e^{i\theta}\omega)$, we define $F_i^\theta = F_i \circ S_\theta$, $R^\theta = R \circ S_\theta$, and $K^\theta = K \circ S_\theta$.

Lemma 1. *For any point $(z_0, \zeta_0) \in b\Omega$ and for any θ, the formal power series of F_i^θ and F_i at (z_0, ζ_0) are identical, $i = 1, 2$.*

Proof. Since $R^\theta = R$, it follows that

$$\frac{F_1^\theta}{(1-R)^{n+2}} + F_2^\theta \log(1-R) = \frac{F_1}{(1-R)^{n+2}} + F_2 \log(1-R). \tag{2.2}$$

Multiplying (2.2) by $(1-R)^{n+2}$, it follows that for any $(z_0, \zeta_0) \in b\Omega$, the Taylor expansions of F_1^θ and F_1 of order $n+1$ at (z_0, ζ_0) are equal. After cancelling out these terms, we instead divide by $\log(1-R)$. We then conclude that $F_2(z_0, e^{i\theta}\zeta_0) = F_2(z_0, \zeta_0)$. Cancelling these terms in (2.2), we now alternately apply this procedure to the terms of degree $n+2+\ell$ in F_1^θ and F_1 at (z_0, ζ_0) and then the terms of degree $\ell+1$ in the expansions of F_2^θ and F_2, for $\ell = 0, 1, 2, \ldots$. We conclude that the full Taylor expansions of F_1^θ and F_1 at (z_0, ζ_0) are equal and also that those of F_2^θ and F_2 are equal, which proves the lemma.

If we now define

$$\widetilde{F}_1(z, \zeta, w, \omega) = \frac{1}{2\pi}\int_0^{2\pi} F_1(z, \zeta e^{i\theta}, w, \omega e^{i\theta})d\theta,$$

then $\widetilde{F}_1 - F_1$ vanishes to infinite order at (z_0, ζ_0). Doing the same for F_2, we conclude that there is a smooth matrix E vanishing to infinite order along the boundary diagonal so that $K = \widetilde{F}_1(1-R)^{-n-2} + \widetilde{F}_2 \log(1-R) + E$. Note that since each term K, $\widetilde{F}_1$, $\widetilde{F}_2$ is invariant under S_θ, the same is true for E. Setting $\widetilde{E} = E(1-R)^{n+2}$, we conclude that if $\widetilde{F}_1 + \widetilde{E}$ is for notational simplicity denoted by F_1 and $\widetilde{F}_2$ by F_2, then we have proved the following.

Proposition 2. *There exist smooth sections F_1 and F_2 of $\mathcal{A}_\Delta(\overline{\Omega}^2, Hom)$ so that $F_i^\theta = F_i$, $i = 1, 2$ for all θ and so that the Bergman kernel K for the projection $L^2(\Omega, \widetilde{E}) \to \mathcal{A}^2(\Omega, \widetilde{E})$ satisfies*

$$K = \frac{F_1}{(1-R)^{n+2}} + F_2 \log(1-R). \tag{2.3}$$

We now show that the sections F_1 and F_2 can be expanded in a Taylor series in powers of $(1-R)$ along the boundary diagonal.

Lemma 3. *For each integer $N = 1, 2, \ldots$, there exist sections C_k^i in $\mathcal{A}_\Delta^*(\overline{\Omega}^2, Hom)$, $k = 0, 1, \ldots, N$ and E_N^i in $\mathcal{A}_\Delta(\overline{\Omega}^2, Hom)$, such that*

$$F_i = \sum_{k=0}^{N} C_k^i (1-R)^k + E_N^i (1-R)^{N+1}. \tag{2.4}$$

To prove Lemma 3, we use the following lemma:

Lemma 4. *Suppose a smooth function $f(x, y, \zeta)$ is defined on a bounded convex open subset W about the origin in $\mathbb{R}^m \times \mathbb{R}^n \times \mathbb{C}$ and that for every pair of integers j, k with $j \geq 0$, $k \geq 1$, f satisfies*

$$\left| \frac{\partial^{j+k}}{\partial \zeta^j \partial \overline{\zeta}^k} f(x, y, \zeta) \right| \leq C_{j,k}(|x| + |\zeta|)^{j+k}. \tag{2.5}$$

Let $T = W \cap \{(0, y, z);\ y \in \mathbb{R}^n,\ z \in \mathbb{C}\}$. Then for each integer $N \geq 0$, there exist smooth functions $E_N(x, y, \zeta)$ and $E^\infty(x, y, \zeta)$ such that

$$f(x, y, \zeta) = \sum_{j=0}^{N} C_j(x, y)\zeta^j + \zeta^{N+1} E_N(x, y, \zeta) + E^\infty(x, y, \zeta), \tag{2.6}$$

and where E^∞ vanishes to infinite order in x and ζ along the set T.

Proof. It follows from Taylor's theorem that

$$f(x,y,\zeta) = \sum_{j+k\le N} \frac{\partial^{j+k} f}{\partial\zeta^j \partial\overline{\zeta}^k}(x,y,0)\frac{\zeta^j\overline{\zeta}^k}{j!k!} + \int_0^1 \sum_{j+k=N+1} \frac{\partial^{N+1} f}{\partial\zeta^j \partial\overline{\zeta}^k}(x,y,\zeta t)\,\frac{\zeta^j\overline{\zeta}^k}{j!k!}\,(1-t)^N dt.$$

Note that if $k>0$, then $\left(\frac{\partial}{\partial\zeta}\right)^j \left(\frac{\partial}{\partial\overline{\zeta}}\right)^k f(x,y,\zeta)$ vanishes to infinite order in x and ζ along the set T. Hence the two terms above corresponding to that pair j,k both yield functions of the form E^∞. The terms with $k=0$ give rise to the sum $\sum C_j(x,y)\zeta^j$ and $\zeta^{N+1}E_N(x,y,\zeta)$.

Proof of Lemma 3. We let (z,ζ) and (w,ω) denote two points in a common coordinate patch of Ω. If we set $r(w) = (R_1(w))^{-\frac{1}{2}}$, then $(w, r(w)) \in b\Omega$. Also set $a(z,w) = (r(w)R_1(z,w))^{-1}$, so that $R(z, a(z,w),\ w, r(w)) = 1$. When $z=w$, then $a(w,w)=r(w)$, so that $|a(z,w)-r(w)| \le C|z-w|$. We let $x=z-w$ and then apply Lemma 4 to the function $f(x,w,\zeta) = F(x+w,\ a(x+w,w)+\zeta,\ w, r(w))$, where $F=$ either $F_1A(w)^{-1}$ on $F_2A(w)^{-1}$. Since F is almost-analytic, we conclude that f satisfies (2.5). Hence the lemma implies we can write

$$\begin{aligned} F(z,\zeta,w,r(w)) &= \sum_{k=0}^{N} C_k(z,w)(\zeta - a(z,w))^k \\ &\quad + e_N(z,\zeta,w,r(w))(\zeta-a(z,w))^{N+1} \\ &\quad + e^\infty(z,\zeta,w,r(w)). \end{aligned}$$

Since $1-R(z,\zeta,w,r(w)) = -r(w)R_1(z,w)(\zeta-a(z,w))$, we can rewrite this as

$$F = \sum_{k=0}^{N} c_k(1-R)^k + e_N(1-R)^{N+1} + e^\infty, \tag{2.7}$$

which holds at all points of the form $(z,\zeta,w,r(w))$. Since $R^\theta = R$, and c_k depends only on (z,w) and not on ζ or ω, the sum in (2.7) is invariant under the map S_θ. If we extend e_N and e^∞ to be invariant under S_θ, then the right-hand side is invariant under S_θ. Since $F_\theta = F$, we conclude that (2.7) holds at all points of the form (z,ζ,w,ω) with $|\omega| = r(w)$. We now extend e_N and e^∞ so that the formal power series in ω of e_N and e^∞

have no holomorphic terms at any point of the form $(z_0, \zeta_0, w_0, \omega_0)$ with $|\omega_0| = r(w_0)$. Since F and each term $c_k(1-R)^k$, $k = 0, \ldots, N$, have this property, this implies that (2.7) now holds modulo an error which vanishes to infinite order on the set where $|\omega_0| = r(w_0)$. But such an error can be absorbed in the term e^∞, so we can actually assume that (2.7) holds identically on $\overline{\Omega} \times \overline{\Omega}$. Finally, note that $(1-R)^{-N-1}$ blows up at only a finite rate as $(z, \zeta) \to (w, \overline{w}) \in b\Omega$. Hence if we define $\tilde{e}^\infty = (1-R)^{-N-1} e^\infty$, then $\tilde{e}^\infty$ is still smooth. Thus we can replace e_N by $e_N + \tilde{e}^\infty$. This proves the local existence of $\tilde{C}_k$.

If we construct $\tilde{C}_k$ and $\widetilde{E}_N$ in a collection of neighborhoods that cover $\overline{\Omega}$, then we would like to argue that the functions $\tilde{C}_k$ and $\widetilde{E}_N$ are unique. But this is not true, since we can always perturb $\tilde{C}_k$ and $\widetilde{E}_N$ by a section vanishing to infinite order on the boundary diagonal of $\overline{\Omega}$. However, it is true that the formal power series of $\tilde{C}_k$ and $\widetilde{E}_N$ are determined all along the boundary diagonal. This makes it easy to patch the various sections together. This completes the proof of Proposition 4.

We are now ready to prove Theorem 2 as stated in the Introduction.

Proof of Theorem 2. If we apply Lemma 3 with $N = n+1$ to the section F_1 in (2.3) and also to F_2 with N an arbitrary positive integer, then it follows that we can write

$$F_1 = \sum_{j=0}^{n+1} c_j(1-R)^j + e_{n+1}(1-R)^{n+2}$$

$$F_2 = \sum_{j=0}^{N} C_j(1-R)^j + e_N(1-R)^{N+1}.$$

Thus we obtain from (2.4) that

$$K = \sum_{j=0}^{n+1} c_j(1-R)^{j-n-2} + \sum_{j=0}^{N} C_j(1-R)^j \log(1-R) + e, \qquad (2.8)$$

where

$$e = e_{n+1} + e_N(1-R)^{N+1}\log(1-R). \qquad (2.9)$$

If we use the identity $(1-R)^{-k-1} = \sum_{d=0}^{\infty} \binom{d+k}{k} R^d$, then the first sum in (2.8) becomes

$$\sum_{j=0}^{n+1} c_j(1-R)^j = \sum_{j=0}^{n+1}\sum_{d=0}^{\infty} c_j \binom{d+n+1-j}{n+1-j} R^d = \sum_{d=0}^{\infty}\sum_{\ell=0}^{n+1} d^{n+1-\ell} a_\ell R^d, \qquad (2.10)$$

for suitable sections a_ℓ in $\mathcal{A}^*_\Delta(\overline{\Omega}^2,\ \mathrm{Hom})$ that are obtained by expanding $c_j \binom{d+n+1-j}{n+1-j}$ in powers of d.

In order to handle the second sum in (2.8), we need the following Taylor expansion: there exist numbers $b_{j,d}$, for $d = 1, \dots, 2N-1$, so

$$(1-t)^j \log(1-t) = \sum_{d=1}^{2N-1} b_{j,d} t^d + \sum_{d=2N}^{\infty} \frac{j!}{d(d-1)\dots(d-j)}\, t^d, \tag{2.11}$$

which holds for all complex numbers t with $|t| \leq 1,\ r \neq 1$. This formula follows by integrating the formula $\log(1-t) = \sum_{d=1}^{\infty} t^d/d$ j times. Note that if (z,ζ) and (w,ω) are in $\overline{\Omega}$, $|R(z,\zeta,w,\omega)| \leq 1$. Hence we can replace t by $R(z,\zeta,w,\omega)$ in (2.11) and the second sum in (2.8) becomes

$$\begin{aligned}&\sum_{j=0}^{N} C_j(1-R)^j \log(1-R)\\ &\quad= \sum_{j=0}^{N} \sum_{d=1}^{2N-1} b_{j,d} C_j R^d + \sum_{j=0}^{N} \sum_{d=2N}^{\infty} \frac{j!}{d(d-1)\dots(d-j)}\, C_j R^d.\end{aligned} \tag{2.12}$$

Since $d \geq 2N$ and $j \leq N$ in the second sum of (2.11), it follows that there exist numbers $B_{j,k}$ for $j = 0, \dots, N,\ k = 1, 2, \dots,$ with $|B_{j,k}| \leq M_N$ so that

$$\frac{j!}{d(d-1)\dots(d-j)} = \frac{j!}{d^{j+1}} \prod_{\ell=1}^{j} \left(1 - \frac{\ell}{d}\right)^{-1} = \frac{1}{d^{j+1}} \sum_{k=0}^{\infty} B_{j,k}\, d^{-k}.$$

Hence the second sum in (2.12) becomes

$$\sum_{d=2N}^{\infty} \sum_{j=0}^{N} \sum_{k=0}^{\infty} B_{j,k} d^{-j-k-1}\, C_j R^d = \sum_{d=2N}^{\infty} \left(\sum_{\ell=n+2}^{\infty} a_\ell d^{-\ell+n+1} \right) R^d, \tag{2.13}$$

where we have defined $a_\ell = \sum_{j+k+2=\ell-n} B_{j,k} C_j$, which is in $\mathcal{A}^*_\Delta(\overline{\Omega}^2,\ \mathrm{Hom})$. Note that the sum defining a_ℓ really only involves $N+1$ terms since $0 \leq j \leq N$. Thus for any integer $m \geq 0$, there is a constant $C_{q,N}$ so that

$$\|a_\ell\|_{C^q(\overline{\Omega}\times\overline{\Omega})} \leq C_{q,N},\ \ell = n+2, n+3, \dots,$$

which implies that

$$\|a_\ell R^d\|_{C^q} \leq C'_{q,N} d^q. \tag{2.14}$$

It follows from (2.8), (2.10), and (2.12)–(2.14) that K_d satisfies

$$\|K_d - \sum_{\ell=0}^{N} d^{n+1-\ell} a_\ell R^d\|_{C^q} \leq \| \sum_{\ell=N+1}^{\infty} d^{n+1-\ell} a_\ell R^d + P_d e\|_{C^q}$$
$$\leq C''_{p,N} d^{n+q-N} + \|P_d e\|_{C^q},$$

where $e = \sum_{d=0}^{\infty} P_d e$ and $P_d e$ is homogeneous of order d in ζ and $\overline{w}$.

In order to estimate $P_d e$, we shall work with slightly different coordinates. In terms of local coordinates (z, ζ, w, ω), define $E(z, \theta, w, \tau) = e(z, r(z)\theta, w, r(w)\tau)$. Note that locally $\overline{\Omega}$ now corresponds to $|\theta| \leq 1$ and $|\tau| \leq 1$. We can represent E by

$$E(z, \theta, \omega, \tau) = \sum_{k=0}^{\infty} g_k(z, \omega) \theta^k \overline{\tau}^k.$$

Let $\rho(z, \theta, w, \tau) = R(z, r(z)\theta, w, r(w)\tau)$. If $D^q E$ is a mixed partial derivative of ρ with $q \leq N+1$, then the only possible singular term occurs if no derivative is applied to $\log(1-\rho)$. Hence $D^q E$ is continuous if $q \leq N$. Moreover,

$$|D^{N+1}E| \leq C_N(1 + |\log|1-\rho||). \tag{2.15}$$

But R satisfies $|1 - R(z, \zeta, w, \omega)| \geq a(|x-w|^2 + |\zeta - \omega|)$ for some constant $a > 0$, so ρ satisfies $|1 - \rho(z, \theta, w, \tau)| \geq c|\rho(z)\theta - \rho(w)\tau|$. Together with (2.15) this shows that along the circle $\theta = e^{i\phi}$,

$$\int_0^{2\pi} |(D^{N+1}E)(z, e^{i\phi}, w, 1)| d\phi \leq C_N,$$

where C_N is independent of z and w. In particular, this implies that

$$\int_0^{2\pi} \left| \left(\frac{\partial}{\partial\phi}\right)^{N+1-q} (D^p_{z,w} E)(z, e^{i\phi}, w, 1)) \right| d\phi \leq C_N.$$

Since the inequality $\int_0^{2\pi} |f^{(m)}(\phi)| d\phi \leq C$ implies that the Fourier series $\sum f_d e^{id\phi}$ of f satisfies $|f_d| \leq C'|d|^{-q}$, $d \neq 0$, we conclude that the Fourier series in ϕ of $D^q_{z,w}E$ satisfies $\|E_d\|_{C^q} \leq C''_N d^{q-N-1}$. If we define $\chi(z, \zeta, w, \omega) = (z, r(z)\zeta, w, r(w)\omega)$, then $E = e \circ \chi$. Since χ preserves the space of sections that are homogeneous of order d, it follows that we can write $e = \sum P_d e$, where $P_d e = E_d \circ \chi^{-1}$. Hence there is a constant T_q so that

$$\|P_d e\|_{C^q} \leq T_q \|E_d\|_{C^q} \leq T_p C'_N d^{q-N-1}.$$

This completes the proof of (0.4).

To prove (0.5), note that (2.10) implies that $a_0(z,z) = \dfrac{c_0(z,z)}{(n+1)!}$. Since the definition of $\tilde{g}$ in (2.1) implies that $|\partial R(\xi)| = 1$ at any boundary point, we obtain $|dR(\xi)| = \sqrt{2}$. Moreover, if $\pi(\xi) = z$, then $\frac{1}{2\pi}(\partial\overline{\partial}R)_{|S_\xi} = \frac{1}{2\pi}\,\partial\overline{\partial}\log R_1(z)$, which is the curvature of R, so we obtain

$$(2\pi)^{-n}\det(\partial\overline{\partial}R_{|S_\xi}) = |\lambda_1(z)\ldots\lambda_n(z)|.$$

Combining these facts with (1.7), we obtain (0.5), which completes the proof of Theorem 2.

We now show how Theorem 2 can be used to compute an asymptotic formula for the projection of $L^2(M, E\otimes L^{*d})$ onto $H^0(M, E\otimes L^{*d})$. If η is the section of L^* that is dual to e in a neighborhood W, then we can write a local section of $E\otimes L^{*d}$ as $\varphi = \varphi'\otimes\eta^{\otimes d}$, where φ' is a local section of E. Using the standard rules for metrics of dual spaces and tensor products, we have that $|\varphi(z)|^2 = |\varphi'(z)|^2|\eta(z)|^{2d} = |\varphi'(z)|^2 R_1(z)^{-d}$, where $|\varphi'(z)|^2 = G(\varphi'(z),\ \varphi'(z))$. If φ is supported in W, then

$$\|\varphi\|^2 = \int_W |\varphi|^2 \operatorname{vol}_g = \int_W |\varphi'|^2 R_1^{-d} \operatorname{vol}_g.$$

It is well known that each section φ of $L^2(M, E\otimes L^{*d})$ can be uniquely identified with a section $I_d(\varphi) \in L^2_d(\Omega, \widetilde{E})$, consisting of the subspace of elements of $L^2(\Omega, \widetilde{E})$ that are homogeneous of order d on each fiber L_z. For the section φ above, if $\varphi' = \sum_{k=1}^{p}\varphi_k(z)e_k$, then $I_d(\varphi)(e, e(z)) = \sum_{k=1}^{p}\zeta^d\varphi_k(z)\tilde{e}_k(z,\zeta)$.

Now consider the metric $g_0 = \pi^* g + \partial R\wedge\overline{\partial}R$, which is degenerate when $R = 0$. If we let $z = (z_1,\ldots,z_n)$ and $z_{n+1} = \zeta$, and if $r_j = \dfrac{\partial}{\partial z_j}(R_1(z)|z_{n+1}|^2)$, then the matrix of g_0 can be written as $g_0 = p^t g_1\overline{p}$, where g_1 is the Hessian matrix of $\sum_{j,k=1}^{n} g_{jk}z_j\overline{z}_k + |z_{n+1}|^2$ and where p is the Jacobian matrix of $P(z) = \left(z_1,\ldots,z_n,\ \sum_{j=1}^{n+1} r_j z_j\right)$. Hence

$\det g_0 = |\det p|^2 \det g_1 = |r_{n+1}|^2 \det g$. Since the volume form of the metric $\sum_{j,k=1}^{n} h_{jk} dz_j \otimes d\bar{z}_k$ is $2^n \det h \text{ vol}_0$, where vol_0 is the Euclidean volume form, we conclude that

$$\text{vol}_{g_0} = 2^{n+1} R_1(z)^2 |\zeta|^2 \det g \text{ vol}_0 = 2R_1(z)^2|\zeta|^2 \text{ vol}_g \wedge dx \wedge dy.$$

It follows that if $\Phi = I_d(\varphi)$, then

$$\begin{aligned}\int_{\pi^{-1}(W)} |\Phi|^2 \text{ vol}_{g_0} &= 2\int_W \int_{|\zeta|^2 < R_1(z)^{-1}} |\varphi'(z)|^2 |R_1(z)|^2 |\zeta|^{2d+2} \text{ vol}_g \wedge dx \wedge dy \\ &= \frac{2\pi}{d+2} \int_W |\varphi'|^2 R_1^{-d} \text{vol}_g = \frac{2\pi}{d+2} \int_W |\varphi|^2 \text{ vol}_g.\end{aligned}$$

By using a partition of unity, we see that if $\Phi = I_d(\varphi)$, where $\varphi \in L^2(M, E \otimes L^{*d})$, then

$$\frac{d+2}{2\pi} \int_\Omega |\Phi|^2 \text{ vol}_{g_0} = \|\varphi\|^2. \tag{2.16}$$

In view of (2.16), we can study the Bergman projection $P_{M,d}$ of $L^2_d(\Omega, \widetilde{E})$ onto $H_d(\Omega, \widetilde{E})$ with respect to the norm given by the left–hand side of (2.16). The kernel of $P_{M,d}$ is described by Theorem 3 of the Introduction.

Proof of Theorem 3. From (2.1), we see that if $R(\xi) \geq \frac{1}{2}$, then $\tilde{g} = g_0$. Hence in the open set $\pi^{-1}(W)$ there is a function $v(z, \zeta)$ such that $v = 0$ if $R_1(z)|\zeta|^2 \geq \frac{1}{2}$ and such that $\text{vol}_{\tilde{g}} = \text{vol}_{g_0} + v(z,\zeta)\text{vol}_g \wedge dx \wedge dy$. If $\Phi_1, \Phi_2 \in L^2_d(\Omega, \widetilde{E})$ are supported in $\pi^{-1}(W)$, then this implies that

$$\int_\Omega (\Phi_1, \Phi_2)\text{vol}_{\tilde{g}} = \int_\Omega (\Phi_1, \Phi_2)\text{vol}_{g_0} + \int_\Omega (\Phi_1, \Phi_2) v(z,\zeta) dx \wedge dy \wedge \text{ vol}_g.$$

Since Φ_1, Φ_2 are homogeneous of degree d and v is supported in the set $R_1(z)|\zeta|^2 \leq \dfrac{1}{2}$, we conclude by applying the Cauchy–Schwarz inequality to the final term that

$$\int_\Omega (\Phi_1, \Phi_2)\text{vol}_{\tilde{g}} = \int_\Omega (\Phi_1, \Phi_2)\text{vol}_{g_0} + \mathcal{O}(\in^{-\lceil} \|\Phi_\infty\| \|\Phi_\in\|). \tag{2.17}$$

By using a partition of unity, this holds for any pair $\Phi_1, \Phi_2 \in L^2_d(\Omega, \widetilde{E})$. In particular, let $\Phi_\mu, \mu = 1, 2, \ldots, N_d$ be an orthonormal basis of $H_d(\Omega, \widetilde{E})$. Then (2.17) implies

$$\delta_{\mu\nu} = \int_\Omega (\Phi_\mu, \Phi_\nu)\text{vol}_{g_0} + \mathcal{O}(\in^{-\lceil}), \ \mu, \nu = \infty, \ldots, \mathcal{N}_\lceil. \tag{2.18}$$

It is a consequence of the Hirzebruch–Riemann–Roch Theorem and the Kodaira Vanishing Theorem that for large d, $\dim H^0(M, E \otimes L^{*d}) = N_d$ is a polynomial in d of degree n. By using (2.18) and the Gram–Schmidt process, we conclude that there is a matrix $C^d = [C^d_{\mu\nu}]$, $1 \leq \mu,\ \nu \leq N_d$ such that for $\mu = 1, \dots, N_d$, $\psi_\mu = \Phi_\mu + \sum_\nu C^d_{\mu\nu}\Phi_\nu$ is an orthonormal basis of $\mathcal{A}^2_d(\Omega, \widetilde{E})$ with respect to the inner product $\langle \Phi, \psi \rangle_{g_0} = \int (\Phi, \psi)\mathrm{vol}_{g_0}$, and also such that $\max_{\mu,\nu} |C^d_{\mu,\nu}| \leq A_m d^{-m}$ for any integer $m \geq 0$.

We now compute $P_{0,d}$, which we define to be the projection onto $\mathcal{A}_d(\Omega, \widetilde{E})$ with respect to the norm $\int_\Omega |F|^2 \mathrm{vol}_{g_0}$. We let $\mathcal{E}_d$ denote any sequence of operators from $L^2(\Omega, \widetilde{E})$ to $C^\infty(\Omega, \widetilde{E})$ such that $\|\mathcal{E}_d(f)\|_m \leq C_m d^{-m}$ for all d holds for any Sobolev norm $\|\quad\|_m$, $m \geq 0$. If $f \in L^2_d(\Omega, \widetilde{E})$, then

$$P_{0,d}f = \sum_{\mu=1}^{N_d} \langle f, \psi_\mu \rangle_{g_0}\ \psi_\mu = \sum_{\mu=1}^{N_d} \langle f, \Phi_\mu \rangle_{\tilde{g}} \Phi_\mu + \mathcal{E}_d(f),$$

where we have used in the second equality that if $\psi \in \mathcal{A}_d(\Omega, \widetilde{E})$, then $\|\psi\|_m \leq C_m d^m \|\psi\|$. By (0.2), we have

$$\sum_{\mu=1}^{N_d} \langle f, \Phi_\mu \rangle_{\tilde{g}} = P_d f + \mathcal{E}_d f = \int_\Omega K_d f \ \mathrm{vol}_{\tilde{g}} = \int K_d f \ \mathrm{vol}_{g_0} + \mathcal{E}_d f,$$

where again in the last equality, since $K_d(\xi, \theta)$ is homogeneous in each fiber in θ, an argument similar to (2.17) shows that the error is of the form $\mathcal{E}_d f$. Finally, note that if we use the left side of (2.16) for the L^2-norm, then we must replace ψ_μ by $\left(\frac{2\pi}{d+2}\right)^{\frac{1}{2}} \psi_\mu$. This gives (0.6). To prove (0.7) and (0.8), note that we can write $\frac{1}{d+2} = \sum_{\ell=0}^{\infty} \frac{(-2)^\ell}{d^{\ell+1}}$.

Proof of the Corollary. We let (z, ζ) denote a point in a frame neighborhood of Ω and we define $\tilde{B}(z, \zeta) = \sum_{\nu=1}^{N_d} |\Phi_\nu(z, \zeta)|^2_{\widetilde{E}}$, where $\Phi_\nu, \nu = 1, \dots, N_d$ is an orthonormal basis of $\mathcal{A}_d(\Omega, \widetilde{E})$ with respect to the norm $\frac{d+2}{2\pi} \int |\Phi|^2 \mathrm{vol}_{g_0}$. By combining (0.7) and (1.2), we see that

$$\tilde{B}(z, \zeta) \sim R_1(z) |\zeta|^{2d} \sum_{\ell=0}^{\infty} \mathrm{tr}\ A_\ell(z) d^{n-\ell}. \tag{2.19}$$

It follows from (2.16) and (2.17) that if $\Phi = I_d(\varphi)$, then

$$|\varphi(z)|^2_{E\otimes L^{*d}} = \frac{d+2}{2\pi}\int_{D_z} 2R_1(z)|\zeta|^2|\Phi(z,\zeta)|^2 dxdy,$$

where $D_z = \{\zeta; |\zeta|^2 < R_1(z)^{-1}\}$. Hence

$$B(z) = \frac{d+2}{2\pi}\int_{D_z} 2R_1(z)|\zeta|^2\tilde{B}(z,\zeta)dxdy. \tag{2.20}$$

The formula (0.9) now follows by combining (2.19) and (2.20) and integrating.

3. Application to a theorem of Tian

We now give the proof of Theorem 4, as stated in the introduction.

Proof of Theorem 4. Given the metric G on E and R on L, we obtain a metric on $(E\otimes L^d)^*$. Let $\varphi_1,\dots,\varphi_N$ be an orthonormal basis of $H^0(M,(E\otimes L^d)^*)$. It is well known that since L^* is positive for large d the map

$$\varphi_*(v\otimes\xi^d) = (\varphi_1(v\otimes\xi^{\otimes d}),\dots,\varphi_N(v\otimes\xi^{\otimes d}))v\in E_z,\ \xi\in L_z,$$

gives rise to an embedding φ of M into the Grassmanian $G_{p,N}$ which is the set of p-dimensional planes S in $\mathbb{C}^N$. If $U_{p,N}$ is the bundle over $G_{p,N}$ consisting of the set of pairs (S,w), where $w\in S$, then we obtain a metric g_{euc} on $U_{p,N}$ by setting $g_{euc}((S,v),\ (S,w)) = v\cdot\overline{w}$. We can use this metric to define a metric on $G_{p,N}$ by

$$g_{Gr} = -\operatorname{Ric}(g_{euc}) = \frac{1}{2\pi}\partial\overline{\partial}\log\det[g^{ij}_{euc}],$$

where $g^{ij}_{euc} = g_{euc}(s_i,s_j)$ and where s_i, $i=1,\dots,p$ is a local frame for $U_{p,N}$. If we view φ_* as a map from $E\otimes L^d = \varphi^* U_{p,N}$ to $U_{p,N}$, then we can pull g_{euc} back to a metric on $E\otimes L^d$ defined by

$$(\varphi^* g_{euc})(v,w) = g_{euc}(\varphi_* v,\varphi_* w) = \sum_{h=1}^{N}\varphi_k(v)\overline{\varphi_k(w)},\ \ v,w\in E\otimes L^d.$$

We now define a metric g_d on M by $g_d = \frac{1}{d}\varphi^* g_{Gr}$. By functoriality,

$$g_d = \frac{1}{d}\varphi^*(-\operatorname{Ric}(g_{euc})) = -\frac{1}{d}\operatorname{Ric}(\varphi^* g_{euc}).$$

If e_i, $i = 1, \ldots, p$ and e are local frames for E and L respectively, then $e_i \otimes e^{\otimes d}$, $i = 1, \ldots, p$, is a local frame for $E \otimes L^d$, so that if we set $M_{ij}(z) = \sum_{k=1}^{N} \varphi_k(e_i \otimes e^{\otimes d})\overline{\varphi_k(e_j \otimes e^{\otimes d})}$, then $g_d = \frac{1}{2\pi d}\partial\bar{\partial}\log\det[M_{ij}]$.

We show how Theorem 3 can be used to compute M_{ij}. Given the metric G on E, we obtain a metric G' on E^* which can be pulled back to Ω to give a metric $\widetilde{G}'$ on $\widetilde{E^*} = \pi^* E^*$. Since $H^0(M, (E \otimes L^d)^*)$ is isometrically equal to $H^0(M, E^* \otimes L^{*d})$, we conclude as in the proof of Theorem 3 that if Φ_k is the element of $\mathcal{A}_d(\Omega, \widetilde{E^*})$ associated with φ_k, then $\Phi_1, \ldots, \Phi_N$ is an orthonormal basis of $\mathcal{A}_d(\Omega, \widetilde{E^*})$ with respect to the norm $\frac{2\pi}{d+2}\int_\Omega |\Phi|^2 \mathrm{vol}_{g_0}$. Let $B(\xi, \theta) = \sum_{k=1}^{N} \Phi_k(\xi) \otimes \overline{\Phi_k(\theta)}$. We now compute an expression for the projection P_d onto $\mathcal{A}_d(\Omega, \widetilde{E^*})$. If $f \in L^2(\Omega, \widetilde{E^*})$, then

$$
\begin{aligned}
\int_\Omega (f(\theta), \overline{B(\xi,\theta)})\mathrm{vol}_{g_0}(\theta) &= \int_\Omega \sum_{k=1}^{N} (f(\theta), \overline{\Phi_k(\xi)} \otimes \Phi_k(\theta))\mathrm{vol}_{g_0}(\theta) \\
&= \int_\Omega \sum_{k=1}^{N} (f(\theta), \Phi_k(\theta))\Phi_k(\xi)\mathrm{vol}_{g_0}(\theta) \qquad (3.1) \\
&= (P_d f)(\xi) = \int_\Omega K_d(\xi,\theta) f(\theta)\mathrm{vol}_{g_0}(\theta).
\end{aligned}
$$

Now let $\eta_1, \ldots, \eta_p$ be the local frame of E^* in W that is dual to $e_1, \ldots, e_p$ and let $\tilde{\eta}_1, \ldots, \tilde{\eta}_p$ denote the pullbacks. We write $B = \sum_{\mu,\nu=1}^{p} B_{\mu,\nu}(\xi,\theta)\tilde{\eta}_\mu(\xi) \otimes \overline{\tilde{\eta}_\nu(\theta)}$, $f = \sum_{\ell=1}^{p} f_\ell(\theta)\tilde{\eta}_\ell(\theta)$, and $G_{jk} = (\tilde{e}_j, \tilde{e}_k)$, so that $G^{-1}_{\nu,\mu} = (\tilde{\eta}_\mu, \tilde{\eta}_\nu)$. Let f_W denote the column vector with entries f_ℓ, $\ell = 1, \ldots, p$, and similarly define matrices B_W and G_W^{-1}. A computation shows that

$$
\int_\Omega (f(\theta), \overline{B(\xi,\theta)})\mathrm{vol}_{g_0}(\theta) = \int_\Omega B_W(\xi,\theta)G_W^{-1}(\theta)f_W(\theta)\mathrm{vol}_{g_0}(\theta). \qquad (3.2)
$$

Since $K_d(\xi,\theta) \in \mathrm{Hom}\,((\widetilde{E^*})_\theta, (\widetilde{E^*})_\xi)$, we can define a matrix $K_{W,d}(\xi,\theta)$ representing $K_d(\xi,\theta)$ with respect to the bases $\tilde{\eta}_1(\theta), \ldots, \tilde{\eta}_p(\theta)$ and $\tilde{\eta}_1(\xi), \ldots, \tilde{\eta}_p(\xi)$. We conclude from (3.1) and (3.2) that

$$
K_{W,d}(\xi,\theta) = B_W(\xi,\theta)G_W^{-1}(\theta). \qquad (3.3)
$$

Note that the matrix $G_W^{-1}(\theta)$ is actually constant on the fibers of L since the metrics $\tilde{G}$ and $\tilde{G}'$ are pullbacks. If we set $\theta = \xi$ and apply Theorem 3, we conclude from (3.3) that there are $p \times p$ matrices $M_1(z)$, $M_2(z), \dots,$ where $z = \pi(\xi)$, so that

$$B_W(\xi, \xi) \sim d^n R(\xi, \xi)^d \lambda(z) \left[G_W(z) + \sum_{\ell=1}^{\infty} d^{-\ell} M_\ell(z) \right], \tag{3.4}$$

and where $\lambda(z) = \prod_{k=1}^{n} |\lambda_k(z)|$.

We claim that $B_W(e, e) = [M_{ij}]$. To see this, note that $\Phi_k(e) = \sum_{i=1}^{p} \varphi_k(e_i \otimes e^{\otimes d}) \tilde{\eta}_i$. It follows that

$$B_W(e, e) = \sum_{k=1}^{N} \sum_{i,j=1}^{p} \varphi_k(e_i \otimes e^{\otimes d}) \overline{\varphi_k(e_j \otimes e^{\otimes d})}\ \tilde{\eta}_i \otimes \overline{\tilde{\eta}_j}$$

which implies that $B_W(e, e) = [M_{ij}]$. Setting $\xi = e(z)$ in (3.4) and $R_1 = R(e, e)$, we obtain

$$[M_{ij}] \sim d^n R_1^d \lambda \left(G_W + \sum_{\ell=1}^{\infty} d^{-\ell} M_\ell \right).$$

Applying $\frac{1}{2\pi d}\, \partial\bar{\partial} \log$, it follows that

$$g_d = \frac{p}{2\pi}\, \partial\bar{\partial} \log R_1 + \frac{1}{d} \left(\frac{p}{2\pi}\, \partial\bar{\partial} \log \lambda + \frac{1}{2\pi}\, \partial\bar{\partial} \log \det G_W \right) + \sum_{\ell=2}^{\infty} d^{-\ell} m_\ell \tag{3.5}$$

where m_ℓ are smooth $(1,1)$–forms in W. Hence (0.10) follows from the fact that $\mathrm{Ric}(R) = -\frac{1}{2\pi}\, \partial\bar{\partial} \log R_1$.

Remark. As noted by Zelditch [5], the coefficient m_1 in (0.10) can also be easily computed. In fact (3.5) implies that $m_1 = -p\, \mathrm{Ric}(-\, \mathrm{Ric} R) - \mathrm{Ric}\,(G)$.

References

[1] L. Boutet de Monvel, J. Sjöstrand, *Sur la singularité des noyaux de Bergman et de Szego,* Journées: Equations aux Dérivées Partielles de Rennes 1975, Astérisque 34–35.

[2] C. Fefferman, *The Bergman kernel and biholomorphic mappings of pseudoconvex domains,* Inventiones Math. **26** (1974) 1–66.

[3] A. Melin, J. Sjöstrand, *Fourier integral operators with complex valued phase functions, Fourier Differential Equations,* Lecture Notes, Springer No. 459, 120–223.

[4] G. Tian, *On a set of polarized Kaehler metrics on algebraic manifolds,* J. Differential Geom. **32** (1990) 99–130.

[5] S. Zelditch, *Szego kernels and a theorem of Tian*, International Math Res. Notices 1998, No.5.

Department of Mathematics
Purdue University
West Lafayette, IN 47907, USA

CHAPTER II

Some Involutive Structures in Analysis and Geometry

Michael Eastwood†

An *involutive structure* on a smooth manifold is a complex subbundle $T^{0,1}$ of the complexified tangent bundle, closed under Lie bracket: $[T^{0,1}, T^{0,1}] \subseteq T^{0,1}$. For the general theory see [6, 16, 25]. Here are some examples.

Foliations: Here $T^{0,1}$ is real: $T^{0,1} = \overline{T^{0,1}}$. Then $T^{0,1}$ is the complexification of a real Frobenius integrable distribution.

Complex structures: Here $T^{0,1} \cap \overline{T^{0,1}} = 0$ and together these bundles span the complexified tangent bundle.

CR structures: Here $T^{0,1} \cap \overline{T^{0,1}} = 0$, but $T^{0,1} \oplus \overline{T^{0,1}}$ may be a strict subbundle of the complexified tangent bundle. It is of codimension one when the CR structure is of hypersurface type.

Lewy: Let $T^{0,1}$ be spanned by the complex vector field

$$L = \frac{\partial}{\partial \bar{z}} + iz\frac{\partial}{\partial t} \qquad \text{where} \qquad \frac{\partial}{\partial \bar{z}} = \frac{1}{2}\left(\frac{\partial}{\partial x} + i\frac{\partial}{\partial y}\right)$$

on $\mathbb{R}^3$. This may also be viewed as defining a CR structure. Lewy showed [21] that there are smooth u for which the equation $Lf = u$ has no local solutions.

Mizohata: Let $T^{0,1}$ be spanned by

$$M = \frac{\partial}{\partial x} + ix\frac{\partial}{\partial y}$$

on $\mathbb{R}^2$. Here, $T^{0,1} \cap \overline{T^{0,1}}$ is not of constant rank. It is spanned by $\partial/\partial x$

†ARC Senior Research Fellow, University of Adelaide.

along the y-axis and is otherwise 0. Defining $\mathbb{R}^2 \to \mathbb{C}$ by $(x,y) \mapsto \zeta = x^2 + 2iy$, we find that $M = 4x\partial/\partial\overline{\zeta}$ when $x \neq 0$. Hence, off the y-axis, M is the Cauchy–Riemann operator in disguise. This geometric interpretation was used by Garabedian [14] to give a very simple example of an unsovable partial differential equation, as follows. Suppose $Mf = i|xy|$ in a neighbourhood of the origin. Then M annihilates $f(x,y) - f(-x,y) - y|y|$ for $x > 0$. As a function of ζ, this is holomorphic to the right of the imaginary axis with continuous real boundary values along this axis. However, the Schwarz reflection principle implies that such boundary values should be real-analytic.

Blowing-up: We shall now consider in detail, $T^{0,1}$ spanned by

$$\frac{\partial}{\partial \overline{z}} \quad \text{and} \quad \frac{\partial}{\partial t} - z\frac{\partial}{\partial s} \tag{1}$$

on $\mathbb{R}^4 = \mathbb{C} \times \mathbb{R}^2$ with coordinates $(x,y,s,t) = (z,s,t)$. These vector fields commute so the formal integrability condition $[T^{0,1}, T^{0,1}] \subseteq T^{0,1}$ is clear. Like the Mizohata example, $T^{0,1} \cap \overline{T^{0,1}} = 0$ except on a hypersurface Σ. In this case $\Sigma = \{y = 0\}$. Along this hypersurface $T^{0,1} \cap \overline{T^{0,1}}$ is spanned by $\partial/\partial t - x\partial/\partial s$. There is a similar geometric interpretation of this structure. Define $\eta : \mathbb{C} \times \mathbb{R}^2 \to \mathbb{C}^2$ by $(z,s,t) \overset{\eta}{\mapsto} (z,w) = (z, s+zt)$. This is a diffeomorphism away from Σ and

$$\frac{\partial}{\partial t} - z\frac{\partial}{\partial s} = -2iy\frac{\partial}{\partial \overline{w}}$$

there. This involutive structure, therefore, is obtained by 'pulling back' the usual complex structure on $\mathbb{C}^2$. Though η drops rank along Σ, the complex structure extends smoothly across but only as an involutive structure.

A *solution* of an involutive structure is a function or distribution which is annihilated by all smooth sections of $T^{0,1}$.

Theorem *A local smooth solution of* (1) *near the origin has the form*

$$h(z, s+zt), \quad \text{for } h(z,w) \textit{holomorphic.}$$

Sketch of proof: For details of the following argument, see [12]. The first of the differential operators says that $f(z,s,t)$ is holomorphic in z. It may, therefore, be expanded as a power series

$$f(z,s,t) = \sum_{k=0}^{\infty} a_k(s,t)z^k$$

for smooth functions $a_k(s,t)$ defined for s and t sufficiently small. Moreover, there are Cauchy estimates for these coefficients, uniform in s and t. The second differential operator implies

$$\frac{\partial a_0(s,t)}{\partial t} = 0 \qquad \text{and} \qquad \frac{\partial a_k(s,t)}{\partial t} = \frac{\partial a_{k-1}(s,t)}{\partial s} \quad \text{for } k \geq 1.$$

It follows that $a_0(s,t)$ is a function, say $b_0(s)$, of s alone. Then

$$\frac{\partial a_1(s,t)}{\partial t} = \frac{\partial a_0(s,t)}{\partial s} = b_0'(s) \quad \Longrightarrow \quad a_1(s,t) = b_1(s) + b_0(s)t,$$

for some smooth function $b_1(s)$. More generally, $a_k(s,t)$ is a polynomial in t:

$$a_k(s,t) = b_k(s) + b_{k-1}'(s)t + \tfrac{1}{2}b_{k-2}'(s)t^2 + \cdots + \tfrac{1}{k!}b_0^{(k)}(s)t^k.$$

We have bounds (the Cauchy estimates) for these polynomials and Lagrange interpolation with $k+1$ equally spaced points enables one to establish bounds on their coefficients, namely the functions $b_k(s)$ and their derivatives. These bounds are sufficient to prove that $f(z,s,t)$ is real-analytic. It therefore extends as a holomorphic function of (z,s,t) as *complex* variables. Now, the second differential operator of (1) may be viewed as a holomorphic vector field on this space of complex variables. The conclusion of the theorem is immediate from this interpretation. □

In fact, it is shown in [12] that f need only be a *distribution* solution of (1) in order to draw the same conclusion.

The *blow-up* of $\mathbb{C}^2$ along $\mathbb{R}^2$ is given by

$$\{((z,w),[a,b]) = ((x+iy, u+iv),[a,b]) \in \mathbb{C}^2 \times \mathbb{RP}_1 \text{ s.t. } ay + bv = 0\}$$

equipped with the obvious projection mapping to $\mathbb{C}^2$. (This is the blow-up of $\mathbb{C}^2$ as a *real* manifold along a real submanifold.) Locally, we can change coordinates

$$(z,s,t) \mapsto ((z, s+zt), [-t,1])$$

to identify this projection with η. The *exceptional variety* $\eta^{-1}(\mathbb{R}^2)$ is Σ with $T^{0,1} \cap \overline{T^{0,1}}$ tangent to the fibres. In any case, it is easy to see that the image of a neighbourhood of $(0,0,0)$ under η is a *double wedge*:

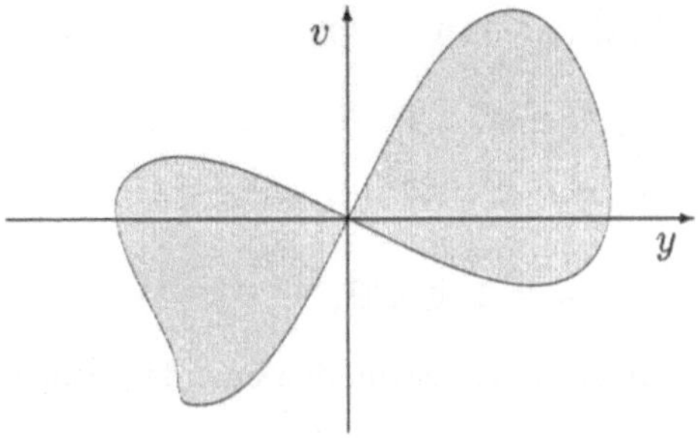

A view of the imaginary plane for fixed (x,u) near the origin in $\mathbb{R}^2$.

The classical edge-of-the-wedge theorem (see [27] for a review) says that a holomorphic function h defined in the interior of such a double wedge and continuous[♯] across the real boundary (the edge), extends as a holomorphic function to a neighbourhood of this boundary in $\mathbb{C}^2$. Indeed, more generally, one can take as edge any totally real submanifold of $\mathbb{C}^n$ and the real blow-up is already useful as giving a manifestly coordinate-free definition of a double wedge. If the edge is real-analytic then one can make a local change of coordinates so that it becomes $\mathbb{R}^n \hookrightarrow \mathbb{C}^n$. However, the edge-of-the-wedge theorem is still true along smooth or even C^1 edges [24].

In the real-analytic case, solving (1) quickly leads to a proof of the edge-of-the-wedge theorem—it only remains to show that $h \circ \eta$ solves (1) in the sense of distributions. The details are in [12] (given in arbitrary dimensions). It is worth noting that, on the blown-up space, the given function satisfies a system of partial differential equations outside a hypersurface Σ and the conclusion is that the function continues to satisfy these equations across the hypersurface. From this point of view, the edge-of-the-wedge theorem more resembles its one-dimensional specialisation (due to Painlevé) which says that a holomorphic function of one complex variable, defined on either side of the real axis and having a continuous extension across this axis, is actually holomorphic there.

The key point concerning (1) is that every distribution solution is a holomorphic function of a basic set of solutions, in this case z and w. In general, such an involutive structure is said to be *hypocomplex*. Thus, hypocomplexity of the involutive structure on the blow-up of $\mathbb{C}^n$ along $\mathbb{R}^n$ is essentially equivalent to the edge-of-the-wedge theorem. As remarked by F. Treves, hypocomplexity of this particular involutive structure may, alternatively, be deduced from microlocal criteria as given, for example, in [4]. C.R. Graham [15] has checked these criteria for the case of a smooth edge too.

Joint work with Graham [13] embarks on a similar programme for CR manifolds. If M is a $(2n+1)$-dimensional CR manifold of hypersurface type and $X \subset M$ is an $(n+1)$-dimensional totally real submanifold, then the CR structure on $M \setminus X$ extends as an involutive structure to the blow-up of M along X. Consider the standard indefinite hypersurface in $\mathbb{C}^3$

$$M = \{(z_1, z_2, w) = (x_1 + iy_1, x_2 + iy_2, u + iv) \text{ s.t. } v = x_2 y_1 - x_1 y_2\}$$

and $X = \mathbb{R}^3$. In an affine coordinate chart

$$\mathbb{C}^3 \ni (z_1, z_2, w) \longmapsto [z_1, z_2, w, 1] \in \mathbb{CP}_3$$

[♯]It is sufficient to assume existence and agreement of distribution boundary values from the two wedges.

this is the indefinite hyperquadric

$$\{[Z_1, Z_2, Z_3, Z_4] \in \mathbb{CP}_3 \text{ s.t. } \operatorname{Im}(Z_1\overline{Z}_2) = \operatorname{Im}(Z_3\overline{Z}_4)\} \tag{2}$$

with $\mathbb{RP}_3$ as totally real submanifold. Microlocal criteria [5] show that the induced involutive structure is hypocomplex. The corresponding edge-of-the-wedge theorem for CR functions follows:

Theorem *Let M be the standard indefinite hypersurface in $\mathbb{C}^3$, as above, and suppose W is a double wedge in M with edge in $\mathbb{R}^3$. Then every CR function defined on W and continuous across the edge extends as a CR function to a neighbourhood of the edge.*

This theorem may also be proved by standard techniques of complex analysis:

Sketch of proof: Firstly, use Lewy extension [17, 21] to obtain an ambient holomorphic extension from W. It may be given by the Cauchy integral formula on an analytic family of discs from which it is easy to obtain an estimate of how far this extension goes. Specifically, if a small ball of radius r, centred on $w \in W \setminus \mathbb{R}^3$, intersects M within $W \setminus \mathbb{R}^3$, then it is possible to extend to distance on the order of r^2 from w. Were we able to extend to distance r, then we would obtain a double wedge in $\mathbb{C}^3$ and the result would follow from the classical edge-of-the-wedge theorem. The rest of the proof is concerned with achieving this further extension. All normal directions to $\mathbb{R}^3$ in M are on an equal footing, so we may as well suppose that W contains a small cone pointing in the y_1-direction:

$$C \equiv \{z \in M \text{ s.t. } \|z\| < \epsilon \text{ and } |y_2| < \epsilon y_1\}$$

for some $\epsilon > 0$. For small $\alpha, \tau \in \mathbb{R}$, consider the embedding

$$\begin{array}{ccc} \mathbb{C}^2 & \hookrightarrow & \mathbb{C}^3 \\ \text{\rotatebox{90}{$\in$}} & & \text{\rotatebox{90}{$\in$}} \\ (\zeta_1, \zeta_2) & \longmapsto & (0, \alpha, 0) + \zeta_1(1, \tau, \alpha) + \zeta_2(0, 0, 1) \\ \| & & \| \\ (\xi_1 + i\eta_1, \xi_2 + i\eta_2) & \longmapsto & (\xi_1, \alpha + \tau\xi_1, \alpha\xi_1 + \xi_2) + i(\eta_1, \tau\eta_1, \alpha\eta_1 + \eta_2). \end{array}$$

On the image Π of this embedding,

$$v - x_2 y_1 + x_1 y_2 = \alpha\eta_1 + \eta_2 - (\alpha + \tau\xi_1)\eta_1 + \xi_1\tau\eta_1 = \eta_2,$$

so $\Pi \cap M$ is given by $\eta_2 = 0$. Thus, for fixed ξ_1 and ξ_2, the η_1-axis lies in M and approaches the plane $\{y_2 = 0\}$ as $\tau \to 0$, while the η_2-axis is nearly perpendicular to M. Thus, by taking τ sufficiently small, we can force the η_1-axis to start off well within the cone C. The r^2-estimate for Lewy extension gives a holomorphic function on the truncated tube over a 'parabolic spike' within Π:

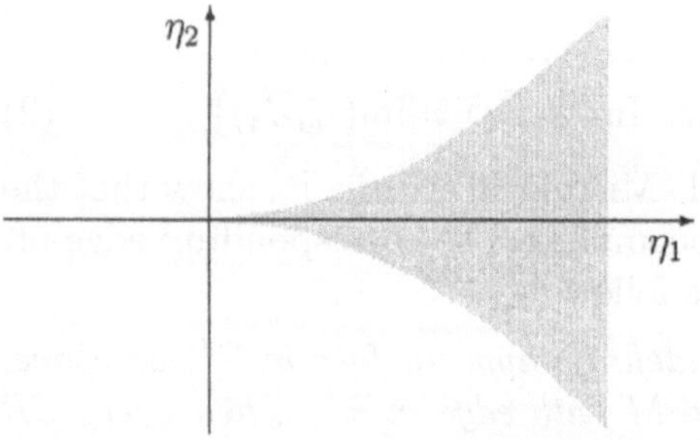

For (ξ_1, ξ_2) small.

Were this a full tube over this base, the Bochner tube theorem would give a holomorphic extension to the tube over the convex hull of the base. In [17], Hörmander proved Bochner's tube theorem by a geometrical arrangement dubbed a 'folding screen' by Komatsu [19]. A careful inspection of this proof reveals that for a truncated tube, such as we have, one can still extend to a more severely truncated tube over a small cone:

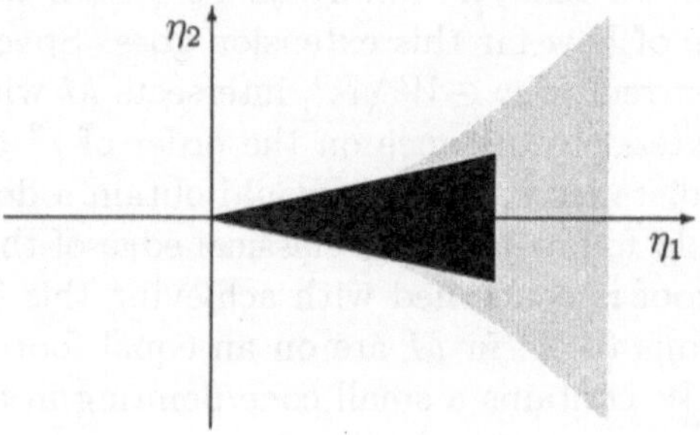

Furthermore, this extension is achieved by pushing analytic discs so continuity up to the edge is preserved. A similar extension for the opposing spike gives a pair of opposing cones. Finally, by allowing α and τ to vary near 0, we sweep out an ambient double wedge and the classical edge-of-the-wedge theorem finishes the proof. □

Such a straightforward generalisation to CR manifolds is not generally valid even for embedded hypersurfaces of indefinite signature. Additional restrictions must be placed on the double wedges including that they contain a null vector for the Levi form. For example, if f is a holomorphic function depending only on the variable w, defined on the half-plane $\{v > 0\}$ and continuous up to the boundary but with no local holomorphic extension through the origin, then restricting f to

$$M = \{(z_1, z_2, w) = (x_1 + iy_1, x_2 + iy_2, u + iv) \text{ s.t. } v = {y_1}^2 - {y_2}^2\}$$

gives a CR function defined on a double wedge but with no CR extension across the edge, $\mathbb{R}^3$. To obtain a positive result in this case, double wedges at the origin should contain a null direction $y_1 = \pm y_2$. Proofs may be effected either by microlocal methods or by folding screen techniques. Here is a typical example from [13]:

Theorem *Suppose M is a locally embeddable CR manifold of hypersurface type whose Levi form everywhere has at least one positive and one negative eigenvalue. Let $X \subset M$ be a totally real submanifold of maximal dimension and W a double wedge in M with edge X. We may identify the normal bundle to X in M with the subspace $J(TX \cap H)$ of H where $H \subset TM$ is the holomorphic distribution and $J \in \mathrm{End}(H)$ is its complex structure. Having done so, suppose that W contains a smooth family of vectors emanating from X which are null for the Levi form. Then every CR function defined on W and continuous across X extends as a CR function to a neighbourhood of X.*

Compact examples

Compact manifolds with an involutive structure are especially interesting (see [16]). A particular example arises geometrically as follows. Let

$$F = \{(L, P) \in \mathbb{CP}_n \times \mathrm{Gr}_2(\mathbb{R}^{n+1}) \text{ s.t. } L \subset \mathbb{C}P\}$$

where $\mathbb{C}P$ denotes the two-dimensional complex linear subspace of $\mathbb{C}^{n+1}$ generated by the two-dimensional real linear subspace P of $\mathbb{R}^{n+1}$. There are forgetful maps η and τ

$$\begin{array}{ccc} & F & \\ {}^{\eta}\swarrow & & \searrow^{\tau} \\ \mathbb{CP}_n & & \mathrm{Gr}_2(\mathbb{R}^{n+1}) \end{array}$$

and it is easy to verify that $\eta : F \to \mathbb{CP}_n$ is the blow-up of $\mathbb{CP}_n$ along $\mathbb{RP}_n$. It acquires, therefore, an involutive structure extending the holomorphic structure on $\mathbb{CP}_n \setminus \mathbb{RP}_n$.

An involutive structure is precisely what is needed to define a sequence of $\bar{d}$-operators in the usual way and, even though the first local cohomology of the involutive structure on F is infinite-dimensional [12, Theorem 2.6], the global cohomology is quite tractable.

Theorem *For any holomorphic vector bundle V on $\mathbb{CP}_n$, there is a canon-*

ical exact sequence

$$0 \to \Gamma(\mathbb{CP}_n, \mathcal{O}(V)) \to \Gamma(\mathbb{RP}_n, \mathcal{E}(V)) \to H^1_{\overline{d}}(F, \widetilde{V}) \to H^1(\mathbb{CP}_n, \mathcal{O}(V)) \to 0$$

$$\begin{array}{cc} \| & \| \\ \textit{holomorphic sections} & \textit{smooth sections} \end{array}$$

where $H^1_{\overline{d}}(F, \widetilde{V})$ is the first involutive cohomology of the pull-back of V to F twisted by the divisor bundle of $\Sigma = \eta^{-1}(\mathbb{RP}_n)$.

Sketch of Proof: The proof (developed jointly with T.N. Bailey and Graham (see [1, 10])) is on the level of formal power series invoking Borel's theorem to realise the necessary smooth functions. For example, the connecting homomorphism $\Gamma(\mathbb{RP}_n, \mathcal{E}(V)) \to H^1_{\overline{d}}(F, \widetilde{V})$ is constructed as follows. The bundle V is just a passenger and may be safely ignored. Now, for f a smooth function on $\mathbb{RP}_n$ choose an 'almost analytic extension' $\hat{f}$ to $\mathbb{CP}_n$. This is a smooth function such that $\overline{\partial}\hat{f}$ vanishes to infinite order along $\mathbb{RP}_n$. The existence of such an extension is a simple application of Borel's theorem as indicated. (If f were real-analytic, then we could take $\hat{f}$ to be its holomorphic extension near $\mathbb{RP}_n$ and $\overline{\partial}\hat{f}$ would vanish in an entire neighbourhood.) The divisor bundle of Σ comes equipped with a canonical 'Heaviside' section H, smooth except along $\mathbb{RP}_n$ where it enjoys a jump discontinuity. Then $H\eta^*\overline{\partial}\hat{f}$ is smooth on F and annihilated by $\overline{d}$. It represents the involutive cohomology class which is the image of f. □

For homogeneous V, the spaces $\Gamma(\mathbb{CP}_n, \mathcal{O}(V))$ and $H^1(\mathbb{CP}_n, \mathcal{O}(V))$ are explicitly computable. In any case, they are finite-dimensional, so the involutive cohomology $H^1_{\overline{d}}(F, \widetilde{V})$ is not much different from the smooth sections of V over $\mathbb{RP}_n$.

It is also possible to investigate the involutive cohomology on F via the mapping τ. Notice that this mapping is simply a fibration with $\mathbb{CP}_1$ fibres. Moreover, the involutive structure on F induces the usual complex structure on each fibre. In particular, the Dolbeault cohomology spaces along these fibres are finite-dimensional. Thus, we may take $\omega \in H^1_{\overline{d}}(F, \widetilde{V})$ and consider $\omega|_{\tau^{-1}(x)}$ as a smooth section of an appropriate vector bundle on $\mathrm{Gr}_2(\mathbb{R}^{n+1})$. In this way, the spaces $H^1_{\overline{d}}(F, \widetilde{V})$ act as intermediaries between spaces of smooth sections of vector bundles over $\mathbb{RP}_n$ and $\mathrm{Gr}_2(\mathbb{R}^{n+1})$, respectively. With $H^1_{\overline{d}}(F, \widetilde{V})$ removed, there results a Radon-like integral transform with $\mathrm{Gr}_2(\mathbb{R}^{n+1})$ regarded as the space of geodesics in $\mathbb{RP}_n$. The advantage of passing through $H^1_{\overline{d}}(F, \widetilde{V})$, however, is that it may be interpreted *precisely* down on $\mathrm{Gr}_2(\mathbb{R}^{n+1})$. (The machinery for doing this is known as the *Penrose transform.*) This allows the kernels and ranges of these integral transforms to be explicitly identified.

For example, if $n = 2$ and we take $V = \mathcal{O}(-2)$, then we may deduce that the *Funk transform* from smooth functions on $\mathbb{RP}_2$ to smooth functions on $\mathbb{RP}_2^*$ is an isomorphism (see [2]). The case $n = 3$ and $V = \mathcal{O}(-2)$ is explained in [10]. The conclusion is that the range of the X-ray transform is smooth solutions of the *ultrahyperbolic wave equation*, a result essentially due to John [18]. By taking V to be the holomorphic cotangent bundle on $\mathbb{CP}_n$, it follows that a smooth 1-form on $\mathbb{RP}_n$ whose integral over every geodesic vanishes is exact. This result is due to Michel [23]. It is generalised considerably in [1]. In order to explain this generalisation it is necessary to discuss the invariant linear differential operators on $\mathbb{RP}_n$. Following [7], let $(a|b, c, \ldots, d, e)$ denote the vector bundle on $\mathbb{RP}_n$ whose fibre over the line L is

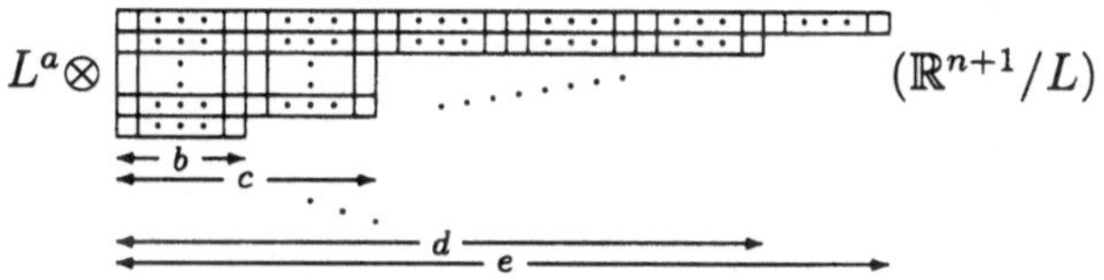

for integers $a, b, c, \ldots, d, e$ with $b \leq c \leq \cdots \leq d \leq e$. (If b is negative, then the Young diagram aquires virtual boxes. Equivalently, a suitable power of $\det(\mathbb{R}^{n+1}/L)$ is included.) The deRham sequence on $\mathbb{RP}_n$

$$\begin{array}{ccccccccc} 0 \to \mathbb{R} \to & \Lambda^0 & \xrightarrow{d} & \Lambda^1 & \xrightarrow{d} & \Lambda^2 & \xrightarrow{d} \cdots \quad \cdots \xrightarrow{d} & \Lambda^n & \to 0 \\ & \| & & \| & & \| & & \| & \\ & (0|0,\cdots,0) & & (1|-1,0,\cdots,0) & & (2|-1,-1,0,\cdots,0) & & (n|-1,\cdots,-1) & \end{array}$$

consists of invariant differential operators. This means that they intertwine the action of $\mathrm{GL}(n+1, \mathbb{R})$ on $\mathbb{RP}_n$ and these bundles. One can take any finite-dimensional representation of $\mathrm{GL}(n+1, \mathbb{R})$ and tensor the deRham sequence with the corresponding trivial bundle over $\mathbb{RP}_n$. The result is a resolution of this representation by invariant differential operators. For example,

$$\begin{aligned} (-1, 0, \ldots, 0) \xrightarrow{d} (1|-1, 0, \ldots, 0) \otimes (-1, 0, \ldots, 0) \\ \xrightarrow{d} (2|-1, -1, 0, \ldots, 0) \otimes (-1, 0, \ldots, 0) \end{aligned}$$

are the first three terms in such a resolution. The bundle $(-1, 0, \ldots, 0)$ is decomposable—there is an exact sequence

$$0 \to (0|-1, 0, \ldots, 0) \to (-1, 0, \ldots, 0) \to (-1|0, \ldots, 0) \to 0.$$

This induces exact sequences

$$0\to\begin{array}{c}(1|-2,0,0,\cdots,0)\\ \oplus\\ (1|-1,-1,0,\cdots,0)\end{array}\to(1|-1,0,\ldots,0)\otimes(-1,0,\ldots,0)$$
$$\to(0|-1,0,\cdots,0)\to 0$$

and

$$0\to\begin{array}{c}(2|-2,-1,0,0,\cdots,0)\\ \oplus\\ (2|-1,-1,-1,0,\cdots,0)\end{array}\to(2|-1,-1,0,\ldots,0)\otimes(-1,0,\ldots,0)$$
$$\to(1|-1,-1,0,\cdots,0)\to 0.$$

It turns out that the induced mappings

$$\begin{array}{c}(0|-1,0\ldots,0)\to(0|-1,0,\ldots,0)\\ \text{and}\\ (1|-1,-1,0,\ldots,0)\to(1|-1,-1,0,\ldots,0)\end{array}\tag{3}$$

are simply the identity and an elementary diagram chase gives

$$(-1|0,\ldots,0)\xrightarrow{\nabla^2}(1|-2,0,\ldots,0)\xrightarrow{\nabla}(2|-2,-1,0,\ldots,0)\tag{4}$$

as the first three terms of a replacement resolution. This is more familiar in its dual form, the *Bernstein–Gelfand–Gelfand* resolution. That (3) are identities is automatic in the dual formulation by looking at central character. A further discussion may be found in [9]. The sequence (4) may be written out explicitly in terms of the flat connection ∇ in any affine coordinate patch:

$$h\longmapsto\nabla_a\nabla_b h\quad\text{and}\quad\psi_{ab}\longmapsto\nabla_a\psi_{bc}-\nabla_b\psi_{ac}.$$

A further iteration of this construction gives

$$(-1|-1,0,\ldots,0)\xrightarrow{\nabla}(0|-2,0,\ldots,0)\xrightarrow{\nabla^2}(2|-2,-2,0,\ldots,0)$$

or, explicitly,

$$\phi_a\mapsto\nabla_a\phi_b+\nabla_b\phi_a\quad\text{and}\quad\theta_{ab}\mapsto\nabla_a\nabla_c\theta_{bd}-\nabla_b\nabla_c\theta_{ad}+\nabla_b\nabla_d\theta_{ac}-\nabla_a\nabla_d\theta_{bc}.$$

This particular resolution was also constructed by Calabi [5]. Just as the deRham sequence implies that a smooth one-form on $\mathbb{RP}_n$ is closed if and only if it is exact, so these two Bernstein–Gelfand–Gelfand resolutions imply that

$$\nabla_a \psi_{bc} = \nabla_b \psi_{ac} \Longleftrightarrow \psi_{ab} = \nabla_a \nabla_b h, \text{ for some } h$$

and

$$\nabla_a \nabla_c \theta_{bd} + \nabla_b \nabla_d \theta_{ac} = \nabla_b \nabla_c \theta_{ad} + \nabla_a \nabla_d \theta_{bc} \Longleftrightarrow$$
$$\theta_{ab} = \nabla_a \phi_b + \nabla_b \phi_a, \text{ for some } \phi_a$$

where ψ_{ab} and θ_{ab} are symmetric forms on $\mathbb{RP}_n$. In [1], these conditions are shown to be further equivalent to ψ_{ab} or θ_{ab} having *zero energy*:

$$\oint_{u \in \gamma} [X^a(u) X^b(u) \psi_{ab}(u)]\, u = 0$$

or

$$\oint_{u \in \gamma} X^a(u) X^b(u) \theta_{ab}(u) = 0$$

for all geodesics γ, where $u \in \mathbb{R}^{n+1}$ is the unit vector representing a point on γ and $X^a(u)$ is the unit tangent vector at u along γ. The second of these is well-known as the condition that an infinitesimal deformation of $\mathbb{RP}_n$ have closed geodesics all of the same length and the conclusion that θ_{ab} may be written as $\nabla_a \phi_b + \nabla_b \phi_a$ is the infinitesimal Blaschke rigidity of $\mathbb{RP}_n$ first proved by Michel [22]. The corresponding statement for $\mathbb{CP}_n$ (first proved by Tsukamoto [26]) is deduced in [11] by using the Penrose transform.

In fact, the Penrose transform fits very well into the global theory of involutive structures. Joint work with Bailey and M.A. Singer [3] develops the Penrose transform from this point of view. The classical Penrose transform (see, for example, [8]) concerns the mapping

$$\tau : \mathbb{CP}_3 \longrightarrow S^4 = \mathbb{HP}_1$$

sending a complex line in $\mathbb{C}^4 = \mathbb{H}^2$ to its quaterionic span. Restricting this to the standard indefinite hyperquadric (2) gives $\tau : Q \to S^3$. Explicitly,

$$[Z_1, Z_2, Z_3, Z_4] \mapsto \begin{pmatrix} \frac{Z_1\overline{Z}_3 + Z_3\overline{Z}_1 + Z_4\overline{Z}_2 + Z_2\overline{Z}_4}{Z_1\overline{Z}_1 + Z_2\overline{Z}_2 + Z_3\overline{Z}_3 + Z_4\overline{Z}_4} \\ \frac{iZ_1\overline{Z}_3 - iZ_3\overline{Z}_1 - iZ_4\overline{Z}_2 + iZ_2\overline{Z}_4}{Z_1\overline{Z}_1 + Z_2\overline{Z}_2 + Z_3\overline{Z}_3 + Z_4\overline{Z}_4} \\ \frac{Z_1\overline{Z}_2 + Z_2\overline{Z}_1 - Z_4\overline{Z}_3 - Z_3\overline{Z}_4}{Z_1\overline{Z}_1 + Z_2\overline{Z}_2 + Z_3\overline{Z}_3 + Z_4\overline{Z}_4} \\ \frac{Z_1\overline{Z}_1 + Z_4\overline{Z}_4 - Z_2\overline{Z}_2 - Z_3\overline{Z}_3}{Z_1\overline{Z}_1 + Z_2\overline{Z}_2 + Z_3\overline{Z}_3 + Z_4\overline{Z}_4} \end{pmatrix}.$$

The Penrose transform may be applied to this fibration. Even though the local $\overline{\partial}_b$-cohomology on Q is quite awkward, the global cohomology is tractable. The machinery of [3] gives, for example,

$$H^1_{\overline{\partial}_b}(Q) \cong \{\text{smooth two-forms on } S^3\}.$$

Again, the point is that the fibres of τ are $\mathbb{CP}_1$'s with complex structure obtained by restricting the CR structure on Q.

References

[1] T.N. Bailey and M.G. Eastwood, *Zero-energy fields on real projective space*, Geom. Dedicata, **67**(1997), 245–258.

[2] T.N. Bailey, M.G. Eastwood, A.R. Gover, and L.J. Mason, *The Funk transform as a Penrose transform*, Math. Proc. Camb. Phil. Soc., **125**(1999), 67–81.

[3] T.N. Bailey, M.G. Eastwood, and M.A. Singer, *The Penrose Transform for Non-holomorphic Correspondences*, in preparation.

[4] M.S. Baouendi and F. Treves, *A microlocal version of Bochner's tube theorem*, Indiana Math. Jour. **31** (1982), 885—895.

[5] C.H. Chang, *Hypoanalyticity with vanishing Levi form*, Bull. Inst. Math. Acad. Sinica **13** (1985), 123–136.

[6] P.D. Cordaro and F. Treves, *Hyperfunctions on Hypo-analytic Manifolds*, Ann. Math. Stud. vol. 136, Princeton University Press, 1994.

[7] M.G. Eastwood, *The generalised Penrose-Ward transform*, Math. Proc. Camb. Phil. Soc. **97** (1985), 165–187.

[8] M.G. Eastwood, *Introduction to Penrose transform*, The Penrose Transform and Analytic Cohomology in Representation Theory, Cont. Math. vol. 154, Amer. Math. Soc. 1993, pp. 71–75.

[9] M.G. Eastwood, *Notes on conformal differential geometry*, Suppl. Rendiconti Circolo Mat. Palermo **43** (1996), 57–76.

[10] M.G. Eastwood, *Complex methods in real integral geometry*, Suppl. Rendiconti Circolo Mat. Palermo **46** (1997), 55–71.

[11] M.G. Eastwood, *Some examples of the Penrose transform*, R.I.M.S. Kokyuroku, Kyoto University, **1058** (1998), 22–28.

[12] M.G. Eastwood and C.R. Graham, *The involutive structure on the blow-up on* $\mathbb{R}^n$ *in* $\mathbb{C}^n$, Comm. Anal. Geom., to appear.

[13] M.G. Eastwood and C.R. Graham, *An edge-of-the-wedge theorem for CR functions*, in preparation.

[14] P.R. Garabedian, *An unsolvable equation*, Proc. A.M.S. **25** (1970), 207–208.

[15] C.R. Graham, *private communication.*

[16] N. Hanges and H. Jacobowitz, *Involutive structures on compact manifolds*, Amer. Jour. Math. **177** (1995), 491–522.

[17] L. Hörmander, *An Introduction to Complex Analysis in Several Variables*, Van Nostrand 1966, North-Holland 1973, 1990.

[18] F. John, *The ultrahyperbolic differential equation with four independent variables*, Duke Math. Jour. **4** (1938), 300–322.

[19] H. Komatsu, *A local version of Bochner's tube theorem*, Jour. Fac. Sci. Univ. Tokyo, Sect. 1A Math. **19** (1972), 201–214.

[20] H. Lewy, *On the local character of the solution of an atypical linear differential equation in three variables and a related theorem for regular functions of two complex variables*, Ann. Math. **64** (1956), 514–522.

[21] H. Lewy, *An example of a smooth linear partial differential equation without solution*, Ann. Math. **66** (1957), 155–158.

[22] R. Michel, *Problèmes d'analyse géométriques liés à la conjecture de Blaschke*, Bull. Soc. Math. France **101** (1973), 17–69.

[23] R. Michel, *Sur quelques problèmes de géométrie globale des géodésiques*, Bol. Soc. Bras. Mat. **9** (1978), 19–38.

[24] J.-P. Rosay, *A propos de "wedges"et d'"edges," et de prolongements holomorphes*, Trans. A.M.S. **297** (1986), 63–72.

[25] F. Treves, *Hypo-analytic Structures*, Princeton University Press, 1992.

[26] C. Tsukamoto, *Infinitesimal Blaschke conjectures on projective spaces*, Ann. Scient. Éc. Norm. Sup. **14** (1981), 339–356.

[27] V.S. Vladimirov, V.V. Zharinov, and A.G. Sergeev, *Bogolyubov's "edge of the wedge" theorem, its development and applications*, Russian Math. Surveys **49:5** (1994), 51–65.

DEPARTMENT OF PURE MATHEMATICS
UNIVERSITY OF ADELAIDE
SOUTH AUSTRALIA 5005

E-mail: meastwoo@spam.maths.adelaide.edu.au

CHAPTER III

The Bergman Kernel in Quantum Mechanics

Charles L. Fefferman

In this article we describe a problem in quantum mechanics, and explain how the elementary properties of the Bergman kernel played a role in its history. We begin by describing the problem and stating the main results. Then we discuss the relevance of the Bergman kernel. (In the sharpest results known at present, the Bergman kernel is out of the picture; but I think the idea is pretty, and I hope it will return.)

The problem in quantum mechanics that I'd like to describe concerns nonrelativistic quantum electrodynamics. This theory concerns nuclei at fixed positions in space, nonrelativistic quantized electrons, and photons. It is a reasonable description of a significant part of the physical world, namely the interaction of light, matter and electricity under ordinary conditions. It accounts in principle for what goes on in a light bulb, but not in a nuclear magnetic resonance machine or a particle accelerator. To construct the theory, let me begin with photons, and bring in the electrons and nuclei later.

Photons arise by quantizing the electromagnetic field. Before quantizing anything, let's discuss classical magnetic fields. In elementary (classical) physics, the magnetic field is given as the curl of a vector potential $A = (A_\mu(x))_{\mu=1,2,3}$ ($x \in \mathbb{R}^3$), where the A_μ are smooth, real-valued functions on $\mathbb{R}^3$. In a suitable gauge (the Coulomb gauge), we may take A to be divergence-free. It is convenient to Fourier analyze A. The Fourier representation of a real, divergence-free vector field on $\mathbb{R}^3$ is as follows:

$$\begin{aligned} A_\mu(x) = \sum_{\lambda=1,2} \int_{\mathbb{R}^3} e_{\lambda,\mu}(k) a_\lambda(k) e^{ik\cdot x} \frac{dk}{\sqrt{|k|}} \\ + \sum_{\lambda=1,2} \int_{\mathbb{R}^3} e_{\lambda,\mu}(k) a^*_\lambda(k) e^{-ik\cdot x} \frac{dk}{\sqrt{|k|}}. \end{aligned} \tag{1}$$

Here, for each nonzero $k \in \mathbb{R}^3$, $e_1(k), e_2(k)$ denote an orthonormal basis for the orthogonal complement of k in $\mathbb{R}^3$; $e_{\lambda,\mu}(k)$ denotes the μth component of $e_\lambda(k)$, for $\lambda = 1,2$ and $\mu = 1,2,3$, $\sqrt{|k|}$ is just a convenient normalizing factor; and $a_\lambda(k)$ is the Fourier coefficient. Equation (1) contains no physics at all, but simply a convenient representation of divergence-free real vector fields.

Next, we allow the $A_\mu(x)$ to evolve in time according to Maxwell's equations in empty space, without charges or currents. This amounts to an infinite-dimensional Hamiltonian system, which becomes very easy to understand in terms of the Fourier representation (1). In fact, the Poisson brackets and Hamiltonian for the mechanical system are given by

$$\{a_\lambda(k), a_{\lambda'}(k')\} = 0, \quad \{a^*_\lambda(k), a^*_{\lambda'}(k')\} = 0, \tag{2}$$

$$\{a_\lambda(k), a^*_{\lambda'}(k')\} = i\delta_{\lambda\lambda'}\delta(k-k'), \tag{3}$$

$$H_{\text{mag}} = \sum_{\lambda=1,2} \frac{1}{2} \int_{\mathbb{R}^3} |k| a^*_\lambda(k) a_\lambda(k)\, dk. \tag{4}$$

(Here, $\delta_{\lambda\lambda'}$ is the Kronecker delta, and $\delta(k-k')$ is the Dirac delta.) From (2), (3), (4), we see that distinct (λ, k) act independently of one another, and that each (λ, k) corresponds to a simple harmonic oscillator. Thus, a classical magnetic field in empty space amounts to an infinite family of uncoupled harmonic oscillators.

Now we can quantize the magnetic field. We simply regard each $a_\lambda(k)$ as an operator, and replace (2), (3) by the commutator relations

$$[a_\lambda(k),\, a_{\lambda'}(k')] = 0, \quad [a^*_\lambda(k),\, a^*_{\lambda'}(k')] = 0, \tag{5}$$

$$[a_\lambda(k),\, a^*_{\lambda'}(k')] = \delta_{\lambda\lambda'}\, \delta(k-k'). \tag{6}$$

The Hamiltonian is still given by (4). Of course, in (4), (5), (6), $a^*_{\lambda'}(k')$ denotes the adjoint of $a_{\lambda'}(k')$. Note that (6) shows that $a_\lambda(k)$ cannot be well-defined for an individual (λ, k), Rather, each a_λ must be regarded as an operator-valued distribution. That is, the smeared operator $\int \theta(k) a_\lambda(k) dk$ makes sense for suitable test functions $\theta(k)$, even though a particular $a_\lambda(k)$ makes no sense.

We would like to relate (4), (5), (6) for $a_\lambda(k)$ back to equation (1), which describes the components $A_\mu(x)$ as operators. As it stands, (1) exhibits $A_\mu(x)$ as another operator-valued distribution, so that $A_\mu(x)$ makes no sense for an individual point x. This will be a very serious difficulty once we put electrons and nuclei into the picture, because we will want to evaluate $A_\mu(x)$ at the positions of the electrons. We will deal with this difficulty in a trivial way, by simply truncating the integral (1). Let $\theta(k)$ be a real, smooth function of compact support on $\mathbb{R}^3$, satisfying $\theta(k) = 1$ for $|k| \leqslant 1$.

For a large parameter Λ (with the dimensions of momentum $|k|$), we crudely replace (1) by

$$\begin{aligned} A_\mu^{(\Lambda)}(x) = \sum_{\lambda=1,2} \int_{\mathbb{R}^3} e_{\lambda,\mu}(k) a_\lambda(k) \theta\left(\frac{k}{\Lambda}\right) e^{ik\cdot x} \frac{dk}{\sqrt{|k|}} \\ + \sum_{\lambda=1,2} \int_{\mathbb{R}^3} e_{\lambda,\mu}(k) a_\lambda^*(k) \theta\left(\frac{k}{\Lambda}\right) e^{-ik\cdot x} \frac{dk}{\sqrt{|k|}}. \end{aligned} \tag{7}$$

Thanks to the cutoff θ, the vector potential $A_\mu^{(\Lambda)}(x)$ makes sense at an individual point x. In fact, $A_\mu^{(\Lambda)}(x)$ is a self-adjoint operator.

The ground-state of the Hamiltonian (4) is the vacuum state Ω, which satisfies

$$a_\lambda(k)\Omega = 0, \tag{8}$$

as in the case of a single harmonic oscillator. Physically, $a_\lambda^*(k)$ creates, and $a_\lambda(k)$ annihilates, a photon of momentum k and polarization λ. Since there are no photons present in the vacuum state Ω, we obtain the zero vector when we apply an annihilation operator to Ω, which accounts for (8). To produce nontrivial states, we let smeared creation operators act on the vacuum, obtaining

$$\sum_{\lambda_1 \cdots \lambda_N} \int g(k_1, \lambda_1; \ldots ; k_N, \lambda_N) a_{\lambda_1}^*(k_1) \cdots a_{\lambda_N}^*(k_N)\, dk_1 \cdots dk_N \Omega \tag{9}$$

for suitable test functions g. We assume that linear combinations of states of the form (9) are dense in the basic Hilbert space on which $a_\lambda(k)$, H_{mag}, $A_\mu^{(\Lambda)}(x)$ act.

This completes the description of the quantization of magnetic fields without electrons or nuclei. The formal rules (4)–(8) make it easy to compute any operator of interest on states of the form (9), as well as to compute inner products of any two states of the form (9). It is then a trivial exercise to prove rigorously that there exist a Hilbert space $\mathcal{H}^{\text{ph}}$, a state $\Omega \in \mathcal{H}^{\text{ph}}$, and operator-valued-distributions $a_\lambda(k)$ on $\mathcal{H}^{\text{ph}}$, unique up to isomorphism, so that equations (4)–(8) hold, and H_{mag}, $A_\mu^{(\Lambda)}(x)$ are self-adjoint operators. To make $A_\mu^{(\Lambda)}(x)$ well-defined at an individual point x, we have paid the price of introducing the cutoff $\theta(k/\Lambda)$ into (7). This removes the effect of photons having energy $|k| \gg \Lambda$ on the quantized magnetic field. At short distances $\ll \Lambda^{-1}$, our theory disagrees with physical reality. However, if we take Λ large compared to the binding energy of an electron in an atom, then photons with $|k| \gg \Lambda$ make only small contributions to quantities of physical interest in the everyday world. For more details, the reader may consult any relevant physics textbook.

Next, let us put photons aside, and discuss electrons and nuclei. Since the nuclei are much heavier than the electrons, we may assume that they do not move. Thus, we take the nuclei to lie at fixed positions $y_1, y_2, \ldots, y_M \in \mathbb{R}^3$, and to have atomic numbers $Z_1, Z_2, \ldots, Z_M \geqslant 1$. We assume all the Z_k are bounded above by $Z \geqslant 1$. (In nature, we can take $Z \sim 100$.) If N electrons are located at $x_1, x_2, \ldots, x_N \in \mathbb{R}^3$, then the potential energy is

$$V_{\text{Coulomb}} = \sum_{1 \leqslant j < k \leqslant N} |x_j - x_k|^{-1} + \sum_{1 \leqslant j < k \leqslant M} Z_j Z_k |y_j - y_k|^{-1} - \sum_{j=1}^{N} \sum_{k=1}^{M} Z_k |x_j - y_k|^{-1}. \tag{10}$$

However, the electrons are not fixed in space. Rather, they are quantized. We suppose that they move in a background magnetic field with a given vector potential $A_\mu(x)$. To see how to quantize the electrons, let us begin with just one electron. A single electron has wave function $\psi(x) \in L^2(\mathbb{R}^3, \mathbb{C}^2)$. The probability density for the position of the electron is $|\psi(x)|^2$, so we assume that $\|\psi\| = 1$. To give the kinetic energy of the electron, we need the Dirac operator on $L^2(\mathbb{R}^3, \mathbb{C}^2)$. To define the Dirac operator, we introduce the Pauli matrices

$$\sigma^1 = \begin{pmatrix} 0 & 1 \\ 1 & 0 \end{pmatrix}, \quad \sigma^2 = \begin{pmatrix} 0 & i \\ -i & 0 \end{pmatrix}, \quad \sigma^3 = \begin{pmatrix} 1 & 0 \\ 0 & -1 \end{pmatrix}$$

acting on $\mathbb{C}^2$. The Dirac operator is

$$\sigma \cdot (i\nabla - A) = \sum_{\mu=1,2,3} \sigma^\mu \left(i \frac{\partial}{\partial x_\mu} - A_\mu(x) \right),$$

acting on $L^2(\mathbb{R}^3, \mathbb{C}^2)$. (Here, $(x_\mu)_{\mu=1,2,3}$ are the rectangular coordinates of $x \in \mathbb{R}^3$.) Then the (Pauli) kinetic energy of the electron is given by $\langle [\sigma \cdot (i\nabla - A)]^2 \psi, \psi \rangle$. If the background vector potential A is set equal to zero, then $[\sigma \cdot (i\nabla - A)]^2 = -\Delta$ by the algebra of the σ matrices, so the kinetic energy reduces to the familiar $\langle -\Delta\psi, \psi \rangle$. The dependence of the Pauli kinetic energy on the magnetic field reflects the fact that electrons have spin 1/2.

Let us now pass from one to N quantized electrons. The state of the system is a wave function $\psi(x_1 \cdots x_N) \in L^2(\mathbb{R}^{3N}, (\mathbb{C}^2)^{\otimes N})$. Thus,

$$\psi(x_1 \cdots x_N) = \left(\psi_{q_1 q_2 \cdots q_N}(x_1 \cdots x_N) \right)_{q_1, \ldots, q_N \in \{1,2\}}.$$

We require that ψ have norm 1, and satisfy the antisymmetry condition

$$\psi_{q_{\sigma_1}\cdots q_{\sigma_N}}(x_{\sigma_1}\cdots x_{\sigma_N}) = (\mathrm{sgn}\sigma)\psi_{q_1\cdots q_N}(x_1\cdots x_N) \tag{11}$$

for permutations σ. (Equation (11) means that electrons are fermions.) Regarding $L^2(\mathbb{R}^{3N},(\mathbb{C}^2)^{\otimes N})$ as the N-fold tensor power of $L^2(\mathbb{R}^3,\mathbb{C}^2)$, we may define the Dirac operator $\sigma_j\cdot(i\nabla_{x_j}-A(x_j))$ as the tensor product $I\otimes\cdots\otimes I\otimes[\sigma\cdot(i\,\nabla-A)]\otimes I\otimes\cdots\otimes I$, where $I=$ identity, and where $\sigma\cdot(i\nabla-A)$ occurs in the jth position. Then the operator $[\sigma_j\cdot(i\nabla_{x_j}-A(x_j))]^2$ represents the kinetic energy of the jth electron, and the total energy of the N electrons and M nuclei is given by $\langle H_{\mathrm{el}}\psi,\psi\rangle$, where

$$H_{\mathrm{el}} = \sum_{j=1}^{N}[\sigma_j\cdot(i\nabla_{x_j}-A(x_j))]^2 + V_{\mathrm{Coulomb}}(x_1\cdots x_N, y_1\cdots y_M) \tag{12}$$

and V_{Coulomb} is given by (10). The electron Hamiltonian H_{el} acts on the Hilbert space $\mathcal{H}_{\mathrm{el}}=\{\psi\in L^2(\mathbb{R}^{3N},(\mathbb{C}^2)^{\otimes N})|(11)\text{ holds}\}$. This completes our description of electrons and nuclei in a fixed, background vector potential $(A_\mu(x))$.

We can now combine our separate discussions of electrons and photons, to define the Hamiltonian for the complete system. Again, we fix nuclei with atomic numbers $Z_1,\ldots,Z_M$ at positions $y_1,\ldots,y_M\in\mathbb{R}^3$, and assume $1\leqslant Z_k\leqslant Z$ for all k. Our basic Hilbert space is $\mathcal{H}_{\mathrm{QED}}=\mathcal{H}_{\mathrm{el}}\otimes\mathcal{H}^{\mathrm{ph}}$. We may think of $\psi\in\mathcal{H}_{\mathrm{QED}}$ as consisting of $\psi=(\psi_{q_1\cdots q_N}(x_1\cdots x_N))_{q_1,\ldots,q_N\in\{1,2\}}$, with each $\psi_{q_1\cdots q_N}$ taking values in $\mathcal{H}^{\mathrm{ph}}$. The total Hamiltonian on $\mathcal{H}_{\mathrm{QED}}$ is obtained by simply adding the two Hamiltonians (4) and (12) for the photons and electrons, using $A_\mu^{(\Lambda)}$ from (7) as the vector potential in (12). Thus, the vector potential in which the electrons live is now the operator-valued vector potential produced by the photons. More precisely, we set

$$\begin{aligned} H_{\mathrm{QED}} &= \Gamma\cdot H_{\mathrm{mag}} + H_{\mathrm{el}} \\ &= \frac{\Gamma}{2}\sum_{\lambda=1,2}\int_{\mathbb{R}^3}|k|a_\lambda^\star(k)a_\lambda(k)\,dk + \sum_{j=1}^{N}[\sigma_j\cdot(i\nabla_{x_j}-A^{(\Lambda)}(x_j)]^2 \\ &\quad + V_{\mathrm{Coulomb}}(x_1\cdots x_N, y_1\cdots y_M) \end{aligned} \tag{13}$$

where Γ is a positive constant, and $A^{(\Lambda)}(x)=(A_\mu^{(\Lambda)}(x))_{\mu=1,2,3}$ is given by (7). Non-relativistic quantum electrodynamics (QED) is the study of the Hamiltonian (13). In principle, it describes a significant part of the physics of the everyday world.

The role of the constant Γ in (13) is worth some attention. In our discussion of H_{mag} and H_{el}, we have not written such constants as the mass of the electron, or Planck's constant. Those constants are not essential, because we may simply pick convenient units in which they are equal to 1. However, when we combine the two Hamiltonians H_{mag} and H_{el} into the total Hamiltonian H_{QED}, then we find that a single dimensionless constant Γ remains, which we cannot affect by our choice of units. Therefore, Γ appears explicitly in (13). The constant Γ is equivalent by a trivial formula to the fine structure constant α. In nature, one has $\alpha \approx \frac{1}{137}$, hence Γ in (13) is approximately 746. The question of whether the properties of the Hamiltonian (13) are significantly dependent on the value of Γ is an important issue in nonrelativistic QED.

The problem we study for nonrelativistic QED is called "H-stability." H-stability is the lower bound

$$\langle H_{\text{QED}}\psi, \psi\rangle \geqslant -C \cdot (N + M) \tag{14}$$

for $\psi \in \mathcal{H}_{\text{QED}}$ of norm 1. Recall that N and M are the number of electrons and nuclei respectively. The constant C in (14) is allowed to depend only on Γ in (13), Λ in (7), and on Z, the upper bound for the atomic numbers $Z_1, Z_2, \dots, Z_M$. In particular, C must be independent of the numbers N, M and the positions $y_1, y_2, \dots, y_M$. H-stability is a very natural and important requirement for a physical system. It arises for two reasons. First and most dramatically, suppose (14) were false. That is, suppose the ground-state energy goes to $-\infty$ faster than linearly in the number of particles. Then the ground state energy of $\sim 2N$ particles would be far lower than that of two isolated systems of $\sim N$ particles. Hence, by bringing together any two macroscopic objects, we would liberate a huge amount of energy. As F. Dyson put it, without (14), any two ordinary objects brought into contact would function as an atomic bomb. A second reason for the basic importance of H-stability is its role in thermodynamics. To explain this, we start by fixing a large cube $\Omega \subset \mathbb{R}^3$. In the same spirit as our discussion of electrons, nuclei and photons on $\mathbb{R}^3$, we can form a Hamiltonian $H^{\Omega}_{N,M}$ for N quantized electrons, M quantized nuclei, and a photon field, confined inside the box Ω.

If we fix two positive parameters ρ, T (the density and temperature), then the basic quantity in thermodynamics is the free energy, defined as

$$F(\rho, T) = \lim_{\substack{\text{vol}\,\Omega \to \infty \\ \frac{N}{\text{vol}\,\Omega}, \frac{M}{\text{vol}\,\Omega} \to \rho}} \frac{1}{\text{vol}\,\Omega} \ln \text{ Trace } e^{-(1/T)H^{\Omega}_{N,M}}. \tag{15}$$

Here, for simplicity, we are taking all the atomic numbers Z_k to be 1.

Other thermodynamic quantities such as pressure and entropy are given by trivial formulas from $F(\rho, T)$ and its derivatives. Therefore, one wants to know that the limit (15) exists, and then to understand its dependence on ρ, T.

The connection of H-stability to thermodynamics is that if (14) fails then, for large enough Ω, already the ground state eigenvalue $E^{\Omega}_{N,M}$ of $H^{\Omega}_{N,M}$ should lie below $-C'(N+M)$ for any a-priori constant C'. That would imply Trace $e^{-(1/T)H^{\Omega}_{N,M}} \geqslant e^{-(1/T)E^{\Omega}_{N,M}} \geqslant e^{+(C'/T)(N+M)}$, so that $\frac{1}{\mathrm{vol}\,\Omega} \ln \mathrm{Trace}\, e^{-(1/T)H^{\Omega}_{N,M}} \geqslant \frac{C'}{T} \cdot \frac{N+M}{\mathrm{vol}\,\Omega}$. Hence,

$$\liminf_{\substack{\mathrm{vol}\,\Omega \to \infty \\ \frac{N}{\mathrm{vol}\,\Omega}, \frac{M}{\mathrm{vol}\,\Omega} \to \rho}} \frac{1}{\mathrm{vol}\,\Omega} \ln \mathrm{Trace}\, e^{-(1/T)H^{\Omega}_{N,M}} \geqslant \frac{C'}{T} \cdot 2\rho.$$

Since C' was an arbitrary constant, this shows that the limit in (15) is infinite. Thus, without H-stability, the fundamental quantity in thermodynamics becomes infinite. Conversely, H-stability is expected to play a part in proving the existence of the limit (15). (See Lieb [6] for an in-depth discussion of H-stability questions.)

After some important papers [5,7,9] on closely related problems, it was shown [2] that H-stability (14) holds, provided the constant Γ is large enough. More precisely, we have the following result.

Theorem 1. *If* $\Gamma > CZ$ *for a universal constant* C, *then* $\langle H_{\mathrm{QED}}\psi, \psi\rangle \geqslant -C'(Z, \Gamma, \Lambda) \cdot M$ *for* $\psi \in \mathcal{H}_{\mathrm{QED}}$ *of norm* 1.

Here, of course, $C'(Z, \Gamma, \Lambda)$ denotes a constant depending only on Z, Γ, Λ. The proof of Theorem 1 is built on a crucial idea of J. Fröhlich.

Theorem 1 leaves open what happens in the physically relevant regime $\Gamma \sim 746$, $1 \leqslant Z \leqslant 100$. The Bergman kernel came in through an attempt to reach that regime. I got as far as the following result.

Theorem 2. *If* $Z = 1$ *and* $\Gamma = 975$, *then* $\langle H_{\mathrm{QED}}\psi, \psi\rangle \geqslant -C'(\Lambda) \cdot M$ *for* $\psi \in \mathcal{H}_{\mathrm{QED}}$ *of norm* 1.

Thus, a system made of electrons, protons and photons would be known to the H-stable if the fine structure constant were $\sim \frac{1}{165}$ instead of its actual value $\sim \frac{1}{137}$.

About the same time, Lieb-Loss-Solovej [8] obtained (by other methods) a closely related inequality with very good constants. Adapting the method of [8], Bugaglaro-Fröhlich-Graf [1] proved the following result.

Theorem 3. *If* $\Gamma \geqslant 746$ *and* $Z \leqslant 6$, *then* $\langle H_{\mathrm{QED}}\psi, \psi\rangle \geqslant -C'(\Lambda)M$ *for every* $\psi \in \mathcal{H}_{\mathrm{QED}}$ *of norm* 1.

So the Bergman kernel was out of the picture.

Finally, returning to the method of Theorem 1, but giving a new twist to Fröhlich's initial idea, Fröhlich, Graf and I proved [3,4] the following theorem, which settles completely the problem of H-stability of nonrelativistic QED.

Theorem 4. *For any* Γ, Z *one has* $\langle H_{\text{QED}}\psi, \psi\rangle \geqslant -C(\Gamma, Z)\Lambda M$ *for every* $\psi \in \mathcal{H}_{\text{QED}}$ *of norm* 1.

Our purpose here is not to explain the proofs of Theorems 1–4, but rather to point out why the Bergman kernel comes into the game. I am fond of this idea, even though it has no part in Theorems 3 and 4. I hope it will prove useful in the future.

The Bergman kernel enters in an attempt to show that the total kinetic energy in a ball or cube Q controls the total number of particles in Q. If we can control the total number of particles in a given cube, then we can prevent the particles from coming too close together, which in turn prevents the Coulomb potential from becoming highly negative. This is a crucial issue in proving a lower bound for the total energy. We give no further explanation here of how to use our estimates in the proof of Theorem 2, but simply concentrate on how the kinetic energy in Q controls the number of particles in Q. We work with a fixed, background vector potential $A = (A_\mu(x))_{\mu=1,2,3}$. So far, we know that the total kinetic energy for a wave function $\psi \in L^2(\mathbb{R}^{3N}, (\mathbb{C}^2)^{\otimes N})$ is given by

$$T = \left\langle \sum_{j=1}^{N} [\sigma_j \cdot (i\nabla_{x_j} - A(x_j))]^2 \psi, \psi \right\rangle,$$

but we have not yet defined the kinetic energy in Q. A trivial integration by parts gives $T = \sum_{j=1}^{N} \int_{\mathbb{R}^{3N}} |\sigma_j \cdot (\nabla_{x_j} - A(x_j))\psi|^2 \, dx_1 \cdots dx_N$, so a natural definition of the kinetic energy in Q is

$$T(Q) = \sum_{j=1}^{N} \int_{\mathbb{R}^{3N}} |\sigma_j \cdot (i\nabla_{x_j} - A(x_j))\psi|^2 \chi_Q(x_j) \, dx_1 \cdots dx_N, \tag{16}$$

where χ_Q denotes the characteristic function of Q. If the electrons are at locations $x_1, x_2, \ldots, x_N \in \mathbb{R}^3$, then the number of electrons in Q is simply $\mathcal{N}(x_1 \cdots x_N) = \sum_{j=1}^{N} \chi_Q(x_j)$.

For a quantum state $\psi \in L^2(\mathbb{R}^{3N}, (\mathbb{C}^2)^{\otimes N})$, the expected number of electrons in Q is $\langle \mathcal{N}_Q\psi, \psi\rangle$, so our problem is to control $\langle \mathcal{N}_Q\psi, \psi\rangle$ in terms of $T(Q)$. In a moment, we will restrict attention to the case where Q is the unit ball or unit cube, and the magnetic field is weak, i.e.

$$\int_Q |\text{curl } A(x)|^2 \, dx \leqslant \eta \quad \text{for a small universal constant } \eta. \tag{17}$$

(Recall that curl $A(x)$ is the magnetic field; the integral in (17) is just the magnetic field energy in Q.) We can get away with assuming (17), because if we rescale (17) from the unit cube to a small cube Q, we obtain the condition

$$\int_Q |\text{curl } A(x)|^2 \, dx \leqslant \frac{\eta}{\text{side } Q}. \tag{18}$$

As Q shrinks to a point, the left side of (18) shrinks to zero, while the right side blows up to ∞. Hence, (18) holds automatically on a small enough cube Q, the size of Q depending on the strength of the magnetic field.

Therefore, for the application to Theorem 2, it is enough to deal with the case $Q =$ unit cube or unit ball, with A satisfying (17). In that case, we want to control $\langle \mathcal{N}_Q \psi, \psi \rangle$ in terms of $T(Q)$.

A discouraging elementary fact is that we can have $T(Q) = 0$ but $\langle \mathcal{N}_Q \psi, \psi \rangle$ arbitrarily large, even when the vector potential is zero. This is easy to see from the fact that the 1-electron Dirac operator $\sigma \cdot \nabla$ on $L^2(Q, \mathbb{C}^2)$ has an infinite-dimensional nullspace. If $\varphi_1, \varphi_2, \ldots$ is an infinite list of orthonormal elements of nullspace $(\sigma \cdot \nabla)$ in $L^2(Q, \mathbb{C}^2)$, then for N arbitrary, we just take $\psi(x_1 \cdots x_N)$ to be an antisymmetrized tensor product of $\varphi_1(x_1), \varphi_2(x_2), \ldots, \varphi_N(x_N)$. (If $x_j \notin Q$, we set $\varphi_j(x_j) = 0$.) So, we cannot hope to control $\langle \mathcal{N}_Q \psi, \psi \rangle$ in terms of $T(Q)$, because we can put arbitrarily many electrons in Q with zero kinetic energy.

What saves the day is that among the arbitrarily many zero-energy electrons we can place in Q, all but a few stay very close to the boundary ∂Q. Hence, even though the kinetic energy in Q does not control the number of electrons in Q, it *does* control the number of electrons in the middle half of Q. Moreover, this is made possible by the elementary properties of the Bergman kernel.

To see the role of the Bergman kernel, let us look at a simple 2-dimensional model problem, with the 3-dimensional Dirac operator replaced by $\bar{\partial}$ on the unit disc $\mathcal{D}$ in one complex variable. We can study antisymmetric wave functions $\psi \in L^2(\mathcal{D}^N)$, with kinetic energy

$$T(\psi) = \sum_{k=1}^{N} \int_{\mathcal{D}^N} \left| \frac{\partial \psi}{\partial \bar{z}_k}(z_1 \cdots z_N) \right|^2 d\,\text{area}\,(z_1) \cdots d\,\text{area}\,(z_N)$$

in place of (16). Like the Dirac operator, $\overline{\partial}$ on the unit disc has an infinite-dimensional nullspace, so again we can have $T(\psi) = 0$ with N arbitrarily large. In fact, an orthonormal basis for the nullspace of $\overline{\partial}$ on $\mathcal{D}$ is of course $\{C_k z^k\}_{k \geqslant 0}$ for suitable normalizing constants C_k. (We note that $C_k z^k$ is strongly concentrated near $\partial\mathcal{D}$ for k large.) The worst case is therefore the antisymmetrized tensor product $\psi(x_0 \cdots z_N) = \frac{1}{\sqrt{(N+1)!}} \sum_\sigma$ $(\mathrm{sgn}\, \sigma)\varphi_{\sigma_0}(z_0) \cdots \varphi_{\sigma_N}(z_N)$ for $N \gg 1$, with $\varphi_k(z) = C_k z^k$. The density of "electrons" in $\mathcal{D}$ is given in this case by

$$\rho(z) = \sum_{k=0}^{N} |\varphi_k(z)|^2. \tag{19}$$

(In general, if $\psi(x_1 \cdots x_N)$ is a many-electron wave function, the electron density is defined as

$$\rho(x) = \sum_{k=1}^{N} \int_{\mathbb{R}^{3N-3}} |\psi(x_1 \cdots x_{k-1}\, x_{k+1} \cdots x_N)|^2\, dx_1 \cdots dx_{k-1} dx_{k+1} \cdots dx_N.) \tag{20}$$

However, the Bergman kernel on the diagonal is $K(z,z) = \sum_{k=0}^{\infty} |\varphi_k(z)|^2$. Comparing this with (19), we see that $\rho(z) \leqslant K(z,z)$. That is, *for any N-electron wave function with zero kinetic energy, the electron density is less than the Bergman kernel on the diagonal.* In particular, the expected number of electrons in the middle half of $\mathcal{D}$ is at most $\int_{|z|<\frac{1}{2}} K(z,z)\, d\,\mathrm{area}\,(z)$, which is a universal constant.

The same idea works for the Dirac operator in place of its $\overline{\partial}$ model. There is a 2×2 matrix valued Bergman kernel $K_A(z,w)$, so that the operator $\mathcal{P}\psi(z) = \int_Q K_A(z,w)\psi(w)dw$ is the orthogonal projection from $L^2(Q, \mathbb{C}^2)$ onto the nullspace of $\sigma \cdot (i\nabla - A)$. The electron density ρ (defined by (20)) for any N-electron wave function with zero kinetic energy satisfies

$$\rho(x) \leqslant \text{ trace } K_A(x,x) \quad \text{for all } x \in Q. \tag{21}$$

Since A is any vector potential satisfying (17), we cannot hope to compute $K_A(x,x)$ exactly. Instead, we can use the extremal property of the Bergman kernel to estimate $K_A(x,x)$. We will prove the following simple estimate.

Lemma 1. *If $\sigma \cdot (i\nabla - A)\psi = 0$ on $Q =$ unit ball, with A satisfying*

(17), *then we have*

$$|\psi(x)|^2 \leqslant \frac{C_0}{(1-|x|)^3} \int\limits_B |\psi(y)|^2 dy \quad \textit{for a universal constant } C_0. \tag{22}$$

Estimate (22) and the extremal property of the Bergman kernel give $|K_A(x,x)| \leqslant \frac{C_0}{(1-|x|)^3}$, where $|K_A|$ denotes the norm of the 2×2 matrix $K_A(x,x)$. Hence, (21) yields

$$\rho(x) \leqslant \text{ trace } K_A(x,x) \leqslant 2|K_A(x,x)| \leqslant \frac{2C_0}{(1-|x|)^3} \tag{23}$$

for zero kinetic energy N-electron wave functions. In particular, if $\mathcal{N}_E(x_1 \cdots x_N) = \sum\limits_{k=1}^{N} \chi_E(x_k)$ is the number of electrons in a subset $E \subset Q$, then

$$\langle \mathcal{N}_E \psi, \psi \rangle \leqslant \int\limits_{x \in E} \frac{2C_0}{(1-|x|)^3}\, dx \quad \text{when } T(Q) = 0. \tag{24}$$

The proof of Lemma 1 is so simple that it can be included here. Applying $\sigma \cdot (i\nabla - A)$ to the equation $\sigma \cdot (i\nabla - A)\psi = 0$, we find that

$$(i\nabla - A)^*(i\nabla - A)\psi = \pm(\sigma \cdot \operatorname{curl} A)\psi, \tag{25}$$

thanks to the algebra of the σ-matrices. Here, $(i\nabla - A)^*(i\nabla - A)$ is a slight variant of the Laplacian. Since curl A is small by (17), equation (25) is not so far from saying that ψ is harmonic. We will correct ψ to lie in the nullspace of $(i\nabla - A)^*(i\nabla - A)$. To do so, we solve the Dirichlet problem

$$(i\nabla - A)^*(i\nabla - A)U = \pm(\sigma \cdot \operatorname{curl} A)\psi \quad \text{on} \quad Q \tag{26}$$

$$U = 0 \quad \text{on} \quad \partial Q. \tag{27}$$

Then $\Phi = \psi - U$ lies in the nullspace of $(i\nabla - A)^*(i\nabla - A)$. It follows that $|\Phi|^2$ is subharmonic, $\Delta|\Phi|^2 \geqslant 0$. Hence, for any $x \in Q$, we have

$$\begin{aligned} |\Phi(x)|^2 &\leqslant \frac{1}{\text{vol } B(x, 1-|x|)} \int\limits_{B(x,1-|x|)} |\Phi(y)|^2\, dy \\ &\leqslant \frac{C}{(1-|x|)^3} \int\limits_Q |\Phi(y)|^2\, dy, \end{aligned} \tag{28}$$

where $B(x,r)$ denotes the ball with center x and radius r.

On the other hand, the solution of the Dirichlet problem (26), (27) is given by

$$U(x) = \pm \int_Q G_A(x,y)(\sigma \cdot \operatorname{curl} A(y))\psi(y)\,dy, \tag{29}$$

where the Green's function G_A satisfies

$$|G_A(x,y)| \leqslant G(x,y), \tag{30}$$

$G(x,y)$ = Green's function for the ordinary Laplacian on Q. Inequality (30) is an immediate consequence of the elementary fact that $(i\nabla - A)^*(i\nabla - A)u = 0$ implies $\Delta|u| \geqslant 0$. From (29) and (30) we get

$$|U(x)| \leqslant \int_Q G(x,y)|\operatorname{curl} A(y)\,||\psi(y)|dy. \tag{31}$$

This lets us estimate the L^2-norm of U using (17). In fact,

$$\begin{aligned} \|U\|_{L^2(Q)} &\leqslant \sup_{y\in Q} \|G(\cdot,y)\|_{L^2(Q)} \int_Q |\operatorname{curl} A(y)||\psi(y)|\,dy \\ &\leqslant C \int_Q |\operatorname{curl} A||\psi|\,dy \\ &\leqslant C\eta^{1/2}\|\psi\|_{L^2(Q)}. \end{aligned} \tag{32}$$

Then since $\Phi = \psi - U$, it follows that $\|\Phi\|_{L^2(Q)} \leqslant (1 + C\eta^{1/2})\|\psi\|_{L^2(Q)}$. Hence, (28) implies

$$|\Phi(x)|^2 \leqslant \frac{C}{(1-|x|)^3} \int_Q |\psi(y)|^2\,dy. \tag{33}$$

So we have a pointwise estimate for Φ. It remains to use (31) to derive a pointwise estimate for U. Since we only know so far that curl A and ψ belong to L^2, and $G(x,y)$ is not L^∞, the most straightforward approach fails. We proceed as follows.

Define a norm $|||\psi||| = \sup_{x\in Q}(1-|x|)^{3/2}\,|\psi(x)|$. Also, write $G(x,y)$ as the sum of two nonnegative functions $G = G_{\mathrm{lo}} + G_{\mathrm{hi}}$, to be picked later. Then (31) gives

$$\begin{aligned} |U(x)| \leqslant &\int_Q G_{\mathrm{lo}}(x,y)|\operatorname{curl} A(y)||\psi(y)|\,dy \\ &+ \int_Q G_{\mathrm{hi}}(x,y)|\operatorname{curl} A(y)||\psi(y)|\,dy. \end{aligned} \tag{34}$$

We can estimate the first term on the right in (34) by

$$\begin{aligned}\int_Q G_{\mathrm{lo}}(x,y)|\operatorname{curl} A(y)||\psi(y)|\,dy \\ &\leqslant \left[\sup_{y\in Q} G_{\mathrm{lo}}(x,y)\right] \|A\|_{L^2(Q)}\|\psi\|_{L^2(Q)} \\ &\leqslant \left[\sup_{y\in Q} G_{\mathrm{lo}}(x,y)\right] \eta^{1/2}\|\psi\|_{L^2(Q)},\end{aligned} \tag{35}$$

thanks to (17). To handle the second term on the right in (34), we use the norm $|||\cdot|||$ as follows.

$$\begin{aligned}\int_Q G_{\mathrm{hi}}(x,y)|\operatorname{curl} A(y)||\psi(y)|\,dy \\ &\leqslant \int_Q G_{\mathrm{hi}}(x,y)|\operatorname{curl} A(y)||\psi(y)|\cdot(1-|y|)^{-3/2}|||\psi|||\,dy \\ &\leqslant \left(\int_Q G_{\mathrm{hi}}^2(x,y)(1-|y|)^{-3}dy\right)^{1/2} \eta^{1/2}|||\psi|||,\end{aligned} \tag{36}$$

by Cauchy-Schwartz and another application of (17). Combining (34), (35), (36), we see that

$$\begin{aligned}|||U||| &= \sup_{x\in Q}(1-|x|)^{3/2}|U(x)| \\ &\leqslant \left[\sup_{x\in Q, y\in Q}(1-|x|)^{3/2}G_{\mathrm{lo}}(x,y)\right]\eta^{1/2}\|\psi\|_{L^2}(Q) \\ &\quad + \left[\sup_{x\in Q}(1-|x|)^3\int_Q G_{\mathrm{hi}}^2(x,y)(1-|y|)^{-3}dy\right]\eta^{1/2}|||\psi|||.\end{aligned} \tag{37}$$

This lets us guess the correct decomposition of $G(x,y)$ into $G_{\mathrm{lo}}+G_{\mathrm{hi}}$. For instance, if we take $G_{\mathrm{lo}}(x,y) = \min\left\{G(x,y), \frac{C_1}{(1-|x|)^3}\right\}$ for a large constant C_1, then the first quantity in brackets in (37) is bounded by C_1. One checks easily that the second quantity in brackets is also bounded, so that (37) implies

$$|||U||| \leqslant C_1\eta^{1/2}\|\psi\|_{L^2(Q)} + C_2\eta^{1/2}|||\psi|||, \tag{38}$$

for universal constants C_1, C_2. On the other hand, (33) means that

$$|||\Phi||| \leqslant C\|\psi\|_{L^2(Q)}. \tag{39}$$

Since $\psi = \Phi + U$, estimates (38) and (39) imply

$$|||\psi||| \leqslant C\|\psi\|_{L^2(Q)} + C_2\eta^{1/2}|||\psi|||. \tag{40}$$

If η is small enough, then $C_2\eta^{1/2} < \frac{1}{2}$ and we can absorb the last term in (40) into the left-hand side, we obtain $|||\psi||| \leqslant C\|\psi\|_{L^2(Q)}$, which is the conclusion of Lemma 1. Here, we assumed that $|||\psi|||$ is finite, in order to absorb a term into the left-hand side of (40). However, we may easily repeat our proof with Q replaced by a ball Q_ϵ of radius $(1-\epsilon)$, and let $\epsilon \to 0$. Since ψ satisfies an elliptic equation on Q, it must be bounded on Q_ϵ for any $\epsilon > 0$, so the triple norm on Q_ϵ is clearly finite. This completes the proof of Lemma 1.

Now that Lemma 1 is proved, we have good control of zero-kinetic energy wave functions, thanks to (24). It remains to pass to wave functions of nonzero kinetic energy. The main tool is following watered-down version of a standard $\overline{\partial}$-agreement.

Lemma 2. *Let $\psi \in L^2(Q, \mathbb{C}^2)$, and suppose ψ is orthogonal to the nullspace of $\sigma\cdot(i\nabla - A)$ with A satisfying* (17). *Then $\|\sigma\cdot(i\nabla - A)\psi\|^2_{L^2(Q)} \geqslant c\|\psi\|^2_{L^2(Q)}$, with c a universal constant.*

Again, the proof is so simple that we can give it here. The main idea is to solve the equation

$$\sigma\cdot(i\nabla - A)\psi = f \tag{41}$$

with

$$\psi \perp C^\infty(\overline{Q}, \mathbb{C}^2) \cap \text{ Nullspace } \sigma\cdot(i\nabla - A). \tag{42}$$

To achieve (42), we take $\psi = \sigma\cdot(i\nabla - A)\phi$ with $\phi\big|_{\partial Q} = 0$. For $g \in C^\infty(\overline{Q}, \mathbb{C}^2) \cap$ Nullspace $\sigma\cdot(i\nabla - A)$ we have

$$\int_Q \psi\cdot\overline{g}\,dx = \int_Q [\sigma\cdot(i\nabla - A)\phi]\cdot\overline{g}\,dx = \int_q \phi\cdot\overline{[\sigma\cdot(i\nabla - A)g]}\,dx = 0,$$

by integration by parts. (The boundary term vanishes, since $\phi\big|_{\partial Q} = 0$.) Thus, (42) holds atuomatically when ψ is given in terms of ϕ as above. It remains to satisfy (41). Thus, we must pick ϕ to satisfy $[\sigma\cdot(i\nabla - A)]^2\phi = f$ on Q, with $\phi\big|_{\partial Q} = 0$. That is,

$$(i\nabla - A)^*(i\nabla - A)\phi = \pm(\sigma\cdot\operatorname{curl} A)\phi + f \quad \text{on} \quad Q \tag{43}$$

with

$$\phi = 0 \quad \text{on} \quad \partial Q. \tag{44}$$

Suppose we can solve (43), (44). Then, by taking inner products with ϕ and integrating by parts, we find that

$$\begin{aligned} \|(i\nabla - A)\phi\|^2 &= \pm\langle(\sigma\cdot\operatorname{curl} A)\phi,\phi\rangle + \langle f,\phi\rangle \\ &\leqslant \int_Q |\operatorname{curl} A||\phi|^2\,dx + \epsilon\|\phi\|^2 + \frac{1}{\epsilon}\|f\|^2 \end{aligned} \tag{45}$$

for a small constant ϵ to be picked later. Since $\big|\nabla|\phi|\big| \leqslant |(i\nabla - A)\phi|$ and $\phi\big|_{\partial Q} = 0$, the Sobolev inequality on Q implies

$$\|\phi\|^2_{L^4(Q)} \leqslant C\|\nabla|\phi|\|^2 \leqslant C\|(i\nabla - A)\phi\|^2. \tag{46}$$

(All norms are in L^2 unless otherwise indicated.) From (17) and (46), we get

$$\int_Q |\operatorname{curl} A||\psi|^2\,dx \leqslant \|\operatorname{curl} A\|\|\psi\|^2_{L^4(Q)} \leqslant C\eta^{1/2}\|(i\nabla - A)\psi\|^2.$$

Putting this into (45), we find that

$$\|(i\nabla - A)\phi\|^2 \leqslant C\eta^{1/2}\|(i\nabla - A)\phi\|^2 + \epsilon\|\phi\|^2 + \frac{1}{\epsilon}\|f\|^2.$$

If η is small enough, then we can absorb the first term on the right into the left-hand side. Thus,

$$\|(i\nabla - A)\phi\|^2 \leqslant 2\epsilon\|\phi\|^2 + \frac{2}{\epsilon}\|f\|^2. \tag{47}$$

This and (46) imply

$$\|\phi\|^2 \leqslant C\epsilon\|\phi\|^2 + \frac{C}{\epsilon}\|f\|^2.$$

Taking ϵ small enough, we can absorb the term $C\epsilon\|\phi\|^2$ into the left-hand side. Hence, we obtain the estimate

$$\|\phi\|^2 \leqslant C\|f\|^2, \tag{48}$$

where we no longer bother to keep track of the constant ϵ. In particular, (48) shows that $f = 0$ forces $\phi = 0$. That is, the boundary value problem (43), (44) has a trivial nullspace. On the other hand, (43), (44) is a self-adjoint elliptic problem. It follows that (43), (44) has a trivial cokernel, and thus may be solved for any $f \in L^2$. Moreover, our solution ϕ may be estimated by (47) and (48); we obtain

$$\|(i\nabla - A)\psi\|^2 \leqslant C\|f\|^2 \tag{49}$$

where again we ignore the dependence on the constant ϵ. Since $\psi = \sigma \cdot (i\nabla - A)\phi$, (49) implies

$$\|\psi\|^2 \leqslant C\|f\|^2, \tag{50}$$

and we know that

$$\sigma \cdot (i\nabla - A)\psi = f, \tag{51}$$

since that is the content of equation (43).

We have learned that for every $f \in L^2(Q, \mathbb{C}^2)$, there is a solution of (51) that satisfies (50). The solution ψ that we constructed satisfies (42), but we don't know that $\psi \perp$ Nullspace $[\sigma \cdot (i\nabla - A)]$ without proving that $C^\infty(\overline{Q}, \mathbb{C}^2) \cap$ Nullspace $[\sigma \cdot (i\nabla - A)]$ is dense in Nullspace $[\sigma \cdot (i\nabla - A)]$. We can finesse this technical point as follows. We let $\psi^\#$ be the projection of ψ onto the orthogonal complement of Nullspace $[\sigma \cdot (i\nabla - A)]$. Then $\phi - \phi^\# \in$ Nullspace $[\sigma \cdot (i\nabla - A)]$, so $\sigma \cdot (i\nabla - A)\psi^\# = \sigma \cdot (i\nabla - A)\psi = f$, and $\|\psi^\#\| \leqslant \|\psi\| \leqslant C\|f\|$. So now we have learned the following: For every $f \in L^2(Q, \mathbb{C}^2)$, there is a solution $\psi^\# \in L^2(Q, \mathbb{C}^2)$ of the equation $\sigma \cdot (i\nabla - A)\psi^\# = f$ with $\psi^\# \perp$ Nullspace $[\sigma \cdot (i\nabla - A)]$. Moreover, we have $\|\psi^\#\| \leqslant C\|f\|$. To complete the proof of Lemma 2, suppose $\psi \in L^2(Q, \mathbb{C}^2)$. Let $f = \sigma \cdot (i\nabla - A)\psi$. If $f \notin L^2$, then $\|\sigma \cdot (i\nabla - A)\psi\|^2 = \infty$, so the conclusion of Lemma 2 holds trivially. If $f \in L^2$, then we construct $\psi^\#$ as above. Since $\sigma \cdot (i\nabla - A)\psi^\# = f = \sigma \cdot (i\nabla - A)\psi$, we know that $\psi - \psi^\# \in$ Nullspace $[\sigma \cdot (i\nabla - A)]$. On the other hand, $\psi^\#$ and ψ are both orthogonal to Nullspace $[\sigma \cdot (i\nabla - A)]$. Hence, $\psi - \psi^\# = 0$, i.e., $\psi^\# = \psi$. Since $\|\psi^\#\| \leqslant C\|f\|$, we obtain $\|\psi\| \leqslant C\|\sigma \cdot (i\nabla - A)\psi\|$, which is the conclusion of Lemma 2.

The reader will surely agree that Lemmas 1 and 2 are trivial. Once one has the idea to relate these quantum mechanical problems to the Bergman kernel, one meets little resistance.

We are ready to use Lemma 2 to pass from zero-energy to arbitrary wave functions. The argument is based on separation of variables. For simplicity, we pretend that our wave functions are scalar-valued. To correct this oversimplification, we need only introduce a lot of additional indices.

On $L^2(\mathbb{R}^3)$, we introduce three orthogonal projections:

Π_0 projects onto φ supported in Q and belonging to the nullspace of $\sigma \cdot (i\nabla - A)$ on Q.

Π_1 projects onto φ supported in Q and orthogonal to the nullspace of $\sigma \cdot (i\nabla - A)$ on Q.

Π_2 projects onto φ supported in $\mathbb{R}^3 \setminus Q$.

For one-electron wave functions, we have the following simple properties.

$$\|\varphi\|^2 = \|\Pi_0\varphi\|^2 + \|\Pi_1\varphi\|^2 + \|\Pi_2\varphi\|^2 \tag{52}$$

$$\sigma\cdot(i\nabla - A)\varphi(x) = \sigma\cdot(i\nabla - A)\Pi_1\varphi(x) \quad \text{for} \quad x\in Q \tag{53}$$

$$|\varphi(x)|^2 \leqslant 2|\Pi_0\varphi(x)|^2 + 2|\Pi_1\varphi(x)|^2 \quad \text{for} \quad x\in Q. \tag{54}$$

Next, we pass to N-electron wave functions $\psi(x_1\cdots x_N)$. For $i = 1,2,3$ and $k = 1,2,\ldots,N$, let Π_i^k denote the projection Π_i acting on the kth electron. That is, for fixed $(x_1\cdots x_{k-1}\,x_{k+1}\cdots x_N)$, we regard ψ as a function of x_k alone, and apply Π_i. The result is what we mean by Π_i^k.

Let $E\subset Q$ be given. For $i_1,\ldots,i_N\in\{0,1,2\}$, let $\psi[i_1\cdots i_N] = \Pi_{i_1}^1\cdots\Pi_{i_N}^N\psi$. Then simple manipulations using (52), (53), (54) show that

$$\begin{aligned}
\left\langle\sum_{k=1}^N \chi_E(x_k)\psi,\,\psi\right\rangle &\leqslant 2\sum_{i_1\cdots i_n}\left\langle\sum_{k\text{ with }i_k=0}\chi_E(x_k)\psi[i_1\cdots i_N],\,\psi[i_1\cdots i_N]\right\rangle\\
&\quad + 2\sum_{i_1\cdots i_N}\sum_{k\text{ with }i_k=1}\left\langle\chi_E(x_k)\psi[i_1\cdots i_N],\,\psi[i_1\cdots i_N]\right\rangle\\
&\leqslant 2\sum_{i_1\cdots i_N}\left\langle\sum_{k\text{ with }i_k=0}\chi_E(x_k)\psi[i_1\cdots i_N],\,\psi[i_1\cdots i_N]\right\rangle\\
&\quad + 2\sum_{i_1\cdots i_N}\sum_{k\text{ with }i_k=1}\|\psi[i_1\cdots i_N]\|^2.
\end{aligned} \tag{55}$$

For a moment, we fix $(i_1\cdots i_N)$, and then fix all the x_k for which $i_k\neq 0$. Then $\psi[i_1\cdots i_N]$ becomes a zero-kinetic-energy wave function in the x_k for which $i_k = 0$. (However, it needn't have norm 1.) Applying estimate (24) to that wave function, and then integrating the result over all $(x_k)_{i_k\neq 0}$, we find that

$$\left\langle\sum_{k\text{ with }i_k=0}\chi_E(x_k)\psi[i_1\cdots i_N],\psi[i_1\cdots i_N]\right\rangle \leqslant C\int_E\frac{dx}{(1-|x|)^3}\cdot\|\psi[i_1\cdots i_N]\|^2.$$

Summing over all $i_1\cdots i_N$, we obtain

$$\sum_{i_1\cdots i_N}\left\langle\sum_{k\text{ with }i_k=0}\chi_E(x_k)\psi[i_1\cdots i_N],\psi[i_1\cdots i_N]\right\rangle \leqslant C\int_E\frac{dx}{(1-|x|)^3} \tag{56}$$

On the other hand, suppose we again fix $i_1\cdots i_N$, but this time we also fix k with $i_k = 1$, then fix $x_1\cdots x_{k-1}x_{k+1}\cdots x_N$. Then $\psi[i_1\cdots i_N]$ becomes a function of x_k alone. Moreover, that function is orthogonal to the nullspace of $\sigma \cdot (i\nabla - A)$ on Q. Applying Lemma 2 to this function, and integrating over all $x_1 \cdots x_{k-1}\ x_{k+1} \cdots x_N$, we learn that $i_k = 1$ implies

$$\begin{aligned} &\|\psi[i_1 \cdots i_n]\|^2 \\ &\quad \leqslant C\|\sigma_k \cdot (i\nabla_{x_k} - A(x_k))\psi[i_1 \cdots i_N]\|^2 \\ &\quad = C\|\sigma_k \cdot (i\nabla_{x_k} - A(x_k))\Pi^1_{i_1} \cdots \Pi^{k-1}_{i_{k-1}}\Pi^{k+1}_{i_{k+1}} \cdots \Pi^N_{i_N}\psi\|^2 \\ &\quad = C \int\limits_{\mathbb{R}^{3N}} \left|\Pi^1_{i_1} \cdots \Pi^{k-1}_{i_{k-1}}\Pi^{k+1}_{i_{k+1}} \cdots \Pi^N_{i_N}\sigma_k \cdot (i\nabla_{x_k} - A(x_k))\psi\right|^2 \\ &\qquad\qquad \chi_Q(x_k)\, dx_1 \cdots dx_N, \end{aligned} \tag{57}$$

thanks to (53). Fixing k, and summing (57) over all $i_1 \cdots i_N$ with $i_k = 1$, we find that

$$\begin{aligned} &\sum_{i_1 \cdots i_N} \chi_{i_k=1}\|\psi[i_1 \cdots i_N]\|^2 \\ &\qquad \leqslant C \int\limits_{\mathbb{R}^{3N}} |\sigma_k \cdot (i\nabla_{x_k} - A(x_k))\psi|^2 \chi_Q(x_k)\, dx_1 \cdots dx_N. \end{aligned}$$

Summing over k, we obtain

$$\begin{aligned} &\sum_{i_1 \cdots i_N} \sum_{k \text{ with } i_k=1} \|\psi[i_1 \cdots i_N]\|^2 \\ &\qquad \leqslant C \sum_{k=1}^{N} \int |\sigma_k \cdot (i\nabla_{x_k} - A(x_k))\psi|^2 \chi_Q(x_k)\, dx_1 \cdots dx_N \\ &\qquad = C\,T(Q) \end{aligned} \tag{58}$$

(see (16)).

Finally, putting (56) and (58) into (55), we obtain the desired control on the number of particles in E, namely

$$\left\langle \sum_{k=1}^{N} \chi_E(x_k)\psi, \psi \right\rangle \leqslant C \int\limits_E \frac{dx}{(1 - |x|)^3} + CT(Q). \tag{59}$$

Estimate (59) holds for all $E \subset Q$ and all $\psi \in \mathcal{H}_{\text{el}}$ of norm 1, provided only that A satisfies (17).

In order to get a reasonable constant in Theorem 2, it was necessary to introduce many minor changes in the proof of (59). The changes affect no ideas, but they do affect the constants. A typical example is that hypothesis (17) should be replaced by the assumption

$$\int\limits_{|x|<1} |\mathrm{curl}\, A(x)|^2 \cdot \left(1+\frac{|x|^2}{2}\right)^{-2} dx < \eta. \tag{60}$$

Since $\frac{4}{9} \leqslant \left(1+\frac{|x|^2}{2}\right)^{-2} \leqslant 1$ for $|x| < 1$, estimates (17) and (60) are essentially equivalent, but the optimal value of η changes. We spare the reader the details.

Finally, let me express my gratitude to the organizers of the conferences on several complex variables, held at Lake Biwa and at RIMS in the summer of 1997, as well as to the Taniguchi Foundation for supporting these meetings. I am grateful also to Ms. Eileen Olszewski for TEXing this paper.

References

[1] Bugagliaro, L., Fröhlich, J., and Graf, G. M., *Stability of quantum electrodynamics with non-relativistic matter*, Phys. Rev. Lett. (to appear).

[2] Fefferman, C., *Stability of Coulomb systems in a magnetic field,* Proc. Nat. Acad. Sci. U.S.A. **92** (1995), 5006–5007.

[3] Fefferman, C., Fröhlich, J., and Graf, G. M., *Stability of nonrelativistic quantum mechanical matter coupled to the (ultraviolet cutoff) radiation field,* Proc. Nat. Acad. Sci. U.S.A. **93** (1996), 1509–1511.

[4] ———, *Stability of ultraviolet-cutoff quantum electrodynamics with non relativistic matter,* Comm. Math. Phys. (to appear).

[5] Fröhlich, J., Lieb, E., and Loss, M., *Stability of Coulomb systems with magnetic field I: The one electron atom,* Comm. Math. Phys. **104** (1986), 251–270.

[6] Lieb, E., *The stability of Matter: From Atoms to Stars: Selecta of Elliott H. Lieb* (W. Thirring, ed.), Springer, Heidelberg, 1991.

[7] Lieb, E., and Loss, M., *Stability of Coulomb systems with magnetic fields II. The many electron atom and the one electron molecule,* Comm. Math. Phys. **104** (1986), 271–282.

[8] Lieb, E., Loss, M., and Solovej, P., *Stability of matter in magnetic fields,* Phys. Rev. Lett. **75** (1995), 985–989.

[9] Loss, M., and Yau, H. Z., *Stability of Coulomb systems with magnejtic fields III. Zero energy states of the Pauli operator,* Comm. Math. Phys. **104** (1986), 283–290.

DEPARTMENT OF MATHEMATICS
PRINCETON UNIVERSITY
PRINCETON, NJ 08544-1000

CHAPTER IV

WKB and the Periodic Table

Charles L. Fefferman

In this article, I'd like to motivate some work that Luis Seco and I have done on the ground-state energy of an atom of atomic number $Z \gg 1$. If one ignores relativistic effects, then the mathematical problem is as follows. Fix a nucleus of charge $+Z$ at the origin. If N electrons are located at $x_1, x_2, \ldots, x_N \in \mathbb{R}^3$, then their potential energy is given by

$$V_{\text{Coulomb}}^{NZ}(x_1, \ldots, x_N) = -\sum_{k=1}^{N} \frac{Z}{|x_k|} + \sum_{1 \leq j < k \leq N} \frac{1}{|x_j - x_k|}. \tag{1}$$

We regard the electrons as quantized, so that the state of the system is given by a wave function

$$\psi(x_1, \ldots, x_N) \in L^2(\mathbb{R}^{3N}).$$

For simplicity, we neglect spin here. (If we had taken spin into account, then ψ would take values in the N-fold tensor power of $\mathbb{C}^2$. This changes no ideas, but introduces factors of 2 into some key formulas).

Not every $\psi \in L^2(\mathbb{R}^{3N})$ is allowed as the wave function for N electrons. The requirements on ψ are as follows:

- ψ must have norm 1 in $L^2(\mathbb{R}^{3N})$
- $\psi(x_1, \ldots, x_N)$ must be *antisymmetric*, i.e.,

$$\psi(x_{\sigma_1}, x_{\sigma_2}, \ldots, x_{\sigma_N}) = (\operatorname{sgn} \sigma)\, \psi(x_1, \ldots, x_N) \quad \text{for permutations } \sigma. \tag{2}$$

The subspace of $L^2(\mathbb{R}^{3N})$ consisting of all functions satisfying (2) will be called $L^2_{\text{antisymm}}(\mathbb{R}^{3N})$.

The (expected) energy of N electrons in the state ψ is given by $\langle H_{NZ}\psi, \psi\rangle$, where the inner product is taken in L^2, and H_{NZ} is the Hamiltonian, given by

$$H_{NZ} = -\sum_{k=1}^{N} (-\Delta_{x_k}) + V_{\text{Coulomb}}^{NZ}. \tag{3}$$

The ground-state energy $E(N, Z)$ of N electrons and a nucleus of charge Z is defined as the lowest eigenvalue (or, more precisely, the infimum of the spectrum) of H_{NZ} acting on $L^2_{\text{antisymm}}(\mathbb{R}^{3N})$.

We may then consider either a neutral atom by taking $E(Z, Z)$, or we may allow nature a chance to force $N \approx Z$ by forming

$$E(Z) = \inf_N E(N, Z). \tag{4}$$

The results known so far are too crude to distinguish these; let us first restrict attention to (4). The mathematical problem we study is that of understanding the asymptotic behavior of $E(Z)$ for large Z. We want to understand the asymptotics of $E(Z)$ so precisely that we can see small irregularities in its behavior. Let me try to explain why one might care about the small irregularities.

We can get a crude idea of how atoms behave by simply dropping the repulsion term $\sum_{j<k} |x_j - x_k|^{-1}$ from (1), to obtain an easier Hamiltonian $\widetilde{H}_{NZ}$. Without repulsion, (3) simplifies enormously. In fact, we have

$$\widetilde{H}_{NZ} = \sum_{k=1}^{N} \left(-\Delta_{x_k} - \frac{Z}{|x_k|} \right) \quad \text{acting on} \quad L^2_{\text{antisymm}}(\mathbb{R}^{3N}),$$

which reduces by separation of variables to the hydrogen atom

$$-\Delta_x - \frac{Z}{|x|} \quad \text{on} \quad L^2(\mathbb{R}^3). \tag{5}$$

Of course, the eigenvalues and eigenfunctions of (5) are computed explicitly in any quantum mechanics textbook.

If $E_1 \leq E_2 \leq E_3 \leq \cdots$ are the eigenfunctions of (5), counted according to multiplicity, then the lowest eigenvalue of $\widetilde{H}_{NZ}$ is easily seen to be

$$\widetilde{E}(N, Z) = E_1 + E_2 + \cdots + E_N. \tag{6}$$

In fact, the eigenvalues of (5) are $\frac{-Z^2}{4n^2}$, with multiplicity n^2 ($n = 1, 2, 3, \ldots$). If we had taken spin into account, then we would have found that the multiplicity of $\frac{-Z^2}{4n^2}$ is $2n^2$, and for the next few paragraphs we work with $E_1, E_2, E_3, \ldots$ given by

$$E_k = \frac{-Z^2}{4n^2} \quad \text{for} \quad N(n) < k \leq N(n+1), \tag{7}$$

where $N(0) = 0$, $N(n+1) - N(n) = 2n^2$.

With E_k and $\widetilde{E}(N, Z)$ given by (6), (7), one computes easily the energy asymptotics of a neutral atom:

$$\widetilde{E}(Z, Z) \approx -(\text{const})\, Z^{7/3} \quad \text{for large} \quad Z. \tag{8}$$

Note that $\widetilde{E}(Z, Z)$ has small irregularities as a function of Z. In fact, one checks easily that $Z^{-2}\widetilde{E}(Z, Z)$ is a piecewise linear function of Z, whose slope changes at the atomic numbers $Z = N(n)$. Thus, the atomic numbers at which $E(Z, Z)/Z^2$ exhibits a kink are

$$Z = 2, 10, 28, \ldots. \tag{9}$$

Now $Z = 2$ and $Z = 10$ are the atomic numbers of helium and neon, the first two noble gases. It is natural to believe that detailed properties of $E(Z, Z)$ affect chemical reactions, since chemical binding arises from the small difference in ground-state energy between, say, a diatomic molecule and two isolated atoms. Note that $Z = 28$ is the wrong prediction for the next noble gas (argon; $Z = 18$). However, we have done amazingly well, considering that we dropped the electron repulsion in (1).

The realization that an oversimplified atom, without electron repulsion, leads to something like the periodic table goes back to Niels Bohr.

Since the 1920s, physicists and mathematicians have worked to understand the asymptotics of $E(Z, Z)$ and of $E(N, Z)$ with the electron repulsion taken into account. See [FS3] for a brief historical discussion.

The main points that we understand so far are as follows. For large Z, both $E(Z)$ and $E(Z, Z)$ are given by

$$E(Z) = -c_0\, Z^{7/3} + c_1\, Z^2 - c_2\, Z^{5/3} + O(Z^{5/3}) \tag{10}$$

for explicit, positive constants c_0, c_1, c_2. To derive this, one relates (3) to an effective one-electron Hamiltonian

$$H = -\Delta_x + Z^{4/3}\, V_{\text{TF}}\big(Z^{1/3}|x|\big) \quad \text{on} \quad L^2(\mathbb{R}^3), \tag{11}$$

where $V_{\text{TF}}(y)$ is a universal function on $(0, \infty)$, defined as the solution of an explicit ordinary differential equation with boundary conditions. The most painful part of the proof of (10) is to derive a precise asymptotic formula for the sum of the negative eigenvalues of (11) by refined WKB theory.

Seco and I have understood (11) well enough to conjecture an asymptotic formula for the irregular variations in the sum of the negative eigenvalues of (11), as a function of Z. We believe our formula can be proven with enough hard work, using the ideas in our papers [FS1–8]. We believe further that this can be used to understand the irregularities in $E(Z)$ for very large Z. It is, however, far from clear that our asymptotic analysis applies to Z as low as 100, and thus has anything to do with chemistry. The prediction of chemical binding energies from quantum mechanics by rigorous mathematics remains a very hard problem.

I've tried to motivate some mathematics here, but have not even tried to explain the statements of the results. The interested reader may get a good idea of what is involved by reading the old survey paper [L] and the introductions to our papers [FS2–8]. See also [IS] for an important result on the ground-state energy of a molecule.

References

[FS1] C. Fefferman and L. Seco, *The ground-state energy of a large atom,* Bull. A.M.S. **23** (2) (1990), 525–530.

[FS2] ———, *Eigenvalues and eigenfunctions of ordinary differential operators,* Adv. Math. **95** (1992), 145–305.

[FS3] ———, *On the Dirac and Schwinger corrections to the ground-state energy of an atom,* ibid. **107** (1994), 1–185.

[FS4] ———, *The density in a one-dimensional potential,* ibid. **107** (2) (1994), 187–364.

[FS5] ———, *The eigenvalue sum for a one-dimensional potential,* ibid. **108** (2) (1994), 263–335.

[FS6] ———, *Aperiodicity of the Hamiltonian flow in the Thomas–Fermi potential,* Rev. Math. Iberoam. **9** (3) (1993), 409–551.

[FS7] ———, *The density in a three-dimensional radial potential,* Adv. Math. **111** (1) (1995), 88–161.

[FS8] ———, *The eigenvalue sum for three-dimensional radial potential,* ibid. **119** (1) (1996), 26–116.

[IS] V. Ivrii and I. M. Sigal, *Asymptotics of the gound state energies of large Coulomb systems,* Annals of Math. **138** (1993), 243–335.

[L] E. Lieb, *Thomas–Fermi and related theories of atoms and molecules, I,* Rev. Mod. Phys. **53** (4) (1981), 603–641.

Department of Mathematics
Princeton University
Princeton, NJ 08544-1000

CHAPTER V

Local Sobolev–Bergman Kernels of Strictly Pseudoconvex Domains

Kengo Hirachi and Gen Komatsu

Introduction

This article grew out of an attempt to understand analytic aspects of Fefferman's invariant theory [F3] of the Bergman kernel on the diagonal of $\Omega \times \Omega$ for strictly pseudoconvex domains Ω in $\mathbb{C}^n$ with smooth (C^∞ or real analytic) boundary. The framework of his invariant theory applies equally to the Szegö kernel if the surface element on $\partial\Omega$ is appropriately chosen, while the Szegö kernel is regarded as the reproducing kernel of a Hilbert space of holomorphic functions in Ω which belong to the L^2 Sobolev space of order 1/2. This fact is our starting point. For each $s \in \mathbb{R}$, we first globally define the Sobolev–Bergman kernel K^s of order $s/2$ to be the reproducing kernel of the Hilbert space $H^{s/2}(\Omega)$ of holomorphic functions which belong to the L^2 Sobolev space of order $s/2$, where the inner product is specified arbitrarily.

In order to put the Sobolev–Bergman kernel K^s in the invariant theory, it is necessary to assume that K^s has two crucial properties which are satisfied by the Bergman kernel $K^{\mathrm{B}} = K^{\mathrm{B}}_\Omega$ and the (invariantly defined) Szegö kernel $K^{\mathrm{S}} = K^{\mathrm{S}}_\Omega$. The first one is the transformation law of weight $w \in \mathbb{Z}$ under biholomorphic mappings $\Phi\colon \Omega_1 \to \Omega_2$

$$K_{\Omega_1} = (K_{\Omega_2} \circ \Phi)\,|\det \Phi'|^{2w/(n+1)} \tag{0.1}$$

for a kernel (or a domain functional) $K = K_\Omega$, where $\det \Phi'$ denotes the holomorphic Jacobian of Φ. If we write $w = \mathrm{w}^{\mathrm{TL}}(K)$ for w in (0.1), then $\mathrm{w}^{\mathrm{TL}}(K^{\mathrm{B}}) = n+1$ and $\mathrm{w}^{\mathrm{TL}}(K^{\mathrm{S}}) = n$. We require the inner product of $H^{s/2}(\Omega)$ to satisfy

$$\mathrm{w}^{\mathrm{TL}}(K^s) = w(s) \quad \text{with} \quad w(s) = n+1-s \in \mathbb{Z},$$

and say that such K^s is weakly invariant. However, we don't know how to define such an inner product for $s > 0$, except for the Szegö kernel case

$s = 1$. So far, we could have defined weakly invariant Sobolev–Bergman kernels $K^s = K^s_\Omega$ only for $s = 1$ and $s \leq 0$ real (see Section 1). This is a motivation to abandon the global definition via the inner product and consider the local kernels regarded as singularities (i.e. kernels modulo smooth error) near a boundary point of reference.

The second crucial property satisfied by the Bergman kernel and the Szegö kernel is that the singularity is simple holonomic, that is,

$$K = \frac{\varphi}{r^w} + \psi \log r \quad (w > 0), \quad K = \varphi r^{-w} \log r \quad (w \leq 0) \tag{0.2}$$

with $w \in \mathbb{Z}$, where r is a (smooth) defining function of $\partial\Omega$ such that $r > 0$ in Ω, and φ, ψ are smooth functions on $\overline{\Omega}$ (near $\partial\Omega$) such that φ does not vanish on $\partial\Omega$. We have $w = \mathrm{w}^{\mathrm{TL}}(K)$ for w in (0.2) if $K = K^{\mathrm{B}}$, K^{S}. Furthermore, the singularities of K^{B} and K^{S} are localizable to a neighborhood of a reference boundary point. In fact, these are obtained by patching locally defined singularities along the boundary $\partial\Omega$. We thus require $w = w(s)$ in defining local Sobolev–Bergman kernels $K^s = K^s_{\mathrm{loc}}$ with simple holonomic singularity. If in addition K^s is weakly invariant, we say that K^s is strongly invariant. This property is necessary in discussing the invariant theory of K^s.

In order to define local Sobolev–Bergman kernels, we first assume for simplicity that the (local) defining function r of $\partial\Omega$ is real analytic, so that we may write $r = r(z, \overline{z})$. We then use Kashiwara's characterization of the local Bergman kernel $K^{\mathrm{B}} = K^{\mathrm{B}}_{\mathrm{loc}}(z, \overline{z})$. Kashiwara [Kas] wrote down a system of microdifferential equations characterizing K^{B} up to a constant multiple by using another system satisfied by $\log r$. According to Boutet de Monvel [BM1]–[BM3], one can in fact define a transformation $\log r \mapsto K^{\mathrm{B}}$, where the singularity $\log r$ represents the domain Ω locally. In other words, K^{B} is a local domain functional via $\log r$. On the other hand, Sato's hyperfunction theory asserts that any simple holonomic singularity $\widehat{K}$, with respect to r which is fixed, is written as $\widehat{K} = \mathbf{A} \log r$, where $\mathbf{A} = \mathbf{A}[\widehat{K}]$ is a specific linear transformation (a microdifferential operator of finite order) which is holomorphic in z. Then $K = \mathbf{A}^{*-1} K^{\mathrm{B}}$ is again a simple holonomic singularity, where $\mathbf{A}^*$ denotes the formal adjoint of $\mathbf{A}$ defined formally by integration by parts without taking the complex conjugate. The mapping $\widehat{K} \mapsto K$ is consistent with Kashiwara's transformation $\log r \mapsto K^{\mathrm{B}}$, and the Szegö kernel $K = K^{\mathrm{S}}$ is obtained by choosing $\widehat{K}$ to be a constant multiple of $1/r$ with an appropriate choice of r which defines $\partial\Omega$ locally. Taking account of this fact, we first define in Subsection 2.1 the local Sobolev–Bergman kernels $K^s = K^s_{\mathrm{loc}}$ with respect to any (local) defining function r by taking

$$\widehat{K^s} = r^{-s} \quad (s > 0), \qquad \widehat{K^s} = r^{-s} \log r \quad (s \leq 0)$$

for $s \in \mathbb{Z}$, where normalization constants are ignored. We then define in Subsection 2.3 the (invariant) local Sobolev–Bergman kernel $K^s_{\mathrm{loc}} = K^s_\Omega$ for each $s \in \mathbb{Z}$ as a local domain functional by requiring that the defining functions $r = r_\Omega$ are so chosen that K^s_{loc} is strongly invariant. Here, the word "strongly" can be omitted, because the strong invariance of K^s_{loc} is reduced to the weak one, the (local) transformation law. In case $\partial\Omega$ is not real analytic but C^∞, the local Sobolev–Bergman kernels are regarded as formal singularities (see Section 3).

As we prove in Subsection 2.2, the invariance of $K^s = K^s_{\mathrm{loc}}$ is equivalent to that of $\widehat{K^s}$, which obviously comes from the transformation law for the defining function $r = r_\Omega$ as a local domain functional. However, the situation is somewhat complicated because the transformation law for r holds only approximately. In [F2], Fefferman constructed r such that $\mathrm{w}^{\mathrm{TL}}(r) = -1$ modulo $O(r^{n+2})$. This error estimate is optimal (Theorem 2). Consequently, the local Sobolev–Bergman kernel K^s_{loc} which by definition is invariant exists if and only if $0 \leq s \leq n+1$ (Theorem 1). These two theorems are the main results of this paper stated in Subsection 2.3. Theorem 1 suggests that, for $0 > s \in \mathbb{Z}$, weakly invariant Sobolev–Bergman kernels K^s which are globally defined do not have simple holonomic singularities, though we don't know anything about the singularities in this case.

We emphasize that the invariance of K^s_{loc} for $0 \leq s \leq n+1$ holds without error, though that of the best possible r is approximate with error of $O(r^{n+2})$. More precisely, the invariance of K^s_{loc} follows from that of r modulo $O(r^{s+1})$ for $0 \leq s \leq n+1$.

For the local Sobolev–Bergman kernel K^s with $0 \leq s \leq n+1$, we can apply Fefferman's invariant theory to get an approximately invariant asymptotic expansion similar to those for K^{B} and K^{S}. Though there are some technical difficulties to be examined such as the polynomial dependence on Moser's normal form coefficients $A = (A^\ell_{\alpha\overline{\beta}})$, we can verify these by inspecting the construction (see Section 4). In fact, the polynomial dependence on $A = (A^\ell_{\alpha\overline{\beta}})$ is taken into account in the definition of K^s_{loc}. All abstract results as in Fefferman [F3] and Bailey-Eastwood-Graham [BEG] for K^{B} are evidently valid as well for K^s, whereas explicit results for K^s such as the determination of universal constants in Graham [G1] and [HKN1], [HKN2] for K^{B} and K^{S} are obtained by computer-aided calculation. These results are stated in Sections 4 with the method of computation explained in Appendix B.

The first author has recently obtained in [Hi] an invariant asymptotic expansion of the Bergman kernel without error via a special family of defining functions r of $\partial\Omega$, where the family is parametrized formally by $C^\infty(\partial\Omega)$ and the transformation law is made to hold within the family. The method applies in getting similar expansions of the Szegö kernel and the local Sobolev–Bergman kernels in the present paper as well. Though the present paper discusses Fefferman's approach so that the best possible defining functions r has the ambiguity $O(r^{n+2})$, the proof of the optimality of this error estimate (i.e. Theorem 2) is done here by using the theory in [Hi] (see Section 5).

1. Globally defined Sobolev–Bergman kernels

Let Ω be a bounded strictly pseudoconvex domain in $\mathbb{C}^n$ with smooth boundary. For $s \in \mathbb{R}$, we denote by $H^{s/2}_{\mathrm{top}}(\Omega)$ the topological vector space consisting of holomorphic functions in Ω which are contained in the L^2 Sobolev space of order $s/2$. When an inner product $(\,\cdot\,,\,\cdot\,)_{s/2}$ is specified, we write $H^{s/2}_{\mathrm{top}}(\Omega)$ as $H^{s/2}(\Omega)$. Then $H^{s/2}(\Omega)$ is a Hilbert space which admits the reproducing kernel $K^s(z,\overline{w})$ for $z, w \in \Omega$ defined by

$$K^s(z,\overline{w}) = \sum_j h_j(z)\,\overline{h_j(w)},$$

where $\{h_j\}_j$ is an arbitrary complete orthonormal system of $H^{s/2}(\Omega)$. We set $K^s(z) = K^s(z,\overline{z})$.

Definition 1.1. The reproducing kernel $K^s(z,\overline{w})$, or rather $K^s(z)$, is called the *Sobolev–Bergman kernel associated with* $H^{s/2}(\Omega)$.

The simplest case is $s = 0$. If $(\,\cdot\,,\,\cdot\,)_0$ is the standard L^2 inner product

$$(h_1,h_2)_0 = \int_\Omega h_1(z)\,\overline{h_2(z)}\,dV(z), \quad dV = \bigwedge_{j=1}^n \frac{dz_j \wedge d\overline{z}_j}{-2i},$$

then $K^0 = K^{\mathrm{B}}$, where K^{B} denotes the Bergman kernel. When we wish to emphasize the dependence on the domain Ω, we write $K^{\mathrm{B}} = K^{\mathrm{B}}_\Omega$. In fact, the Bergman kernel is a domain functional, and it is elementary that if $\Phi\colon \Omega_1 \to \Omega_2$ is biholomorphic, then

$$K^{\mathrm{B}}_{\Omega_1}(z) = K^{\mathrm{B}}_{\Omega_2}(\Phi(z))\,|\det \Phi'(z)|^2,$$

where $\Phi' = \partial\Phi/\partial z$ and thus $\det \Phi'$ is the holomorphic Jacobian of Φ. More generally, we follow Fefferman and make the following:

Definition 1.2. If a domain functional $K = K_\Omega$ satisfies

$$K_{\Omega_1}(z) = K_{\Omega_2}(\Phi(z))\,|\det \Phi'(z)|^{2w/(n+1)} \tag{1.1}$$

whenever $\Phi\colon \Omega_1 \to \Omega_2$ is biholomorphic, then we say that K satisfies the *transformation law of weight* w, and write $\mathrm{w}^{\mathrm{TL}}(K) = w$.

Another well-known example is the Szegö kernel K^1. Here, we may choose an inner product on $H^{1/2}(\Omega)$ to be given by

$$(h_1, h_2)_{1/2} = \int_{\partial\Omega} h_1(z)\,\overline{h_2(z)}\,\sigma(z),$$

where σ is a surface element on $\partial\Omega$. Thus $H^{1/2}(\Omega)$ depends on σ. It is possible to choose σ in such a way that K^1 satisfies the transformation law, as follows.

Let us take a smooth positive defining function $\rho \in C^\infty(\overline{\Omega})$, and thus

$$\Omega = \{z \in \mathbb{C}^n;\ \rho(z) > 0\}, \qquad d\rho(z) \neq 0 \quad \text{for} \quad z \in \partial\Omega.$$

Let $J[\,\cdot\,]$ denote the *Levi determinant* or the (complex) *Monge-Ampère operator* defined by

$$J[\rho] = (-1)^n \det \begin{pmatrix} \rho & \partial\rho/\partial\overline{z}_k \\ \partial\rho/\partial z_j & \partial^2\rho/\partial z_j\partial\overline{z}_k \end{pmatrix} \quad (j, k = 1, \ldots, n). \tag{1.2}$$

We then have $\mathrm{w}^{\mathrm{TL}}(K^1) = n$, provided the surface element σ is subject to the normalization

$$\sigma \wedge d\rho = J[\rho]^{1/(n+1)}\,dV \quad \text{on } \partial\Omega.$$

In this case, we write K^1 as K^{S} and call it the *invariant Szegö kernel* or just the *Szegö kernel.* Thus

$$\mathrm{w}^{\mathrm{TL}}(K^{\mathrm{B}}) = n+1, \quad \mathrm{w}^{\mathrm{TL}}(K^{\mathrm{S}}) = n.$$

These numbers coincide with the magnitude of the singularities. In fact, according to a celebrated theorem of Fefferman [F1] (see also Boutet de Monvel and Sjöstrand [BS]), there exist functions $\varphi^{\mathrm{B}} = \varphi^{\mathrm{B}}[\rho]$ and $\psi^{\mathrm{B}} = \psi^{\mathrm{B}}[\rho]$ in $C^\infty(\overline{\Omega})$ such that

$$\frac{\pi^n}{n!} K^{\mathrm{B}} = \frac{\varphi^{\mathrm{B}}}{\rho^{n+1}} + \psi^{\mathrm{B}} \log \rho, \quad \left(\varphi^{\mathrm{B}} - J[\rho]\right)\big|_{\partial\Omega} = 0. \tag{1.3}$$

Similarly, there exist $\varphi^{\mathrm{S}} = \varphi^{\mathrm{S}}[\rho]$ and $\psi^{\mathrm{S}} = \psi^{\mathrm{S}}[\rho]$ in $C^\infty(\overline{\Omega})$ such that

$$\frac{\pi^n}{(n-1)!} K^{\mathrm{S}} = \frac{\varphi^{\mathrm{S}}}{\rho^n} + \psi^{\mathrm{S}} \log \rho, \quad \left(\varphi^{\mathrm{S}} - J[\rho]^{n/(n+1)}\right)\Big|_{\partial\Omega} = 0. \tag{1.4}$$

(Note that $J[\rho] > 0$ on $\partial\Omega$ by the strict pseudoconvexity.)

We are interested in the Sobolev–Bergman kernel K^s satisfying

$$\mathrm{w}^{\mathrm{TL}}(K^s) = w(s), \quad \text{where} \quad w(s) = n + 1 - s \in \mathbb{Z}. \tag{1.5}$$

It will be also natural to require the existence of $\varphi^s, \psi^s \in C^\infty(\overline{\Omega})$ such that

$$c_{s,n}^{\mathrm{SB}} K^s = \begin{cases} \varphi^s \rho^{-w(s)} + \psi^s \log \rho & \text{for } w(s) > 0, \\ \varphi^s \rho^{-w(s)} \log \rho & \text{for } w(s) \le 0, \end{cases} \tag{1.6}$$

where $c_{s,n}^{\mathrm{SB}} \neq 0$ are normalization constants so chosen that

$$\left(\varphi^s - J[\rho]^{w(s)/(n+1)}\right)\Big|_{\partial\Omega} = 0.$$

Definition 1.3. A Sobolev–Bergman kernel K^s is said to be *weakly invariant* if the condition (1.5) holds. If in addition the condition (1.6) holds. then K^s is said to be *invariant.*

If the conditions (1.5) and (1.6) are not taken into account, it is easy to give examples of Sobolev–Bergman kernels K^s for any $s \in \mathbb{R}$, by specifying an inner product $(\,\cdot\,,\,\cdot\,)_{s/2}$. For instance, if $s/2 > 0$ is an integer, then we may take, with the usual (commutative) multi-index notation,

$$(h_1, h_2)_{s/2} = \int_\Omega \sum_{|\alpha|+|\beta| \le s/2} \left(\partial_z^\alpha \partial_{\overline{z}}^\beta h_1\right)\left(\partial_{\overline{z}}^\alpha \partial_z^\beta \overline{h}_2\right) dV, \tag{1.7}$$

though the condition (1.5) breaks down.

Remark 1. In case Ω is a ball in $\mathbb{C}^n$ and $H^{s/2}(\Omega)$ is specified by the inner product (1.7) with $s/2 \in \mathbb{N}_0$, Boas [Bo] showed that the reproducing kernel K^s takes the form (1.6), where the logarithmic terms appear even for $0 < s < n$. It is easy to define an inner product of $H^s(\Omega)$, for each domain Ω which is biholomorphic to a ball, in such a way that the transformation law (1.1) with $w = w(s)$ holds for the reproducing kernels $K = K^s$ of such domains. However, such reproducing kernels are not defined for domains which are not biholomorphic to a ball. In other words, a domain functional $K^s = K_\Omega^s$ is not determined as a weakly invariant Sobolev–Bergman kernel.

In case $s < 0$ is a real number, we can define a weakly invariant Sobolev–Bergman kernel as follows. An inner product on $H_{\mathrm{top}}^{s/2}(\Omega)$ is given by

$$(h_1, h_2)_{s/2} = \int_\Omega h_1(z)\, \overline{h_2(z)}\, \rho(z)^{-s}\, dV(z)$$

for any smooth defining function $\rho > 0$ of Ω. Moreover, we may replace ρ by any continuous function $u > 0$ of the same magnitude as ρ to have

$$(h_1, h_2)_{s/2} = \int_\Omega h_1(z)\,\overline{h_2(z)}\,u(z)^{-s}\,dV(z). \tag{1.8}$$

We have:

Proposition 1.1. *Let $0 > s \in \mathbb{R}$. If $u = u_\Omega$ satisfies*

$$\mathrm{w}^{\mathrm{TL}}(u) = -1, \quad 0 < \inf u/\rho \le \sup u/\rho < +\infty, \tag{1.9}$$

then the Sobolev–Bergman kernel K^s defined by (1.8) is weakly invariant.

Proof. If $\Phi\colon \Omega_1 \to \Omega_2$ is biholomorphic, then

$$u_1 = (u_2 \circ \Phi)\,|\det \Phi'|^{-2/(n+1)}, \quad u_\ell = u_{\Omega_\ell} \quad (\ell = 1, 2).$$

It then follows that an isometry $\Phi^*\colon H^{s/2}(\Omega_2) \to H^{s/2}(\Omega_1)$ is given by

$$\Phi^*\widetilde{h} = (\widetilde{h} \circ \Phi)(\det \Phi')^{w(s)/(n+1)}.$$

If $\{\widetilde{h}_j\}$ is a complete orthonormal system of $H^{s/2}(\Omega_2)$, then a complete orthonormal system $\{h_j\}$ of $H^{s/2}(\Omega_1)$ is defined by $h_j = \Phi^*\widetilde{h}_j$. Thus, the transformation law (1.1) for $K^s = K^s_\Omega$ follows from

$$\sum_j |h_j|^2 = \sum_j |\widetilde{h}_j \circ \Phi|^2\,|\det \Phi'|^{2w(s)/(n+1)}.$$

□

Examples of $u = u_\Omega$ satisfying (1.9) are given by

$$u^{\mathrm{B}} = \left(c^{\mathrm{SB}}_{n,n}\,K^{\mathrm{B}}\right)^{-1/(n+1)} \quad \text{or} \quad u^{\mathrm{S}} = \left(c^{\mathrm{SB}}_{n-1,n}\,K^{\mathrm{S}}\right)^{-1/n}.$$

Another important example is given by the solution $u = u^{\mathrm{MA}}$ of the boundary value problem

$$J[u] = 1 \ \text{ and } \ u > 0 \ \text{ in } \Omega; \qquad u = 0 \ \text{ on } \partial\Omega. \tag{1.10}$$

The unique existence of a solution of (1.10) in $C^\infty(\Omega) \cap C^{n+3/2-\varepsilon}(\overline{\Omega})$ was proved by Cheng and Yau [CY]. Thus the first relation in (1.9) follows from the fact (see [F2]) that if $\Phi\colon \Omega_1 \to \Omega_2$ is biholomorphic, then

$$J[u_1] = J[u_2] \circ \Phi, \quad \text{where} \quad u_1 = (u_2 \circ \Phi) \cdot |\det \Phi'|^{-2/(n+1)}.$$

The second relation in (1.9) follows from the asymptotic expansion due to Lee and Melrose [LM]:

$$u^{\mathrm{MA}} \sim \rho \sum_{k=0}^{\infty} \eta_k \cdot (\rho^{n+1} \log \rho)^k, \quad \eta_k \in C^\infty(\overline{\Omega})$$

where $\eta_0|_{\partial\Omega} > 0$. Thus $u = u^{\mathrm{MA}}$ satisfies (1.9).

If Ω is the unit ball Ω_{ball}, then

$$u^{\mathrm{B}}_{\Omega_{\mathrm{ball}}}(z) = K^{\mathrm{S}}_{\Omega_{\mathrm{ball}}}(z) = u^{\mathrm{MA}}_{\Omega_{\mathrm{ball}}} = 1 - |z|^2.$$

In contrast to Boas' result in Remark 1, we have:

Proposition 1.2. *Let $K^s = K^s_\Omega$ be a weakly invariant Sobolev–Bergman kernel of order $s/2$, $0 > s \in \mathbb{R}$, defined by the inner product* (1.8) *with either one of $u = u^{\mathrm{B}}$, u^{S} or u^{MA}. Then*

$$K^s_{\Omega_{\mathrm{ball}}}(z) = \frac{\Gamma(w(s))}{\pi^n \, \Gamma(1-s)} \frac{1}{(1-|z|^2)^{w(s)}}.$$

Proof. For $n = 1$, the result follows by using the fact that monomials form a complete orthogonal system of $H^{s/2}(\Omega_{\mathrm{ball}})$. For $n \geq 2$, we consider

$$K^s_{\mathrm{aux}}(z) = \sum_{\alpha \in \mathbb{N}_0^n} |h_\alpha(z)|^2, \qquad h_\alpha(z) := \frac{z^\alpha}{\|z^\alpha\|_{s/2}},$$

where $\|\cdot\|_{s/2}$ is the norm corresponding to the inner product $(\,\cdot\,,\,\cdot\,)_{s/2}$. It suffices to show that

$$K^s_{\mathrm{aux}}(z) = \frac{\Gamma(w(s))}{\pi^n \, \Gamma(1-s)} \frac{1}{(1-|z|^2)^{w(s)}}, \qquad K^s_{\mathrm{aux}}(z) = K^s_{\Omega_{\mathrm{ball}}}(z).$$

The first equality is obtained by direct computation using the result for $n = 1$. The second one is equivalent to the completeness of the orthonormal system $\{h_\alpha\}$, and the proof of this fact is done by noting that

$$K^s_{\Omega_{\mathrm{ball}}}(z) = \sup \left\{ |h(z)|^2 / \|h\|^2_{s/2}; \ \ 0 \neq h \in H^{s/2}(\Omega_{\mathrm{ball}}) \right\},$$

just as in the proof for $s = 0$ given by Hörmander [Hö]. □

2. Definition of local Sobolev–Bergman kernels

In this section, we consider the local Sobolev–Bergman kernel of order $s/2$ for $s \in \mathbb{Z}$ and the invariance in the sense of (1.5). We begin with the motivation because the definition is somewhat technical.

An important fact is that the singularities of the Bergman kernel $K^{\mathrm{B}}(z)$ and the (invariant) Szegö kernel $K^{\mathrm{S}}(z)$ as in (1.3) and (1.4) can be localized to any boundary point, say $p \in \partial\Omega$. That is, if $\Omega_1 \cap U = \Omega_2 \cap U$ for a neighborhood $U \subset \mathbb{C}^n$ of p, then $K_{\Omega_1} - K_{\Omega_2}$ for $K = K^{\mathrm{B}}$ or K^{S} is *smooth* near $p \in \partial\Omega$, where smooth means C^∞ or C^ω (real analytic) in accordance with the regularity of $\partial\Omega$ near p. Furthermore, one can define local kernels $K_{\mathrm{loc}} = K^{\mathrm{B}}_{\mathrm{loc}}$ and $K^{\mathrm{S}}_{\mathrm{loc}}$ by requiring the following three conditions:

(i) $K_{\mathrm{loc}}(z, \overline{w})$ is holomorphic in z and anti-holomorphic in w for $z, w \in \Omega \cap U$. Two local kernels K_{loc} and $\widetilde{K}_{\mathrm{loc}}$ are identified when the difference is smooth in $\mathbb{C}^n$ near p. Thus U can be shrunk arbitrarily.

(ii) $K_{\mathrm{loc}} = K^{\mathrm{B}}_{\mathrm{loc}}$ and $K^{\mathrm{S}}_{\mathrm{loc}}$ have singularities of the form (1.3) and (1.4), respectively, where $\varphi = \varphi^{\mathrm{B}}$, φ^{S} and $\psi = \psi^{\mathrm{B}}$, ψ^{S} are smooth in $\Omega \cap U$.

(iii) Reproducing properties modulo smooth errors hold, that is,

$$\int_{\Omega\cap U} K^{\mathrm{B}}_{\mathrm{loc}}(z, \overline{w})\, f_1(w)\, dV(w) - f_1(z) \sim 0,$$

$$\int_{\partial\Omega\cap U} K^{\mathrm{S}}_{\mathrm{loc}}(z, \overline{w})\, f_2(w)\, \sigma(w) - f_2(z) \sim 0,$$

for holomorphic functions f_1 and f_2 in U_0, where $U_0 \subset \mathbb{C}^n$ is an open set satisfying $p \in U \Subset U_0$, and each f_2 is regarded as the boundary value. In case $\partial\Omega$ is C^∞ near p, $f_1(z)$ and $f_2(z)$ are required to be of polynomial growth in $1/\mathrm{dist}\,(z, \partial\Omega)$. (If $\partial\Omega$ is C^ω near p, then no restriction on f_1 and f_2 is necessary, provided the pairings are interpreted in the sense of hyperfunctions, cf. Kaneko [Kan].)

The local kernels $K^{\mathrm{B}}_{\mathrm{loc}}$ and $K^{\mathrm{S}}_{\mathrm{loc}}$ are uniquely determined by the requirements (i)–(iii). We wish to define the local Sobolev–Bergman kernel K^s_{loc} for $s \in \mathbb{N}$ in a similar way. Our main concern is the invariance in the sense of (1.5) under local biholomorphic mappings. However, the condition (iii) uses the inner products, and we don't know how to define $(\,\cdot\,,\,\cdot\,)_s$ for $s \in \mathbb{N}$ such that the Sobolev–Bergman kernel K^s is invariant. We thus abandon (iii) and instead adopt Kashiwara's characterization of the local Bergman kernel $K^{\mathrm{B}}_{\mathrm{loc}}$, a method which applies equally to the local Szegö kernel $K^{\mathrm{S}}_{\mathrm{loc}}$.

In this section, we assume that $\partial\Omega$ is C^ω near p. We are only concerned with local kernels $K_{\mathrm{loc}}(z, \overline{w})$ defined near $(z, w) = (p, p)$, and thus the subscript loc will be omitted.

2.1. Kashiwara's transformation. We fix a local defining function r of Ω near $p \in \partial\Omega$ and assume for a moment that r is real analytic. Then r has a holomorphic extension to a neighborhood of $M \times \overline{M}$ in $\mathbb{C}^n \times \mathbb{C}^n$, where $M \subset \partial\Omega$ is a neighborhood of p (or more precisely, a germ of $\partial\Omega$ at p). Denoting it again by r, we set, for $m \in \mathbb{Z}$,

$$\widehat{K}_m[r] = \begin{cases} \dfrac{1}{m!}\, r^m \log r & \text{for } m \geq 0, \\ (-1)^{m+1}(-m-1)!\, \dfrac{1}{r^{-m}} & \text{for } m < 0, \end{cases}$$

and consider singularities of the form

$$K = \begin{cases} \varphi \widehat{K}_{-w}[r] + \psi \widehat{K}_0[r] & \text{if } w > 0, \\ \varphi \widehat{K}_{-w}[r] & \text{if } w \leq 0, \end{cases}$$

where φ an ψ are holomorphic in $(z, \overline{z})$ near $M \times \overline{M}$ for some M. We denote by $\mathcal{C}_p^\times$ the totality of K such that $\varphi \neq 0$ near $M \times \overline{M}$. By [SKK], if $\widehat{K} \in \mathcal{C}_p^\times$ then, for any holomorphic microdifferential operator $P = P(z, \partial_z)$, there exists an antiholomorphic microdifferential operator $Q = Q(\overline{z}, \partial_{\overline{z}})$ such that $P\widehat{K} = Q\widehat{K}$. Furthermore, if $P_j = P_j(z, \partial_z)$ for $j = 1, \ldots, 2n$ are chosen independently then $\widehat{K}$ is determined up to a multiplicative constant by

$$P_j(z, \partial_z)\widehat{K}(z, \overline{z}) = Q_j(\overline{z}, \partial_{\overline{z}})\widehat{K}(z, \overline{z}) \quad \text{for} \quad j = 1, \ldots, 2n. \tag{2.1}$$

(A more rigorous description of [SKK] will be given in Appendix A.) Let us consider another system of microdifferential equations for $K \in \mathcal{C}_p^\times$

$$P_j^*(z, \partial_z)K(z, \overline{z}) = Q_j^*(\overline{z}, \partial_{\overline{z}})K(z, \overline{z}) \quad \text{for} \quad j = 1, \ldots, 2n, \tag{2.2}$$

where P_j^*, Q_j^* are formal adjoints of P_j, Q_j, respectively. The independence of P_j implies that of P_j^*, so that the solution of (2.2), if it exists, is unique up to a multiplicative constant.

Kashiwara's theorem ([Kas]). *If $\widehat{K} = \widehat{K}_0[r] = \log r$ in* (2.1) *then* (2.2) *is satisfied by the local Bergman kernel $K = K^{\mathrm{B}}$.*

By [SKK], if $\widehat{K} \in \mathcal{C}_p^\times$ then there exists a unique invertible holomorphic microdifferential operator $\mathbf{A}[\widehat{K}]$ such that

$$\widehat{K}(z, \overline{z}) = \mathbf{A}(z, \partial_z)\widehat{K}_0(z, \overline{z}) \quad \text{with} \quad \mathbf{A} = \mathbf{A}[\widehat{K}], \quad \widehat{K}_0 = \widehat{K}_0[r]. \tag{2.3}$$

Thus, Kashiwara's theorem yields

Lemma 2.1. *If* (2.1) *and* (2.3) *hold for* $\widehat{K} \in \mathcal{C}_p^\times$, *then* (2.2) *is satisfied by*

$$K(z, \overline{z}) = \mathbf{A}^*(z, \partial_z)^{-1} K^{\mathrm{B}}(z, \overline{z}).$$

Proof. Since $\mathbf{A}$ is a holomorphic operator and thus $Q_j \mathbf{A} = \mathbf{A} Q_j$, it follows that $P_j \mathbf{A} \widehat{K}_0 = Q_j \mathbf{A} \widehat{K}_0 = \mathbf{A} Q_j \widehat{K}_0$, that is, $\mathbf{A}^{-1} P_j \mathbf{A} \widehat{K}_0 = Q_j \widehat{K}_0$, so that Kashiwara's theorem yields $Q_j^* K^{\mathrm{B}} = \mathbf{A}^* P_j^* \mathbf{A}^{*-1} K^{\mathrm{B}} = \mathbf{A}^* P_j^* K$. Using $\mathbf{A}^{*-1} Q_j^* = Q_j^* \mathbf{A}^{*-1}$, we get $P_j^* K = \mathbf{A}^{*-1} Q_j^* K^{\mathrm{B}} = Q_j^* K$. □

Since $\mathbf{A}[K^{\mathrm{B}}]^* = \mathbf{A}[K^{\mathrm{B}}]$, it follows that

$$\mathcal{C}_p^\times \ni \widehat{K} \mapsto K \in \mathcal{C}_p^\times \tag{2.4}$$

given by Lemma 2.1 is an involution. We refer to it as *Kashiwara's transformation.*

Definition 2.1. Let r be a real analytic local defining function of Ω near $p \in \partial\Omega$. For $s \in \mathbb{Z}$, we define $K^s[r] = K$ by $\widehat{K} = \widehat{K}_{-s}[r]$ in (2.4) and call $K^s[r]$ the *local Sobolev–Bergman kernel of order $s/2$ with respect to r.*

By the definition via Kashiwara's theorem, we have $K^0[r] = (\text{const.})\, K^{\mathrm{B}}$ independently of the choice of r. We also have $K^1[r] = (\text{const.})\, K^{\mathrm{S}}$ if $J[r] = 1$ on $\partial\Omega$.

2.2. Biholomorphic transformation law. We wish to define a local Sobolev–Bergman kernel of Sobolev order $s/2$ for $s \in \mathbb{Z}$ as a local domain functional $K^s = (K^s_\Omega)_\Omega$ near the reference points $p_\Omega \in \partial\Omega$, say $p_\Omega = 0 \in \mathbb{C}^n$, where we continue to assume that $\partial\Omega$ is real analytic near 0. In the definition, we require three conditions of which the first two are:

Condition SB1. Each K^s_Ω is of the form $K^s_\Omega = K^s[r_\Omega]$, where r_Ω is a local defining function of Ω near $0 \in \mathbb{C}^n$. That is, K^s_Ω is the local Sobolev–Bergman kernel with respect to r_Ω.

Condition SB2. The family $r = r_\Omega$ is so chosen that $K^s = K^s_\Omega$ satisfies the transformation law of weight $w(s)$

$$K^s_\Omega = (K^s_{\widetilde{\Omega}} \circ \Phi) |\det \Phi'|^{2w(s)/(n+1)} \quad \text{with} \quad w(s) = n + 1 - s \tag{2.5}$$

under local biholomorphic mappings $\Phi \colon \Omega \to \widetilde{\Omega}$ defined near the origin such that $\Phi(0) = 0$.

The third condition is somewhat complicated, and the precise statement is postponed to the next subsection. That condition is motivated by the result of this subsection.

Assume Condition SB1. Then the validity of Condition SB2 depends on the approximate transformation law of weight -1 for the family $r = (r_\Omega)_\Omega$:

$$r_\Omega = (r_{\widetilde{\Omega}} \circ \Phi)|\det \Phi'|^{-2/(n+1)} \mod O(r_\Omega^N), \tag{2.6}$$

where $O(r_\Omega^N)$ stands for terms which are smoothly divisible by r_Ω^N. In fact, we have:

Proposition 2.1. *Assume there exists $N_0 \in \mathbb{N}$ such that $r = r_\Omega$ satisfies the transformation law (2.6) for $N = N_0$ but not for $N = N_0 + 1$. Then the transformation law (2.5) is valid if and only if $0 \leq s \leq N_0 - 1$.*

This is consistent with the independence of K^{B} and the dependence of K^{S} on $r = (r_\Omega)$.

In the proof of Proposition 2.1, we need the following property of Kashiwara's transformation.

Lemma 2.2. *Assume Condition SB1. Then $K_\Omega = K_\Omega^s$ satisfies (2.5) if and only if $\widehat{K}_\Omega$ satisfies*

$$\widehat{K}_\Omega = (\widehat{K}_{\widetilde{\Omega}} \circ \Phi)|\det \Phi'|^{2s/(n+1)}. \tag{2.7}$$

Proof. We shall show that (2.7) implies (2.5). The proof of the converse is similar. What we have to show is that

$$\widehat{K}_\Omega = |f|^{2s}\Phi^*\widehat{K}_{\widetilde{\Omega}} \quad \text{implies} \quad K_\Omega = |f|^{2w(s)}\Phi^* K_{\widetilde{\Omega}},$$

where $f = (\det \Phi')^{1/(n+1)}$ and Φ^* stands for the pull-back by Φ. Let us abbreviate by writing $\mathbf{A}_\Omega = \mathbf{A}[\widehat{K}_\Omega]$, and similarly for $\widetilde{\Omega}$ in place of Ω. Then the assumption (2.7) is further written as

$$\mathbf{A}_\Omega \log r_\Omega = |f|^{2s}\Phi^*\mathbf{A}_{\widetilde{\Omega}}(\Phi^{-1})^*\Phi^* \log r_{\widetilde{\Omega}}.$$

The right side is simplified by setting $\widetilde{\mathbf{A}} = \Phi^*\mathbf{A}_{\widetilde{\Omega}}(\Phi^{-1})^*$, using $\Phi^* \log r_{\widetilde{\Omega}} = \log r_\Omega$, and choosing a holomorphic microdifferential operator $P = P(z, \partial_z)$ such that $P \log r_\Omega = \overline{f}^s \log r_\Omega$. Since $\widetilde{\mathbf{A}}$ is a holomorphic operator, it follows that

$$\mathbf{A}_\Omega \log r_\Omega = f^s \widetilde{\mathbf{A}} P \log r_\Omega, \quad \text{so that} \quad \mathbf{A}_\Omega = f^s \widetilde{A} P$$

(see Appendix A). Using $\Phi^*\mathbf{A}_{\widetilde{\Omega}}^*(\Phi^{-1})^* = f^{-n-1}\widetilde{\mathbf{A}}^* f^{n+1}$, we get

$$(\mathbf{A}_\Omega^*)^{-1} = f^{w(s)}\Phi^*(\mathbf{A}_{\widetilde{\Omega}}^*)^{-1}(\Phi^{-1})^* f^{-n-1}(P^*)^{-1},$$

where f^{-n-1} acts as a multiplication operator. We apply both sides to K^{B}_{Ω}. Noting that $P^* K^{\mathrm{B}}_{\Omega} = \overline{f}^s K^{\mathrm{B}}_{\Omega}$, we have

$$(P^*)^{-1} K^{\mathrm{B}}_{\Omega} = \overline{f}^{-s} K^{\mathrm{B}}_{\Omega} = f^{n+1} \overline{f}^{w(s)} \Phi^* K^{\mathrm{B}}_{\Omega}.$$

Since $\overline{f}^{w(s)}$ commutes with a holomorphic operator, it follows that

$$K_\Omega = |f|^{2w(s)} \Phi^* (\mathbf{A}^*_{\Omega})^{-1} K^{\mathrm{B}}_{\Omega} = |f|^{2w(s)} \Phi^* K_{\Omega},$$

which is the desired conclusion (2.5). □

Proof of Proposition 2.1. Let us abbreviate by writing $\widetilde{r} = r_{\Omega} \circ \Phi$ and $f = (\det \Phi')^{1/(n+1)}$. In case $s \leq 0$, (2.7) is written as

$$r_{\Omega}^{-s} \log r_{\Omega} = \widetilde{r}^{-s} |f|^{2s/(n+1)} \log \widetilde{r}.$$

For $s = 0$, this is always the case. For $s < 0$, this is valid if and only if (2.6) holds for any positive integer m. In case $s > 0$, (2.7) is written as

$$r_{\Omega}^{-s} = \widetilde{r}^{-s} |f|^{2s/(n+1)}.$$

This is valid if and only if (2.6) holds for $N \geq s + 1$. Thus the desired result follows from Lemma 2.2. □

2.3. Definition of local Sobolev–Bergman kernel. We are in a position to state a condition on the family $r = (r_\Omega)_\Omega$, to be called Condition SB3. This consists of the approximate transformation law (2.6) for $m = s + 1$ and the polynomial dependence on Moser's normal form coefficients.

Recall that Moser's normal form is a real hypersurface of the form

$$N(A): \quad \rho_A = 2u - |z'|^2 - \sum_{\ell=0}^{\infty} \sum_{|\alpha|,|\beta| \geq 2} A^{\ell}_{\alpha\overline{\beta}} \, z'_\alpha \, \overline{z'_\beta} \, v^\ell = 0,$$

with normal coordinates $z = (z', z_n) \in \mathbb{C}^{n-1} \times \mathbb{C}$, $u = \operatorname{Re} z_n$, $v = \operatorname{Im} z_n$, such that $A = (A^{\ell}_{\alpha\overline{\beta}})$ is subject to the following conditions:

(N1) Each $A^{\ell}_{p\overline{q}} = (A^{\ell}_{\alpha\overline{\beta}})_{|\alpha|=p,|\beta|=q}$ is a bisymmetric tensor of type (p, q) on $\mathbb{C}^{n-1}$. That is, α, β are ordered multi-indices such as $\alpha = \alpha_1 \ldots \alpha_p$, $1 \leq \alpha_j \leq n - 1$, and $A^{\ell}_{\alpha\overline{\beta}}$ is unchanged under permutation of α and that of β.

(N2) $A^{\ell}_{\alpha\overline{\beta}}$ is Hermitian symmetric, that is $\overline{A^{\ell}_{\alpha\overline{\beta}}} = A^{\ell}_{\beta\overline{\alpha}}$.

(N3) $\operatorname{tr} A^{\ell}_{2\overline{2}} = 0$, $(\operatorname{tr})^2 A^{\ell}_{2\overline{3}} = 0$, $(\operatorname{tr})^3 A^{\ell}_{3\overline{3}} = 0$, where tr stands for the usual tensorial trace taken with respect to $\delta^{j\overline{k}}$.

Some notation is in order. By $\mathcal{N}$, we denote the totality of $A = (A^{\ell}_{\alpha\overline{\beta}})$ satisfying the conditions (N1)–(N3). We define $\mathcal{N}^{\omega}$ to be the set of $A \in \mathcal{N}$ such that $N(A)$ is real analytic. (In general, $N(A)$ is a formal surface.) The strictly pseudoconvex side $\rho_A > 0$ of $N(A)$ is denoted by $\Omega(A)$, which makes sense near the origin. We use the coordinates $(z', \overline{z'}, \rho_A, v)$ for functions on $\Omega \cup N(A)$.

We have assumed that each $\partial\Omega$ is real analytic near the origin, so that we can place it locally in Moser's normal form $N(A)$ with $A \in \mathcal{N}^{\omega}$. More precisely, there exists a local biholomorphic mapping Φ_A such that $\Phi_A(\Omega) = \Omega(A)$ and $\Phi_A(\partial\Omega) = N(A)$ locally. For r_Ω, we set

$$r_A = (r_\Omega \circ \Phi_A^{-1}) |\det \Phi_A'|^{2/(n+1)},$$

and consider the Taylor expansion about the origin

$$r_A = \sum_{k=1}^{N-1} c_k(z', \overline{z'}, v)\, \rho_A^k + O(\rho_A^N). \tag{2.8}$$

More precisely, we require that the family $(r_A)_{r \in \mathcal{N}^{\omega}}$ is well-defined in the sense of (2.8). Now we pose:

Condition SB3. In case $s > 0$, the family $r = (r_\Omega)_\Omega$ satisfies (2.6) for $N = s + 1$. Furthermore, in (2.8) for $N = s + 1$, any coefficient of the Taylor expansion of $c_k(z', \overline{z'}, v)$ about the origin is a universal polynomial in $A \in \mathcal{N}^{\omega}$. In case $s < 0$, the requirements above hold for any $N \in \mathbb{N}$. In case $s = 0$, no requirement is imposed.

Definition 2.2. By a local Sobolev–Bergman kernel of order $s/2$, $s \in \mathbb{Z}$, we mean a local domain functional $K^s = (K^s_\Omega)$ satisfying Conditions SB1–3.

By virtue of Proposition 2.1, the existence of a local Sobolev–Bergman kernel is reduced to that of a family of defining functions $r = (r_\Omega)_\Omega$ satisfying Condition SB3. Our main result of this paper is:

Theorem 1. *A local Sobolev–Bergman kernel of order $s/2$ ($s \in \mathbb{Z}$) exists if and only if $0 \le s \le n + 1$.*

The nonexistence part of Theorem 1 is a consequence of:

Theorem 2. *There does not exist a family of C^∞ local defining functions $r = (r_\Omega)$ satisfying the requirements in Condition SB3 with $N = n + 3$.*

The proof of Theorem 2 is given in Section 5. Let us observe that Theorem 1 follows from Theorem 2. It suffices to show the existence of

$r = (r_\Omega)$ satisfying Condition SB3 with $N = n+2$ in place of $N = s+1$. But this has been done by Fefferman [F2]. He constructed r_Ω satisfying $J[r_\Omega] = 1 + O(r_\Omega^{n+1})$ and (2.6) for $N = n+2$. Specifically, one starts from an arbitrary smooth local defining function ρ of Ω, and defines ρ_s for $s = 1, \ldots, n+1$ successively by

$$\rho_1 = J[\rho]^{-1/(n+1)}\rho, \quad \frac{\rho_s}{\rho_{s-1}} = 1 + \frac{1 - J[\rho_{s-1}]}{c_s}, \quad c_s = s(n+2-s). \tag{2.9}$$

Then $J[\rho_s] = 1 + O(\rho^s)$, and ρ_s satisfies the approximate transformation law (2.6) for $N = s+1$. Thus, we may set $r_\Omega = \rho_{n+1}$. It is clear that r_Ω is real analytic whenever the initial ρ is. The polynomial dependence on $A \in \mathcal{N}$ as in Condition SB3 is examined if we locally place $\partial\Omega$ in normal form $N(A)$ and start from $\rho = \rho_A$. In fact, the universality of the polynomials in Condition SB3 follows from the transformation law (2.6) for $N = n+2$.

3. Local Sobolev–Bergman kernels (the C^∞ case)

3.1. Polynomial dependence in the real analytic case. In order to define local Sobolev–Bergman kernels in the C^∞ category, we rewrite Condition SB3 under Conditions SB1 and SB2. That is, we need to state the polynomial dependence on Moser's normal form coefficients $A = (A^\ell_{\alpha\overline{\beta}})$ more explicitly.

Let us first recall the notion of biweight on $A^\ell_{\alpha\overline{\beta}}$ for $A = (A^\ell_{\alpha\overline{\beta}}) \in \mathcal{N}$ defined by

$$\mathrm{w}_2(A^\ell_{\alpha\overline{\beta}}) = (|\alpha| + \ell - 1, |\beta| + \ell - 1).$$

This comes from the transformation law under dilations

$$\phi_\lambda(z', z_n) = (\lambda z', |\lambda|^2 z_n) \qquad \text{for} \quad \lambda \in \mathbb{C}^*.$$

The notion for polynomials in A to be of (homogeneous) biweight is defined by

$$\mathrm{w}_2(P_1(A)P_2(A)) = \mathrm{w}_2(P_1(A)) + \mathrm{w}_2(P_2(A))$$

for monomials $P_1(A)$ and $P_2(A)$. If $P(A)$ is a polynomial of biweight (w', w''), we write

$$\mathrm{w}_2(P(A)) = (w', w''), \quad \mathrm{w}^{\mathrm{dil}}(P(A)) = \frac{1}{2}(w' + w''),$$

and call $\mathrm{w}^{\mathrm{dil}}(P(A))$ the weight of $P(A)$ with respect to dilations. Then, a polynomial in A is of weight w with respect to dilations if and only if it is a linear combination of polynomials of biweight (w', w'') such that $w' + w'' = 2w$. We have no essential change if we replace $\mathcal{N}$ by $\mathcal{N}^\omega$.

Let $K^s = (K^s_\Omega)$ be the local Sobolev–Bergman kernel of order $s/2$ in Definition 2.2, so that each $\partial\Omega$ is real analytic near the reference point assumed to be the origin $0 \in \mathbb{C}^n$. As in the previous section, we locally place $\partial\Omega$ in normal form $N(A)$, and write $K^s = (K^s_A)_{A\in\mathcal{N}^\omega}$, where each K^s_A corresponds to $\Omega(A)$. In fact, (K^s_A) is a subfamily of (K^s_Ω), but there is no loss of information via the transformation law

$$K^s_A = (K^s_\Omega \circ \Phi_A^{-1})|\det \Phi'_A|^{-2w(s)/(n+1)} \tag{3.1}$$

for Φ_A in Subsection 2.3. Note that (3.1) is consistent with (2.5). As in (2.8), we have

$$K^s_A = \sum_{m=0}^{\infty} \sum_{\alpha,\beta,\ell} P^{\ell m}_{\alpha\overline{\beta}}(A)\, z'_\alpha\, \overline{z'_\beta}\, v^\ell\, \widehat{K}_{m-w(s)}[\rho_A], \tag{3.2}$$

where $P^{\ell m}_{\alpha\overline{\beta}}(A)$ are universal polynomials in $A \in \mathcal{N}$ determined by $K^s = (K^s_A)$. Furthermore,

$$\mathrm{w}^{\mathrm{dil}}(P^{\ell m}_{\alpha\overline{\beta}}(A)) = \frac{1}{2}(|\alpha| + |\beta|) + \ell + m. \tag{3.3}$$

As before, we refer to the universality of the polynomials $P^{\ell m}_{\alpha\overline{\beta}}(A)$ in (3.2) as the polynomial dependence of $K^s = (K^s_\Omega)$ on A. This follows from Condition SB3 and the construction in Subsection 2.1. Here, a crucial fact is the polynomial dependence of the local Bergman kernel $K^0 = K^{\mathrm{B}}$ on A, a fact which has been examined in [HKN1].

Let us restrict ourselves to the half line $z = \gamma_t$ for $t > 0$ small defined by $\gamma_t = (0, t/2) \in \mathbb{C}^{n-1} \times \mathbb{C}$. Then (3.2) implies

$$K^s_A(\gamma_t) = \sum_{m=0}^{\infty} P_m(A)\, \widehat{K}_{m-w(s)}[t], \tag{3.4}$$

where $P_m(A) = P^{0m}_{0\overline{0}}(A)$. Thus (3.3) yields

$$\mathrm{w}^{\mathrm{dil}}(P_m(A)) = m. \tag{3.5}$$

Since $w^{\mathrm{dil}}(A^\ell_{\alpha\overline{\beta}}) > 0$, it follows from (3.5) that:

Lemma 3.1. *Each polynomial $P_m(A)$ in (3.4) depends only on $A^\ell_{\alpha\overline{\beta}}$ such that $\mathrm{w}^{dil}(A^\ell_{\alpha\overline{\beta}}) \le m$.*

A crucial fact is the following.

Proposition 3.1. *The expansion* (3.4) *determines* $K^s = (K^s_A)$.

Proof. We first take a small neighborhood $M \subset \partial\Omega$ of the origin. For any $q \in M$ fixed, we then place M about q in normal form $N(A)$ with some $A \in \mathcal{N}^\omega$. By [CM], we may take the local biholomorphic mappings $\Phi_{q,A}\colon M \to N(A)$ with $\Phi_{q,A}(q) = 0$ to depend on $q \in M$ real analytically. Setting

$$K^s_{q,A} = (K^s_\Omega \circ \Phi^{-1}_{q,A})|\det \Phi'_{q,A}|^{-2w(s)/(n+1)},$$

we have, as in (3.4),

$$K^s_{q,A}(\gamma_t) = \sum_{m=0}^{\infty} P_m(A)\,\widehat{K}_{m-w(s)}[t]. \tag{3.6}$$

The point is that $P_m(A)$ in (3.6) are independent of $q \in M$. This fact follows from the universality of $P^{\ell m}_{\alpha\overline{\beta}}(A)$ in (3.2). The expansion (3.2) about the origin is recovered from (3.6) by varying $q \in M$. Thus (3.4) determines (3.2). □

3.2. Definition of local Sobolev–Bergman kernels in the C^∞ category. Let us define local Sobolev–Bergman kernels $K^s = (K^s_\Omega)$ near $0 \in \partial\Omega$ in case each $\partial\Omega$ is merely C^∞. We regard each K^s_Ω as a formal singularity. In other words, we ignore the difference by flat functions. As before, it suffices to specify $K^s = (K^s_A)_{A\in\mathcal{N}}$ given by the transformation law (3.1). This is done by real analytic approximation. More precisely, we first truncate $A = (A^\ell_{\alpha\overline{\beta}}) \in \mathcal{N}$ by neglecting $A^\ell_{\alpha\overline{\beta}}$ such that $\mathrm{w}^{\mathrm{dil}}(A^\ell_{\alpha\overline{\beta}}) > N$ for $N \in \mathbb{N}$ large, and denote the results by A_N. Then $N(A_N)$ are algebraic real hypersurfaces, for which we can consider an expansion of the form (3.2). By Proposition 3.1, this expansion is determined by an expansion of the form (3.4). In this new expansion, the coefficients $P_m(A)$ for $m \le N$ are determined by A_N, a fact which follows from Lemma 3.1. In other words, these $P_m(A)$ are unchanged if A_N are replaced by A_{N+1}. Consequently, we have the expansions (3.4) and (3.2) for any $A \in \mathcal{N}$ even when $A \notin \mathcal{N}^\omega$. Therefore, $K^s = (K^s_\Omega)$ for $\partial\Omega \in C^\infty$ near $0 \in \mathbb{C}^n$ is well-defined.

Remark 2. Let $s \in \mathbb{Z}$ and $s \notin [0, n+1]$. Then by Theorem 2, there does not exist a local Sobolev–Bergman kernel of order $s/2$. Nevertheless, we can define a similar local domain functional $K^s = (K^s_\Omega)$ with ambiguity. We require $K^s = (K^s_\Omega)$ to satisfy Conditions SB1–3, but the exact transformation law (2.5) in Condition SB2 is replaced by an approximate one. The existence of $K^s = (K^s_\Omega)$ in the real analytic category is proved as in the exact kernels case $s \in [0, n+1]$, though we have to be more careful in inspecting the construction in Subsection 2.1. The ambiguity of K comes from that of $\widehat{K}$ via that of $\mathbf{A}(z, \partial_z)$. The definition of $K^s = (K^s_\Omega)$ in the C^∞ category is also similar to that in the exact kernels case $s \in [0, n+1]$ in the previous subsection. We have (3.1) if each K^s_Ω is regarded as an equivalence class with respect to the ambiguity. For the approximate kernels as above, one can develop Fefferman's invariant theory as in the next section.

4. Invariant expansions of local Sobolev–Bergman kernels

4.1. Ambient metric construction. Let $K^s = (K^s_\Omega)$ be the local Sobolev–Bergman kernel of order $s/2$ in the C^∞ category, so that $s \in \mathbb{Z}$ satisfies $0 \le s \le n+1$. As before, we set $w(s) = n+1-s$. Let $r = (r_\Omega)$ be a family of C^∞ local defining functions satisfying Condition SB3 with $N = n+2$ in place of $N = s+1$. It has been known for $s = 0, 1$ (that is, for the Bergman kernel and the Szegö kernel) that K^s admits an expansion of the form

$$K^s_\Omega = \sum_{m=0}^{n} W^s_m[r_\Omega]\, \widehat{K}_{m-w(s)}[r_\Omega] \qquad \mathrm{mod}\ O(\widehat{K}_s[r_\Omega]), \tag{4.1}$$

where $W^s_m = W^s_m[r_\Omega]$ are Weyl functionals of weight m given by the ambient metric construction (cf. [F3], [BEG], [HKN1], [Hi]). Terminology will be reviewed below in this subsection (Definitions 4.1 and 4.2). If $n = 2$ and $s = 0, 1$, then (4.1) is refined as follows (cf. [G2], [HKN1], [HKN2]):

$$K^s_\Omega = \sum_{m=0}^{5} W^s_m[r_\Omega]\, \widehat{K}_{m-w(s)}[r_\Omega] \qquad \mathrm{mod}\ O(\widehat{K}_{s+3}[r_\Omega]), \tag{4.2}$$

where $W^s_m = W^s_m[r_\Omega]$ ($m \ne 3$) are Weyl–Fefferman functionals of weight m. Here, the case $m = 3$ is exceptional and we explain it at the end of this subsection. The proof of these facts yields the following:

Proposition 4.1. *An expansion of the form* (4.1) *holds in general for* $0 \le s \le n+1$.

Proposition 4.2. *An expansion of the form* (4.2) *for* $n = 2$ *holds in general for* $0 \le s \le 3$.

In fact, we have defined the local Sobolev–Bergman kernel in such a way that Propositions 4.1 and 4.2 are obvious. In order to explain it, we begin by recalling the ambient metric construction. For simplicity of notation, we drop the subscript Ω in r_Ω and write $K^s[r]$ for K^s_Ω. Though our description below looks global near $\partial\Omega$, it is obvious as before how to localize or formalize to a neighborhood of a boundary point of reference.

The ambient metric $g = g[r]$ is defined by the potential $r_\#(z_0, z) = |z_0|^2 r(z)$ on $\mathbb{C}^* \times \overline{\Omega}$, where $z_0 \in \mathbb{C}^* = \mathbb{C} \setminus \{0\}$ is an extra variable. That is, g is a Lorentz-Kähler metric in a neighborhood of $\mathbb{C}^* \times \partial\Omega$, inside Ω. Specifically,

$$g = \sum_{j,k=0}^{n} g_{j\overline{k}}\, dz_j d\overline{z}_k = \sum_{j,k=0}^{n} \frac{\partial^2 r_\#}{\partial z_j \partial \overline{z}_k}\, dz_j d\overline{z}_k.$$

Denoting by $R = R[r]$ the curvature tensor of g, we consider successive covariant derivatives $R^{(p,q)} = \overline{\nabla}^{q-2}\nabla^{p-2}R$. Regarding components of $R^{(p,q)}$ as independent variables, we manufacture complete contractions, with respect to g, of the form

$$W_\# = \operatorname{contr}\Big(R^{(p_1,q_1)} \otimes \cdots \otimes R^{(p_m,q_m)}\Big), \tag{4.3}$$

where $\sum p_\ell = \sum q_\ell = 2(m+w)$, the definition of w called the weight of $W_\#$. By a *Weyl polynomial* $W_\#$ of weight w, we mean a linear combination of complete contractions of the form (4.3) of weight w. Here, $W_\#$ is regarded as a polynomial in components of $R^{(p,q)}$ for all $p, q \geq 2$.

Given a Weyl polynomial $W_\#$ of weight w, we now regard it as a functional of r and write $W_\# = W_\#[r]$. Setting $W[r] = W_\#[r]|_{z_0=1}$, we have

$$W_\#[r](z_0, z) = |z_0|^{2w} W[r](z).$$

Using the terminology in [HKN2], we pose:

Definition 4.1. $W = W[r]$ is called a *Weyl functional* of weight w.

If $W = W[r]$ is a Weyl functional of weight w, then the following transformation law holds under biholomorphic mappings $\Phi : \Omega_1 \to \Omega_2$

$$W[r_1] = (W[r_2] \circ \Phi)|\det \Phi'|^{2w/(n+1)}, \tag{4.4}$$

provided r_j are defining functions of Ω_j, subject to the restriction at the beginning of this section, such that $r_1 = (r_2 \circ \Phi)|\det \Phi'|^{-2/(n+1)}$. Furthermore, (4.4) holds modulo $O(r^{n+1-w})$, without assuming the relation between r_1 and r_2. Consequently, it follows from the construction that if $w \leq n$ then the boundary value of $W[r]$ is a CR invariant of weight w. This is a consequence of the polynomial dependence of $W[r]$ on $A \in \mathcal{N}$ in the sense as before.

Definition 4.2 (cf. [HKN2]). Let $n = 2$. We say that a Weyl functional $W = W[r]$ of weight w is a *Weyl–Fefferman functional* if $W[r]$ modulo $O(r^{6-w})$ is independent of the choice of r.

If $n = 2$ and $W = W[r]$ is a Weyl–Fefferman functional of weight w, then (4.4) holds modulo $O(r^{6-w})$. Hence, if $w \leq 5$ then the boundary value of $W[r]$ is a CR invariant of weight w.

By a CR invariant of weight w, we mean a polynomial $P(A)$ in $A \in \mathcal{N}$ satisfying the transformation law

$$P(A) = P(\widetilde{A})|\det \Phi'(0)|^{2w/(n+1)}$$

under any local biholomorphic mapping $\Phi\colon N(A) \to N(\widetilde{A})$ such that $\Phi(0) = 0$. We denote the totality of these $P(A)$ by I_w^{CR}. Any CR invariant can be regarded as a smooth function on $\partial\Omega$. Propositions 4.1 and 4.2 are consequences of the following fact, except for $W_3^s[r_\Omega]$ in (4.2).

Proposition 4.3. *If $n \geq 3$ and $w \leq n$, then any* CR *invariant of weight w is realized by the boundary value of a Weyl functional of weight w. If $n = 2$, $w \leq 5$ and $w \neq 3$, then any* CR *invariant of weight w is realized by the boundary value of a Weyl–Fefferman functional of weight w.*

For the proof, see [BEG] and [HKN2].

Remark 3. Let us say that a Weyl functional is linear (resp. nonlinear) if the corresponding Weyl polynomial is linear (resp. nonlinear).

(1°) Let $n = 2$ and $w \leq 5$. Then, any nonlinear Weyl functional of weight w is a Weyl–Fefferman functional and any linear Weyl–Fefferman functional of weight w is trivial. Now let $W \neq 0$ be a linear Weyl functional of weight w. If $w \leq 2$ then the boundary value of W is zero, whereas if $w = 3$ then the boundary value of W is nonzero and gives rise to a CR invariant. The vector space of CR invariants of weight 3 is one dimensional, and thus a base is realized by the boundary value of a linear Weyl functional, though the ambiguity estimate is too rough. (Cf. [HKN2] for the detail.)

(2°) Let $n \geq 3$ and $w \leq n+1$. It is plausible that any Weyl functional of weight w has the ambiguity modulo $O(r^{n+2-w})$ and that any CR invariant of weight w is realized by the boundary value of a Weyl functional of weight w. If $w \leq n$ then any linear Weyl functional of weight w is trivial (cf. [F3]). It is desirable to define the notion of Weyl–Fefferman functionals as in the case of $n = 2$ by the optimal ambiguity estimate for nonlinear Weyl functionals.

(3°) According to the theory developed in [Hi] and roughly explained in the next section, the Weyl functionals $W_w = W_w[r]$ of arbitrary weight w make sense as functionals of a special family of defining functions r, where the ambiguity of r is measured by a parameter and its effect on $W_w = W_w[r]$ is taken into account. In this sense, Propositions 4.1 and 4.2 can be refined in such a way that (4.1) and (4.2) are infinite asymptotic series. Here, we don't need a refinement of Proposition 4.3, which is stated in Subsection 5.3.

We conclude this subsection by explaining what is $W_3^s[r]$ in (4.3), where the subscript Ω in r_Ω is dropped. For each s, this is a constant multiple of $\eta_1^{\mathrm{G}} = \eta_1^{\mathrm{G}}[r]$ which appears in Graham's asymptotic solution of $J[u] = 1$:

$$u^{\mathrm{G}} = r \sum_{k=0}^{\infty} \eta_k^{\mathrm{G}} \cdot (r^{n+1} \log r)^k, \quad \eta_k^{\mathrm{G}} \in C^\infty(\overline{\Omega})$$

in the general case of dimension $n \geq 2$. This is a formal series, and the difference of flat functions along $\partial\Omega$ is ignored in determining η_k^{G}. We have

$$\eta_0^{\mathrm{G}} = 1 + ar^{n+1} + O(r^{n+2}) \qquad \text{with} \quad a \in C^\infty(\partial\Omega),$$

and u^{G} is uniquely constructed by specifying a. We have approximate transformation laws

$$\eta_{k,\Omega_1}^{\mathrm{G}} = (\eta_{k,\Omega_2}^{\mathrm{G}} \circ \Phi)|\det \Phi'|^{2k} \mod O(r^{n+1})$$

under (local) biholomorphic mappings $\Phi\colon \Omega_1 \to \Omega_2$. In particular, each η_k^{G} modulo $O(r^{n+1})$ is independent of a and r, as far as r is subject to the condition at the beginning of this subsection. By construction, the polynomial dependence on $A \in \mathcal{N}$ is valid as before. Thus, η_1^{G} for $n = 2$ behaves like a Weyl–Fefferman functional of weight 3.

4.2. Explicit result in dimension ≥ 3. Let $n \geq 3$. It is proved in [G2] that $I_0^{\mathrm{CR}} = \mathbb{C}$, $I_1^{\mathrm{CR}} = \{0\}$ and that I_2^{CR} is generated by

$$\|A_{2\overline{2}}^0\|^2 = \sum_{|\alpha|=|\beta|=2} |A_{\alpha\overline{\beta}}^0|^2.$$

Consequently, we have for $W_m^s = W_m^s[r_\Omega]$ in the expansion (4.1),

$$W_0^s = 1, \quad W_1^s = 0, \quad W_2^s[r_\Omega]\big|_{\partial\Omega} = c^s(n)\|A_{2\overline{2}}^0\|^2, \tag{4.5}$$

where $c^s(n)$ are universal constants. By [HKN1],

$$c^0(n) = \frac{2}{3(n-1)n}, \quad c^1(n) = \frac{2}{3(n-2)(n-1)}.$$

By a similar proof, we have:

Proposition 4.4. *The constants $c^s(n)$ in (4.5) are given by*

$$c^s(n) = \frac{2}{3(n-s-1)(n-s)} \quad \textit{for} \quad s \neq n-1, n,$$

and $c^{n-1}(n) = -2/3$, $c^n(n) = 2/3$.

4.3. Explicit results in dimension two. Let $n = 2$. We first note by [G2] that $I_0^{\mathrm{CR}} = \mathbb{C}$ and that I_1^{CR} and I_2^{CR} are trivial. Consequently, we have for $W_m^s = W_m^s[r_\Omega]$ in the expansion (4.2),

$$W_0^s = 1, \quad W_1^s = 0, \quad W_2^s = 0.$$

It remains to determine $\psi^s = W_3^s + W_4^s r + W_5^s r^2$, where we abbreviated by writing r and W_j^s in place of r_Ω and $W_j^s[r_\Omega]$, respectively. By [G2] and [HKN2], we have

$$\dim I_3^{\mathrm{CR}} = \dim I_4^{\mathrm{CR}} = 1, \quad \dim I_5^{\mathrm{CR}} = 2.$$

More precisely, I_3^{CR} and I_4^{CR} are generated by $A_{4\bar{4}}^0$ and $|A_{2\bar{4}}^2|^2$, respectively: I_5^{CR} is spanned by $F_5^{\mathrm{CR}}(1,0)$ and $F_5^{\mathrm{CR}}(0,1)$, where

$$F_5^{\mathrm{CR}}(a,b) = F(a,b,-2a+(10/9)b,-a+b/3)$$

with $F(a,b,c,d) = a|A_{5\bar{2}}^0|^2 + b|A_{4\bar{3}}^0|^2 + \mathrm{Re}\{(cA_{3\bar{5}}^0 - idA_{2\bar{4}}^1)A_{4\bar{2}}^0\}$. By Graham [G2], the boundary value of η_1^{G} is $4A_{4\bar{4}}^0$. It is proved in [HKN2] that if $p+q-2 = 4, 5$ then $\|R^{(p,q)}\|^2$ is a Weyl–Fefferman functional of weight $w = p+q-2$, where $\|R^{(p,q)}\|^2$ stands for the squared norm of the tensor $R^{(p,q)}$ with respect to the ambient metric g restricted to $z_0 = 1$. (The squared norm need not be nonnegative because g is a Lorentz metric.) Furthermore, the boundary values of $\|R^{(5,2)}\|^2$ and $\|R^{(4,3)}\|^2$ are linearly independent as CR invariants. Consequently, we may set

$$\psi^s = c_0^s\eta_1^{\mathrm{G}} + c_1^s\|R^{(4,2)}\|^2 r + \left(c_2^s\|R^{(5,2)}\|^2 + c_3^s\|R^{(4,3)}\|^2\right)r^2 + O(r^3), \quad (4.6)$$

where c_j^s for $j = 0, \ldots, 3$ are universal constants.

Proposition 4.5. *The constants c_j^s in (4.6) are given by*

$$\begin{array}{llll} c_0^0 = -3, & c_1^0 = 3/1120, & c_2^0 = 61/141120, & c_3^0 = 3/7840, \\ c_0^1 = -2, & c_1^1 = 1/3360, & c_2^1 = 1/23520, & c_3^1 = 1/13230, \\ c_0^2 = -1, & c_1^2 = -1/10080, & c_2^2 = -1/70560, & c_3^2 = -1/169344, \\ c_0^3 = 1, & c_1^3 = 1/4480, & c_2^3 = 1/33075, & c_3^3 = 1/31360. \end{array}$$

The proof of Proposition 4.5 is done by locally placing $\partial\Omega$ in normal form $N(A)$ and restricting both sides of (4.6) to the half line $\gamma_t = (0, t/2)$, $t > 0$. By [HKN2], we have

$$\begin{aligned} \|R^{(4,2)}\|^2(\gamma_t) &= 2^8\, q_1(7,0) + 2^8\, q_2(117, 435, 936, 0, 50, 0)\, t + O(t^2), \\ \|R^{(5,2)}\|^2(\gamma_t) &= 4 \cdot (5!)^2 q_2(5/2, 9, 18, 0, 1, 0) + O(t), \\ \|R^{(4,3)}\|^2(\gamma_t) &= 4 \cdot (5!)^2 q_2(37/30, 5, 57/5, 0, 4/3, 0) + O(t), \end{aligned} \tag{4.7}$$

where

$$\begin{aligned} q_1(d_1, d_2) &= d_1 |A^0_{2\bar{4}}|^2 + d_2 A^0_{5\bar{5}}, \\ q_2(d_1, d_2, d_3, d_4, d_5, d_6) &= \mathrm{Re}\Big(2d_1 i A^1_{2\bar{4}} A^0_{4\bar{2}} + 2d_2 A^0_{3\bar{5}} A^0_{4\bar{2}} + d_3 |A^0_{3\bar{4}}|^2 \\ &\qquad + d_4 A^2_{4\bar{4}} + d_5 |A^0_{2\bar{5}}|^2 + d_6 A^0_{6\bar{6}}\Big). \end{aligned}$$

Though [HKN2] does not give the expansion of $\eta_1^G(\gamma_t)$ for general $N(A)$, an algorithm of computation is provided. Computer-aided calculation yields:

Lemma 4.1.

$$\begin{aligned} \eta_1^G(\gamma_t) &= 4\, A^0_{4\bar{4}} + q_1(368/5, -20)\, t \\ &\quad + q_2(226/15, -312, -1956/5, 2, -680/3, 60)\, t^2 + O(t^3). \end{aligned}$$

A method of computation of $\psi^s(\gamma_t)$ is given in Appendix B. Again, computer-aided calculation yields:

Lemma 4.2. *With q_1 and q_2 as in Lemma 4.1,*

$$\begin{aligned} \psi^0(\gamma_t) &= -12 A^0_{4\bar{4}} + q_1(-216, 60)\, t \\ &\quad + q_2(-36, 900, 1116, -6, 660, -180)\, t^2 + O(t^3), \\ \psi^1(\gamma_t) &= -8 A^0_{4\bar{4}} + q_1(-440/3, 40)\, t \\ &\quad + q_2(-248/9, 1840/3, 760, -4, 4040/9, -120)\, t^2 + O(t^3), \\ \psi^2(\gamma_t) &= -4 A^0_{4\bar{4}} + q_1(-664/9, 20)\, t \\ &\quad + q_2(-131/9, 310, 386, -2, 680/3, -60)\, t^2 + O(t^3), \\ \psi^3(\gamma_t) &= 4 A^0_{4\bar{4}} + q_1(74, -20)\, t \\ &\quad + q_2(15, -312, -390, 2, -228, 60)\, t^2 + O(t^3). \end{aligned}$$

Proposition 4.5 is proved by using Lemmas 4.1 and 4.2, together with (4.7) and the result for $s = 0$ or $s = 1$ given in [HKN2].

4.4. A construction of CR invariants of weight five in dimension two. As an implication of Lemmas 4.1 and 4.2, we now give a linear relation satisfied by $\eta_{\mathrm{I}}^{\mathrm{G}}$ and local Sobolev–Bergman kernels of order $s/2$ for $s = 0, 1, 2, 3$. Let us first normalize by setting

$$\eta_{\mathrm{I}} = \frac{\eta_{\mathrm{I}}^{\mathrm{G}}}{4}, \quad \psi_{\mathrm{I}}^0 = \frac{\psi^0}{4}, \quad \psi_{\mathrm{I}}^1 = -\frac{\psi^1}{4}, \quad \psi_{\mathrm{I}}^2 = -\frac{\psi^2}{8}, \quad \psi_{\mathrm{I}}^3 = -\frac{\psi^3}{12}$$

so that the evaluation at $z = 0$ gives rise to $\eta_{\mathrm{I}} = \psi_{\mathrm{I}}^s = A_{4\bar{4}}^0$ for $s = 0, 1, 2, 3$. To get a CR invariant of weight four, we next set

$$\eta_{\mathrm{II}} = \frac{5}{2}\,\frac{\eta_{\mathrm{I}} - \psi_{\mathrm{I}}^3}{r}, \quad \psi_{\mathrm{II}}^0 = 2\,\frac{\psi_{\mathrm{I}}^0 - \psi_{\mathrm{I}}^3}{r}, \quad \psi_{\mathrm{II}}^1 = \frac{9}{4}\,\frac{\psi_{\mathrm{I}}^1 - \psi_{\mathrm{I}}^3}{r}, \quad \psi_{\mathrm{II}}^2 = 3\,\frac{\psi_{\mathrm{I}}^2 - \psi_{\mathrm{I}}^3}{r}.$$

Then $\eta_{\mathrm{II}} = \psi_{\mathrm{II}}^s = |A_{2\bar{4}}^0|^2$ at $z = 0$ for $s = 0, 1, 2$. We thus set

$$\eta_{\mathrm{III}} = \frac{\eta_{\mathrm{II}} - \psi_{\mathrm{II}}^2}{r}, \quad \psi_{\mathrm{III}}^0 = 6\,\frac{\psi_{\mathrm{II}}^0 - \psi_{\mathrm{II}}^2}{r}, \quad \psi_{\mathrm{III}}^1 = \frac{48}{5}\,\frac{\psi_{\mathrm{II}}^1 - \psi_{\mathrm{II}}^2}{r}.$$

Then

$$\eta_{\mathrm{III}}|_{z=0} = q_2(7/12, -5/2, -6, 0, -5/6, 0),$$
$$\psi_{\mathrm{III}}^0|_{z=0} = \psi_{\mathrm{III}}^1|_{z=0} = q_2(1, -6, -18, 0, -4, 0).$$

The right sides are CR invariants of weight five which are linearly independent. In particular, we see that $\dim I_5^{\mathrm{CR}} \geq 2$. This observation was indeed used as a motivation of getting results in [HKN2] about I_5^{CR}.

5. Proof of Theorem 2

5.1. Nonexistence of exactly invariant defining functions. We prove Theorem 2 stated in Subsection 2.3. This is done by using the nonexistence of a local defining function $r = r_\Omega$ satisfying exact transformation law of weight -1. To state it more precisely, we introduce spaces $\mathcal{F}_{\mathrm{def}}^m$ of local defining functions for $m \geq 3$ ($m \in \mathbb{Z}$) as follows. Recall first that $C_{\mathrm{def}}^\infty(\overline{\Omega})$ is the totality of functions $r \in C^\infty(\overline{\Omega})$ such that $r > 0$ in Ω and $dr \neq 0$ on $\partial\Omega$. Localizing it, we have a sheaf of (smooth) local defining functions $C_{\mathrm{def},\partial\Omega}^\infty(\overline{\Omega}) = (C_{\mathrm{def},p}^\infty(\overline{\Omega}))_{p\in\partial\Omega}$. If $\partial\Omega = N(A)$ with $A \in \mathcal{N}$, we write $C_{\mathrm{def},A}^\infty = C_{\mathrm{def},0}^\infty(\overline{\Omega})$, where we disregard the difference by flat functions at the origin. Then, $C_{\mathrm{def}}^\infty = (C_{\mathrm{def},A}^\infty)_{A\in\mathcal{N}}$ is a space of local domain functionals which represent local defining functions. We denote by

$$\mathcal{F}_{\mathrm{def}}^m = (\mathcal{F}_{\mathrm{def},A}^m)_{A\in\mathcal{N}} \quad \text{for} \quad m \geq 3 \quad (m \in \mathbb{Z}),$$

the totality of $r = (r_A)_{A\in\mathcal{N}} \in C^\infty_{\text{def}}$ such that r satisfies the transformation law of weight -1 modulo $O(r^m)$ and that if

$$r_A(\gamma_t) = \sum_{j=1}^{m-1} P_j(A)t^j + O(t^m)$$

in Moser's normal coordinates, then $P_j(A) \in I_j^{\text{CR}}$. Then we have the following:

Proposition 5.1. $\mathcal{F}^{n+3}_{\text{def}} = \emptyset$.

Postponing the proof for a moment, we first observe that Theorem 2 follows from this.

Proof of Theorem 2. We may assume $w \leq -1$ by considering Kashiwara's transformation. Assume there exists a local Sobolev–Bergman kernel of weight w, $K = \varphi r^{-w}\log r$, where $r \in C^\infty_{\text{def}}$ and $\varphi \in C^\infty$ with $\varphi(0) \neq 0$. Setting $\rho = \varphi^{-1/w}r$, we have $K = \rho^{-w}\log\rho$ and $\rho \in C^\infty_{\text{def}}$. Furthermore, $\rho \in \cap\mathcal{F}^m_{\text{def}}$, but this contradicts Proposition 5.1. □

The proof of Proposition 5.1 requires some results in [Hi]. In [Hi], a subclass $\mathcal{F}$ of $\mathcal{F}^{n+2}_{\text{def}}$ is defined so that

$$\mathcal{F} = (\mathcal{F}_A)_{A\in\mathcal{N}} \not\subset \mathcal{F}^{n+3}_{\text{def}}, \tag{5.1}$$

and that the ambient metric construction gives rise to Weyl functionals $W = W[r]$ of arbitrary weight $w \in \mathbb{N}_0$ on the class $\mathcal{F}$. We have the following two lemmas.

Lemma 5.1. *If $\rho \in \mathcal{F}^m_{\text{def}}$ with $m \geq 3$, then*

$$\rho = cr + \sum_{j=1}^{m-2} W_j[r]r^{j+1} + O(r^m) \quad \textit{for} \quad r \in \mathcal{F},$$

where $c > 0$ is a universal constant and $W_j = W_j[r]$ are Weyl functionals of weight j on $\mathcal{F}$.

Lemma 5.2. *If $W = W[r]$ is a Weyl functional of weight $w \in \mathbb{N}_0$ on $\mathcal{F}$, then $r^w W[r]$ modulo $O(r^{n+3})$ is independent of $r \in \mathcal{F}$.*

In the proof of Proposition 5.1, only these lemmas and (5.1) are used. Even the definition of $\mathcal{F}$ is not necessary.

Proof of Proposition 5.1. Assuming $\mathcal{F}^{n+3}_{\text{def}} \neq \emptyset$, we pick $\rho \in \mathcal{F}^{n+3}_{\text{def}}$. It then follows from Lemma 5.1 that

$$\rho = cr + \sum_{j=1}^{n+1} W_j[r]r^{j+1} + O(r^{n+3}) \quad \text{for} \quad r \in \mathcal{F}. \tag{5.2}$$

We set $\phi[r] = \sum W_j[r]r^j$. It then follows from Lemma 5.2 that $\phi[r]$ modulo $O(r^{n+3})$ is independent of $r \in \mathcal{F}$. This also holds for $r\phi[r]$, because $\mathcal{F} \subset \mathcal{F}_{\mathrm{def}}^{n+2}$ and $\phi[r] = O(r)$. Thus, (5.2) with $\rho \in \mathcal{F}_{\mathrm{def}}^{n+3}$ implies $r \in \mathcal{F}_{\mathrm{def}}^{n+3}$, but this contradicts (5.1). □

5.2. Definition of the class $\mathcal{F}$ and a review of [Hi]. Before proving Lemmas 5.1 and 5.2 with (5.1), let us give the definition of $\mathcal{F}$. It suffices to fix Ω and define a subclass $\mathcal{F}_{\partial\Omega}$ of $C^\infty_{\mathrm{def}}(\overline{\Omega})$ so that the localization of $\mathcal{F}_{\partial\Omega}$ gives rise to $\mathcal{F}$. We begin by considering the boundary value problem

$$J_\#[U] = |z^0|^{2n} \text{ and } U > 0 \text{ in } \mathbb{C}^* \times \Omega, \quad U = 0 \text{ on } \mathbb{C}^* \times \partial\Omega \tag{5.3}$$

for functions $U = U(z^0, z)$, where

$$J_\#[U] = (-1)^n \det(U_{j\bar{k}})_{0\le j,k\le n}, \quad U_{j\bar{k}} = \partial^2 U/\partial z^j \partial \bar{z}^k.$$

This is a lift of the Monge–Ampère operator in the sense that if $U(z^0, z) = |z^0|^2 u(z)$ then $J_\#[U] = |z^0|^{2n} J[u]$. But we are concerned with asymptotic solutions of (5.3) of the form

$$U = r_\# + r_\# \sum_{k=1}^{\infty} \eta_k \cdot (r^{n+1} \log r_\#)^{n+1} \quad \text{with} \quad \eta_k \in C^\infty(\overline{\Omega}), \tag{5.4}$$

where $r_\#(z^0, z) = |z^0|^{2n} r(z)$ with $r \in C^\infty_{\mathrm{def}}(\overline{\Omega})$. Note that r is not prescribed but determined together with U. We call r the smooth part of U and denote the totality of these r by $\mathcal{F}_{\partial\Omega}$. The fact $\mathcal{F}_{\partial\Omega} \neq \emptyset$ is proved by solving a formal initial value problem for (5.3) near $\partial\Omega$ with an extra initial condition

$$X^{n+2} r|_{\partial\Omega} = a \in C^\infty(\partial\Omega),$$

where X is a real vector field which is transversal to $\partial\Omega$. The unique existence of the asymptotic solution U for each data $a \in C^\infty(\partial\Omega)$ is valid and the operation of taking the smooth part $U \mapsto r$ is injective, provided we ignore the difference by flat functions along $\partial\Omega$. Thus $a \mapsto r$ is essentially a bijection $C^\infty(\partial\Omega) \to \mathcal{F}_{\partial\Omega}$. The construction is local near a boundary point, or even formal, as we explain at the end of this subsection.

An important fact is that one can formulate an exact transformation law

$$r = (\widetilde{r} \circ \Phi) |\det \Phi'|^{-2/(n+1)} \tag{5.5}$$

under biholomorphic mappings $\Phi\colon \Omega \to \widetilde{\Omega}$. Specifically, if $\widetilde{r} \in \mathcal{F}_{\partial\widetilde{\Omega}}$ and if r is defined by (5.5) then $r \in \mathcal{F}_{\partial\Omega}$. In this sense, Weyl functionals, $W = W[r]$ for $r \in \mathcal{F}_{\partial\Omega}$, of weight w satisfies the exact transformation law

$$W[r] = (W[\widetilde{r}] \circ \Phi) |\det \Phi'|^{2w/(n+1)}. \tag{5.6}$$

A main result of [Hi] states that if ψ^{B} is regarded as a functional of $r \in \mathcal{F}_{\partial\Omega}$, then

$$\psi^{\mathrm{B}}[r] = \sum_{k=0}^{m-1} W_{k+n+1}[r]r^k + O(r^m) \quad \text{for any} \quad m \in \mathbb{N},$$

where $W_j = W_j[r]$ are Weyl functionals of weight j. The proof of this fact applies without change to Lemma 5.1. We thus regard Lemma 5.1 as proved, where the localization is taken into account as follows.

In the definition of the local space $\mathcal{F} = (\mathcal{F}_A)_{A\in\mathcal{N}}$, we may set $X = \partial/\partial\rho$ for Moser's normal coordinates. Then each $\mathcal{F}_A$ is parametrized by a space of formal power series as follows:

$$(\partial^{n+2}r/\partial\rho^{n+2})\,\big|_{\rho=0} = \sum_{\alpha,\beta,\ell} C^{\ell}_{\alpha\overline{\beta}}\, z'_{\alpha}\, \overline{z'_{\beta}}\, v^{\ell} \quad \text{for} \quad r \in \mathcal{F}_A.$$

We thus have a bijection $\mathcal{C} \ni C \mapsto r = r_{A,C} \in \mathcal{F}_A$ for each $A \in \mathcal{N}$, where $\mathcal{C}$ denotes the totality of $C = (C^{\ell}_{\alpha\overline{\beta}})$. This bijection is the localization of the composition operator $C^{\infty}(\partial\Omega) \to \mathcal{F}_{\partial\Omega}$ given by $a \mapsto U$ and $U \mapsto r$. Consequently, we have a bijection

$$\mathcal{N} \times \mathcal{C} \ni (A, C) \mapsto r_{A,C} \in \mathcal{F}_A, \tag{5.7}$$

where C parametrizes the ambiguity of $r_{A,C}$. Setting $r_C = (r_{A,C})_{A\in\mathcal{N}}$, we denote by $\mathcal{F}$ the totality of r_C for $C \in \mathcal{C}$. Then $\mathcal{F} \subset C^{\infty}_{\mathrm{def}}$. It is easy to see that $\mathcal{F} \subset \mathcal{F}^{n+2}_{\mathrm{def}}$ (see [Hi]), and (5.1) is clear from the definition. Abusing notation, we write r in place of r_C, so that selecting $r \in \mathcal{F}$ is equivalent to specifying $C \in \mathcal{C}$. The point of introducing the class $\mathcal{F}$ is the exact transformation laws (5.5) and (5.6), where $C \in \mathcal{C}$ must vary. It is therefore necessary to regard the space $\mathcal{F}$ itself as a family of local domain functionals parametrized by $C \in \mathcal{C}$.

5.3. Reduction to the boundary. We have justified (5.1) and Lemma 5.1. To prove Lemma 5.2, we need to consider the boundary value of each Weyl functional on $\mathcal{F}$, say $W = W[r]$, where $r = (r_{A,C})_{A\in\mathcal{N}}$ with $r_{A,C}$ in (5.7). More precisely, we take the restriction of $W[r]$ to the origin $0 \in N(A)$. Denoting it by $P_W = P_W(A, C)$, we see by inspecting the construction that P_W is a polynomial in $(A, C) \in \mathcal{N} \times \mathcal{C}$. Let $I^{\mathrm{W}}_w(\mathcal{N} \times \mathcal{C})$ denote the totality of such polynomials which come from Weyl functionals of weight w on $\mathcal{F}$. We define a subspace $I^{\mathrm{W}}_w(\mathcal{N})$ of $I^{\mathrm{W}}_w(\mathcal{N} \times \mathcal{C})$ to be the totality of $P_W(A, C)$ which are independent of $C \in \mathcal{C}$. Then, another main result of [Hi] states that

$$I^{\mathrm{W}}_w(\mathcal{N}) = I^{\mathrm{CR}}_w \quad \text{for} \quad w \in \mathbb{N}_0 \tag{5.8}$$

and that if $n \geq 3$ (resp. $n = 2$) then

$$I_w^{\mathrm{W}}(\mathcal{N} \times \mathcal{C}) = I_w^{\mathrm{W}}(\mathcal{N}) \quad \text{for} \quad w \leq n+2 \quad (\text{resp. } w \leq 5), \tag{5.9}$$

where the weight restriction in (5.9) is optimal. In the following, Lemma 5.2 is proved by using (5.8), while (5.9) shows that the error estimate in Lemma 5.2 is optimal.

Proof of Lemma 5.2. This is a refinement of Fefferman's Ambiguity Lemma in [F3]. As in [F3], the problem is reduced to the case $\partial\Omega = N(A)$ with $A \in \mathcal{N}$, via the transformation law for r and $W = W[r]$. In Moser's normal coordinates, we investigate the behavior of $r^w W[r]$ along the half line $\gamma_t = (0, t) \in \mathbb{C}^{n-1} \times \mathbb{C}$, $t > 0$. We have

$$(r^w W[r])(\gamma_t) = \sum_{j=m}^{n+2} P_j(A, C) t^j + O(t^{n+3}),$$

where $P_j(A, C)$ are polynomials in $(A, C) \in \mathcal{N} \times \mathcal{C}$. Furthermore, $P_j(A, C)$ is of weight j. It suffices to show that $P_j(A, C)$ are independent of $C \in \mathcal{C}$. Assume that $P_j(A, C)$ depends on C. Since

$$\mathrm{w}(A^\ell_{\alpha\overline{\beta}}) \geq 2, \quad \mathrm{w}(C^\ell_{\alpha\overline{\beta}}) \geq n+1$$

for $A = (A^\ell_{\alpha\overline{\beta}}) \in \mathcal{N}$ and $C = (C^\ell_{\alpha\overline{\beta}}) \in \mathcal{C}$, it follows that $P_j(A, C)$ is linear homogeneous. Consequently, the Weyl polynomial $W_\#$ must be linear, so that we may assume $W_\# = \mathrm{tr}(\nabla^{(p,p)} R)$. By the linearity of $P_j(A, C)$, the assumption implies that $P_j(0, C) \neq 0$, so that we are reduced to the case $A = 0 \in \mathcal{N}$. In this case, $N(A)$ is the boundary of a Siegel domain, and any asymptotic solution of (5.3) of the form (5.4) is (formally) smooth. Consequently, any ambient metric is Ricci-flat, so that $W_\#$ must vanish. We thus have $P_j(0, C) = 0$, a contradiction. □

Appendix

Appendix A. Holomorphic microfunctions. Proofs of the facts stated below are found for instance in a textbook by Schapira [S].

Let X be a complex manifold and Y a complex hypersurface. Then Y is locally given by the zeros of a holomorphic function $f(z)$ such that $df \neq 0$. A germ of a holomorphic microfunction at $p \in Y$ is, by definition, an equivalence class modulo $\mathcal{O}_{X,p}$ of a germ of a (multi-valued) holomorphic function in $X \setminus Y$ of the form

$$\varphi f^{-m} + \psi \log f \quad \text{with} \quad m \in \mathbb{Z}, \quad \varphi, \psi \in \mathcal{O}_{X,p}.$$

Let $\mathcal{C}_{Y|X,p}$ denote the vector space of those equivalence classes. Then a sheaf of holomorphic microfunctions is defined by $\mathcal{C}_{X|Y} = (\mathcal{C}_{Y|X,p})_{p\in Y}$. For $L \in \mathcal{C}_{Y|X,p}$, the singular support of L is contained in

$$N = T^*_Y X \setminus 0 = \{(p,\xi) \in T^*X;\ p \in Y,\ \xi = c\, df|_{z=p},\ c \in \mathbb{C}^*\},$$

the conormal bundle of $Y \subset X$. (In [SKK], $\mathcal{C}_{Y|X}$ is defined to be a sheaf on the projective conormal bundle $N/\mathbb{C}^*$, which can be identified with Y.) The sheaf $\mathcal{E}_X$ of microdifferential operators is defined in such a way that a germ $P(z,\partial_z) \in \mathcal{E}_{X,\hat{p}}$ acts on $\mathcal{C}_{Y|X,p}$, where $\hat{p} = (p,\xi) \in N$. Specifically, $\mathcal{E}_{X,\hat{p}}$ is a ring generated by

$$z_1, \ldots, z_n,\ \partial_{z_1}, \ldots, \partial_{z_n} \ \text{ and } \ \partial_{z_n}^{-1},$$

where $z = (z_1, \ldots, z_n)$ is a local coordinate system of X such that $z_n = f$. The action of $\partial_{z_n}^{-1}$ on $L \in \mathcal{C}_{Y|X,p}$ is given by a curvilinear integral

$$\partial_{z_n}^{-1} L(z) = \int_{p'}^{z} L(z)\, dz_n,$$

where $p' \in X \setminus Y$ is chosen so close to p that the right side (modulo $\mathcal{O}_{X,p}$) is independent of the choice of p'.

We say that $L \in \mathcal{C}_{Y|X,p}$ is *nondegenerate* if L is represented by a function of the form

$$\varphi f^{-m} + \psi \log f \ \text{ for } \ m > 0, \ \text{ or } \ \varphi r^{-m} \log f \ \text{ for } \ m \le 0, \qquad (A.1)$$

where φ is nonvanishing. If $L \in \mathcal{C}_{Y|X,p}$ is of the form $L = P \log f$ with $P = P(z,\partial_z) \in \mathcal{E}_{X,\hat{p}}$, then L is nondegenerate if and only if P is elliptic (i.e. invertible).

In what follows, we consider the case $X = \mathbb{C}^n \times \overline{\mathbb{C}^n}$, the complexification of the diagonal $\{(z,w) \in X;\ w = \overline{z}\} = \mathbb{C}^n \cong \mathbb{R}^{2n}$. Let Ω be a domain in $\mathbb{C}^n$ such that the boundary is locally given by a real-analytic defining function $\rho(z,\overline{z})$ near a boundary point of reference. Then the complexification of the boundary $\partial\Omega$ is locally given by $Y = \{(z,w) \in X;\ \rho(z,w) = 0\}$.

Lemma A.1. *If Ω is strictly pseudoconvex locally, then every holomorphic microfunction $L \in \mathcal{C}_{Y|X,(z_0,w_0)}$ is written as*

$$L(z,w) = P(z,\partial_z) \log \rho(z,w) = Q(w,\partial_w) \log \rho(z,w), \qquad (A.2)$$

where $P \in \mathcal{E}_{\mathbb{C}^n,(z_0,d_z\rho)}$ and $Q \in \mathcal{E}_{\overline{\mathbb{C}^n},(w_0,d_w\rho)}$ are microdifferential operators determined uniquely by L.

In this lemma, we may replace $\log\rho$ by any nondegenerate holomorphic microfunction K with support Y. It then follows that for any $P(z,\partial_z) \in \mathcal{E}_{\mathbb{C}^n,(z_0,d_z\rho)}$ there exists a unique $Q(w,\partial_w) \in \mathcal{E}_{\overline{\mathbb{C}^n},(w_0,d_w\rho)}$ such that

$$P(z,\partial_z)K = Q(w,\partial_w)K.$$

Let Q^* denote the formal adjoint of Q. Then the correspondence $P \mapsto Q^*$ gives rise to an isomorphism of rings $\mathcal{E}_{\mathbb{C}^n,(z_0,d_z\rho)} \to \mathcal{E}_{\overline{\mathbb{C}^n},(w_0,-d_w\rho)}$, which is called the *quantized contact transformation with kernel* K. The following is clear from Lemma A.1.

Lemma A.2. *If two kernels $K, \widetilde{K} \in C_{Y|X,(z_0,w_0)}$ give the same quantized contact transformation, then $K = c\widetilde{K}$ with some constant $c \in \mathbb{C}^*$.*

If $K \mapsto \widehat{K}$ is Kashiwara's transformation, then

$$P(z,\partial_z)K = Q(w,\partial_w)K \quad \text{if and only if} \quad P^*(z,\partial_z)\widehat{K} = Q^*(w,\partial_w)\widehat{K}.$$

In particular, the quantized contact transformation $P(z,\partial_z) \mapsto Q^*(w,\partial_w)$ with kernel K is given by the inverse of the quantized contact transformation $Q(w,\partial_w) \mapsto P^*(z,\partial_z)$ with kernel $\widehat{K}$.

The proof of Lemma A.1 (e.g., in Shapira [S]) simply yields the following lemma, which was used in the proof of Lemma 3.2.

Lemma A.3. *If L in* (A.2) *is of the form* (A.1) *with ρ in place of f and with φ nonvanishing, then P and Q are operators of order $\le m$.*

Appendix B. Method of computing the asymptotic expansion.

We here explain the method of computing the expansion of K^s.

Let us first recall the procedure for computing the Bergman kernel K^0 due to Boutet de Monvel. We take a $\mathbb{C}$-valued defining function of the complexification of $\partial\Omega$ of the form $U(z,\overline{z}) = z_n + \overline{z}_n - z' \cdot \overline{z'} - H(z,\overline{z'})$, where

$$H(z,\overline{z'}) = \sum_{|\alpha|,|\beta|\ge 2,\ell\ge 0} B^{\ell}_{\alpha\overline{\beta}}\, z'_\alpha\, \overline{z'_\beta}\, z_n^{\ell}.$$

Then each $B^{\ell}_{\alpha\overline{\beta}}$ is a polynomial in $A = (A^{\ell}_{\alpha\overline{\beta}}) \in \mathcal{N}$. Let $\mathbf{A}_0 = \mathbf{A}_0(z,\partial_z)$ be a microdifferential operator of infinite order given by the total symbol

$$\mathbf{A}_0(z,\zeta) = \exp\left(-H(z,-\zeta'/\zeta_n)\zeta_n\right).$$

We define weight by

$$\mathrm{w}(z_j) = -\mathrm{w}(\partial_{z_j}) = -1/2 \quad (j<n), \quad \mathrm{w}(z_n) = -\mathrm{w}(\partial_{z_n}) = -1.$$

(For more about the notion of weight, see Section 3 of [HKN2].) Then $\mathbf{A}_0$ can be regarded as an asymptotic series as weight tends to $-\infty$. We can verify $\log U = \mathbf{A}_0(z, \partial_z) \log \rho_0$ by using $\partial_{z_j} \partial_{z_n}^{-1} \log \rho_0 = -\overline{z_j} \log \rho_0$. Therefore the Bergman kernel $K^0[r]$ for Ω (up to a constant multiple $(-\pi)^n$) is given by

$$K^0[r] = \mathbf{A}_0^{*-1}(z, \partial_z)\, \widehat{K}_{-n-1}[\rho_0]. \tag{B.1}$$

Here the inverse of $\mathbf{A}_0^*$ is defined by $\mathbf{A}_0^{*-1} = \sum_{k=0}^{\infty}(1 - \mathbf{A}_0^*)^k$, which is an asymptotic series as weight tends to $-\infty$ because each term of $1 - \mathbf{A}_0^*$ has negative weight.

We generalize (B.1) to K^s for $s > 0$. First, write

$$\widehat{K}_s[r] = \sum_{\ell=1}^{s} a_\ell(z, \overline{z'})\, \widehat{K}_\ell[U]$$

and define a microdifferential operator of infinite order by the total symbol

$$\mathbf{A}_s(z, \zeta) = \mathbf{A}_0(z, \zeta) \sum_{\ell=1}^{s} a_\ell(z, -\zeta'/\zeta_n)\zeta_n^\ell.$$

Then we get $\widehat{K}_s[r] = \mathbf{A}_s(z, \partial_z) \log \rho_0$ by using $\mathbf{A}_0(z, \partial_z)\partial_{z_n}^\ell \log \rho_0 = \widehat{K}_\ell[U]$. Thus we have

$$K^s[r] = \mathbf{A}_s^{*-1}(z, \partial_z)\, K^0[\rho_0].$$

Here $\mathbf{A}_s^{*-1}$ is defined by the series

$$\mathbf{A}_s^{*-1} = \partial_{z_n}^{-s} \sum_{k=0}^{\infty} (1 - \mathbf{A}_s^* \partial_{z_n}^{-s})^k,$$

in which each term in $1 - \mathbf{A}_s^* \partial_{z_n}^{-s}$ has negative weight.

Method of proving Lemma 4.2. We only need to know the first five terms in

$$\mathbf{A}_s^{*-1}(z, \zeta)\big|_{z_1=\zeta_1=0} = \sum_{k=-s}^{\infty} c_k\, \zeta_2^{-k},$$

that is, the terms of weight $\geq -s - 5$ in the right-hand side. Such terms can be computed from the the terms of $\mathbf{A}_s$ that have weight $\geq s - 5$. Details of this computation are discussed in [HKN2]. □

Proof of Proposition 4.4. We only need to compute $K^s(\gamma_t)$ for a surface in normal form for which $\|R^{(2,2)}\|^2(0) = \|A_{2\overline{2}}^0\|^2 \neq 0$. We here take the surface $\rho = \rho_0 - F = 0$, where $F = z_1^2\overline{z}_2^2 + z_2^2\overline{z}_1^2$, for which $\|A_{2\overline{2}}^0\|^2 = 2$.

Starting from this ρ, we set ρ_1, ρ_2 and ρ_3 as in Subsection 2.3. Then we have $r = \rho_3 + O(\rho^3)$. Since each term in ρ^3 has weight less than -3, we see that $r = r_3 +$ (terms of weight < -3). Thus we have

$$r = \rho + \left(\frac{16|z_1 z_2|^2 \rho_0}{n+1} - \frac{8(|z_1|^2 + |z_2|^2)\rho_0^2}{(n+1)n} + \frac{16\rho_0^3}{3(n+1)n(n-1)}\right) \qquad (B.2)$$
$$+ \text{(terms of weight} < -3).$$

In particular, we get

$$r(\gamma_t) = t + 2c' t^3 + O(t^4) \quad \text{with} \quad c' = \frac{8}{3(n+1)n(n-1)}. \qquad (B.3)$$

Next we write $\widehat{K}_s[r] = \mathbf{A}_s(z, \partial_z) \log \rho_0$. Then from (B.2) we get

$$\mathbf{A}_s(z,\zeta) = \zeta_n^s - \widetilde{F}\zeta_n^{s-1} - \Big(-\frac{\widetilde{F}^2}{2} + \frac{16s}{n+1} z_1 z_2 \zeta_1 \zeta_2$$
$$+ \frac{8s(s-1)}{(n+1)n}(z_1\zeta_1 + z_2\zeta_2) + \frac{16s(s-1)(s-2)}{3(n+1)n(n-1)}\Big)\zeta_n^{s-2}$$
$$+ \text{(terms of weight} < s-3),$$

where $\widetilde{F} = z_1^2\zeta_2^2 + z_2^2\zeta_1^2$. Thus we have

$$\mathbf{A}_s^{*-1}(z,\zeta)\big|_{z'=\zeta'=0} = \zeta_n^{-s} + \widetilde{c}^s \zeta_n^{-s-2} + \text{(terms of weight} < -s-3),$$

where

$$\widetilde{c}^s = -4 + \frac{16s}{n+1} + \frac{16s(s-1)}{(n+1)n} + \frac{16s(s-1)(s-2)}{3(n+1)n(n-1)}.$$

Therefore we get, for $s = 0, 1, \ldots, n-2$,

$$K^s(\gamma_t) = t^{s-n-1}\left(1 + \frac{\widetilde{c}^s t^2}{(n-s)(n-s-1)} + O(t^3)\right), \qquad (B.4)$$

and

$$\begin{aligned} K^{n-1}(\gamma_t) &= t^{-2} + (-\widetilde{c}^{n-1} + O(t)) \log t, \\ K^n(\gamma_t) &= t^{-1} + (\widetilde{c}^n t + O(t^2)) \log t, \qquad (B.5) \\ K^{n+1}(\gamma_t) &= (1 + \widetilde{c}^{n+1} t^2/2 + O(t^3)) \log t. \end{aligned}$$

Using (B.3) and (B.4), we have $\varphi_s(\gamma_t) = 1 + 2((n-s+1)c' + \widetilde{c}^s)t^2 + O(t^3)$ for $m = 0, 1, \ldots, n-2$. Thus we get

$$c_s = (n-s+1)c' + \widetilde{c}^s = \frac{2}{3(s-n+1)(s-n)}.$$

The constants c_s for $s \geq n-1$ are determined by using (B.5) in the same manner. □

References

[BEG] T. N. Bailey, M. G. Eastwood and C. R. Graham, *Invariant theory for conformal and CR geometry,* Ann. of Math. **139** (1994), 491–552.

[Bo] H. P. Boas, *Holomorphic reproducing kernels in Reinhardt domains,* Pacific J. Math. **112** (1984), 273–292.

[BM1] L. Boutet de Monvel, *Complément sur le noyau de Bergman,* Sém. EDP, École Polytech. Exposé n° XX, 1985–86.

[BM2] L. Boutet de Monvel, *Le noyau de Bergman en dimension* 2, Sém. EDP, École Polytech. Exposé n° XXII, 1987–88.

[BM3] L. Boutet de Monvel, Singularity of the Bergman kernel, *in* "Complex Geometry", Lecture Notes in Pure and Appl. Math. 143, pp. 13–29, Dekker, 1993.

[BS] L. Boutet de Monvel et J. Sjöstrand, *Sur la singularité des noyaux de Bergman et de Szegö*, Soc. Math. de France, Astérisque **34–35** (1976), 123–164.

[CY] S.-Y. Cheng and S.-T. Yau, *On the existence of a complete Kähler metric on non-compact complex manifolds and the regularity of Fefferman's equation,* Comm. Pure Appl. Math. **33** (1980), 507–544.

[CM] S. S. Chern and J. K. Moser, *Real hypersurfaces in complex manifolds,* Acta Math. **133** (1974), 219–271.

[F1] C. Fefferman, *The Bergman kernel and biholomorphic mappings of pseudoconvex domains,* Invent. Math. **26** (1974), 1–65.

[F2] C. Fefferman, *Monge-Ampère equations, the Bergman kernel, and geometry of pseudoconvex domains*, Ann. of Math. **103** (1976), 395–416; *Correction*, ibid., **104** (1976), 393–394.

[F3] C. Fefferman, *Parabolic invariant theory in complex analysis*, Adv. in Math. **31** (1979), 131–262.

[G1] C. R. Graham, Scalar boundary invariants and the Bergman kernel, *in* "Complex Analysis II", Lecture Notes in Math. 1276, pp. 108–135, Springer, 1987.

[G2] C. R. Graham, *Higher asymptotics of the complex Monge-Ampère equation*, Compositio Math. **64** (1987), 133–155.

[Hi] K. Hirachi, *Construction of boundary invariants and the logarithmic singularity of the Bergman kernel*, to appear in *Ann. of Math.*.

[HKN1] K. Hirachi, G. Komatsu and N. Nakazawa, Two methods of determining local invariants in the Szegö kernel, *in* "Complex Geometry", Lecture Notes in Pure and Appl. Math. 143, pp. 77–96, Dekker, 1993.

[HKN2] K. Hirachi, G. Komatsu and N. Nakazawa, *CR invariants of weight five in the Bergman kernel*, to appear in *Adv. in Math.*.

[Hö] L. Hörmander L^2 *estimates and existence theorems for the* $\overline{\partial}$ *operator,* Acta Math. **113** (1965), 89–152.

[Kan] A. Kaneko, Introduction to Hyperfunctions, Kluwer, 1988.

[Kas] M. Kashiwara, *Analyse micro-locale du noyau de Bergman* Sém. Goulaouic-Schwartz, École Polytech., Exposé n° VIII, 1976–77.

[LM] J. Lee and R. Melrose, *Boundary behaviour of the complex Monge-Ampère equation*, Acta Math. **148** (1982), 159–192.

[M] J. K. Moser, *Holomorphic equivalence and normal forms of hypersurfaces*, Proc. Sympos. Pure Math., 27, 2, pp. 109–112, Amer. Math. Soc., 1975.

[SKK] M. Sato, T. Kawai and M. Kashiwara, Microfunctions and pseudo-differential equations, *in* "Hyperfunctions and Pseudo-Differential Equations," Lecture Notes in Math. 287, pp. 265–529, Springer, 1973.

[Sch] P. Schapira, "Microdifferential Systems in the Complex Domain," Grundlehren Math. Wiss. 269, Springer, 1985.

DEPARTMENT OF MATHEMATICS
OSAKA UNIVERSITY,
TOYONAKA 560, JAPAN

E-mail: hirachi@math.sci.osaka-u.ac.jp
komatsu@math.sci.osaka-u.ac.jp

CHAPTER VI

Quantitative Estimates for Global Regularity

J. J. Kohn

Introduction

Let $\Omega \subset \mathbb{C}^n$ be a bounded pseudoconvex domain with a smooth boundary. We denote by $L_2(\Omega)$ the space of square-integrable functions on Ω and by $\mathcal{H}(\Omega)$ the space of square-integrable holomorphic functions on Ω. Let $B\colon L_2(\Omega) \to \mathcal{H}(\Omega)$ denote the Bergman projection operator, which is the orthogonal projection of $L_2(\Omega)$ onto $\mathcal{H}(\Omega)$. Here we will be concerned with the global regularity of B in terms of Sobolev norms, that is, the question of when $B(H^s(\Omega)) \subset H^s(\Omega)$ where $H^s(\Omega)$ denotes the Sobolev space of order s. Of course, if B preserves $H^s(\Omega)$ locally (i.e., if $B(H^s_{\text{loc}}(\Omega)) \subset H^s_{\text{loc}}(\Omega)$), then B also preserves $H^s(\Omega)$ globally. Aspects of the local question are very well understood, in particular when Ω is of finite D'Angelo type (see [Ca1] and [D'A]). Local regularity can still occur when the D'Angelo type is infinite, as in the examples given in [Chr2] and [K2]. Local regularity fails whenever there is a complex curve V in the boundary of Ω. In that case, if $P \in V$, then for given s there exists an $f \in L_2(\Omega)$ such that $\zeta f \in H^s(\Omega)$ for every smooth function ζ with support in a fixed small neighborhood of P and such that $\zeta B(f) \notin H^s(\Omega)$ whenever $\zeta = 1$ in some neighborhood of P. In contrast, global regularity always holds for small s. That is, if Ω is pseudoconvex, then there exists $\eta > 0$ such that $B(H^s(\Omega)) \subset H^s(\Omega)$ for $s \leqslant \eta$. Furthermore, there is a series of results showing global regularity under a variety of conditions (see [Ca2], [BC], [Ch], [BS1], and [BS2]).

The interest in global regularity came in the early 1970's (see [K1]) and by the end of the 1980's there was a general impression that global regularity was always valid on pseudoconvex domains. In 1984 Barrett (see [Ba2]) found a domain Ω with a smooth boundary but not pseudoconvex, such that there is an $f \in C_0^\infty(\Omega)$ for which Bf is not bounded. It came as a great surprise when Kiselman (see [Ki]) and Barrett (see [Ba1]) found that global regularity does not always hold in the pseudoconvex case. Kiselman showed that on a modified Diederich–Fornaess worm domain W (see [DF1]), there exists a function $f \in C^\infty(\overline{W})$ such that Bf is not Hölder continuous for any positive Hölder exponent. Kiselman used a worm domain W which does not have a smooth boundary. Then Barrett showed that for smooth worm domains Ω, there exist s such that $B(H^s(\Omega)) \not\subset H^s(\Omega)$. Christ in [Chr1] proved the remarkable result that for any worm domain Ω, there exists an s such that $B(C^\infty(\overline{\Omega})) \not\subset H^s(\Omega)$. Recently, Siu (see [S]) constructed a special worm domain Ω with smooth boundary for which there exists $f \in C^\infty(\overline{\Omega})$ such that Bf is not Hölder continuous for any positive Hölder exponent.

This paper is devoted to a quantitative analysis of a result of Boas and Straube (see [BS1]). Their result states that if Ω has a smooth plurisubharmonic defining function, then $B(H^s(\Omega)) \subset H^s(\Omega)$ for all s. Our starting point is a theorem of Diederich and Fornaess (see [DF2]) which asserts that every pseudoconvex domain Ω has a smooth defining function ρ such that there exists $\delta > 0$ so that $-(-\rho)^\delta$ is plurisubharmonic. Our main result is the following.

Theorem. *Let $\Omega \subset \mathbb{C}^n$ be a bounded pseudoconvex domain with a smooth boundary. Then there exists positive constants η and A with the following property. Let ρ be a smooth defining function of Ω such that $-(-\rho)^\delta$ is plurisubharmonic and let g be defined by $\rho = gr$, where r is a smooth defining function with $\sum |r_{z_i}|^2 = 1$ on $b\Omega$, the boundary of Ω. Then $B(H^s(\Omega)) \subset H^s(\Omega)$ whenever: either $s \leqslant \eta$, or $s > \eta$ and*

$$A^s(1-\delta)\max_{b\Omega}\Big(1+\frac{|g_{z_i}|}{g}\Big)^3 \leqslant 1.$$

We call the readers' attention to the extensions of this result given in section 5. The proof of the above theorem is based on the following two methods which, I believe, will also prove useful in other contexts.

(1) Construction of the pseudodifferential operators $T^{(s)}$ which measure smoothness microlocalized in the "bad" direction.

(2) The use of the weights $|r|^\sigma$ to get precise Sobolev estimates.

1. A-priori estimates for the $\overline{\partial}$-Neumann problem

Let $\Omega \subset \mathbb{C}^n$ be a bounded domain with smooth boundary. Let $L_2^q(\Omega)$ denote the space of $(0,q)$-forms on Ω with square-integrable coefficients. Let $\overset{q}{\mathrm{Dom}}_\Omega(\overline{\partial}) \subset L_2^q(\Omega)$ consist of all $\varphi \in L_2^q(\Omega)$ such that $\overline{\partial}\varphi \in L_2^{q+1}(\Omega)$, where $\overline{\partial}\varphi$ is meant in the sense of distributions. Let $\overline{\partial}^*$ denote the L_2-adjoint of $\overline{\partial}$ with domain denoted by $\overset{q}{\text{Dom}}(\overline{\partial}^*)$. We set $\mathcal{D}^q = \overset{q}{\text{Dom}}(\overline{\partial}) \cap \overset{q}{\text{Dom}}(\overline{\partial}^*)$ and let $Q(\varphi, \psi)$ be defined by

$$Q(\varphi, \psi) = (\overline{\partial}\varphi, \overline{\partial}\psi) + (\overline{\partial}^*\varphi, \overline{\partial}^*\psi) \tag{1.1}$$

for $\varphi, \psi \in \mathcal{D}^q$.

The basic theorem that solves the $\overline{\partial}$-Neumann problem in L_2 on pseudoconvex domains is the following. (For an exposition of this material, see [FK].)

1.2 Theorem. *If $\Omega \subset \mathbb{C}^n$ is a pseudoconvex, bounded domain with a smooth boundary, then given $\alpha \in L_2^{q+1}(\Omega)$, there exists a unique $\varphi \in \mathcal{D}^{q+1}$ such that*

$$Q(\varphi, \psi) = (\alpha, \psi) \tag{1.3}$$

for all $\psi \in \mathcal{D}^{q+1}$. Further, if $\overline{\partial}\alpha = 0$, then $\overline{\partial}\varphi = 0$ and $\overline{\partial}\,\overline{\partial}^*\varphi = \alpha$. Thus $u = \overline{\partial}^*\varphi$ is the unique solution of the equation $\overline{\partial}u = \alpha$ with the property that u is orthogonal to the space of square-integrable $\overline{\partial}$-closed $(0, q)$-forms. When $q = 0$, then u is orthogonal to the space of square-integrable holomorphic functions.

We denote by N_{q+1} the operator defined by $N_{q+1}\alpha = \varphi$. Thus the solution u above can be written as $\overline{\partial}^* N_{q+1}\alpha$. We are really interested in the case $q = 0$ (which corresponds to the Bergman projection) but we need general q because of an induction argument. We define the Bergman projection $B\colon L_2(\Omega) \to \mathcal{H}(\Omega)$. $\mathcal{H}(\Omega)$ denotes the space of square-integrable holomorphic functions and B is the orthogonal projection onto $\mathcal{H}(\Omega)$. By linear algebra, it follows from Theorem 1.2 that

$$Bf = f - \overline{\partial}^* N_1 \overline{\partial} f\,. \tag{1.4}$$

Similarly, if $B_q\colon L_2^q(\Omega) \to \mathcal{H}^q(\Omega)$ denotes the orthogonal projection, where $\mathcal{H}^q(\Omega)$ denotes the space of square-integrable $\overline{\partial}$-closed $(0, q)$-forms, we have

$$B_q f = f - \overline{\partial}^* N_{q+1} \overline{\partial} f\,.$$

1.5 Definition. A defining function for Ω is a function $\rho \in C^\infty(U)$, where U is a neighborhood of $b\Omega$ such that $\rho < 0$ in $U \cap \Omega, \rho = 0$ on $b\Omega$ and $d\rho \neq 0$ on $b\Omega$. We will denote by r a fixed defining function with the further property that $\sum |r_{z_j}|^2 = 1$ on $b\Omega$.

Let $\dot{\mathcal{D}}^q = \mathcal{D}^q \cap C^\infty(\overline{\Omega})$. Then a form $\varphi = \sum \varphi_j d\overline{z}_j$ is in $\dot{\mathcal{D}}^1$ if, and only if $\varphi_j \in C^\infty(\overline{\Omega})$ and satisfies

$$\sum r_{z_j}\varphi_j = 0 \quad \text{on} \quad b\Omega\,. \tag{1.6}$$

The following is a basic a priori estimate for the $\bar\partial$-Neumann problem. We will deal first with $(0,1)$-forms and show how to generalize to $(0,q)$-forms.

1.7 Theorem. *Let $\Omega \subset \mathbb{C}^n$ be a bounded, pseudoconvex domain with a smooth boundary. Let $\lambda \in C^\infty(\overline{\Omega})$ be a function such that $\lambda \geqslant 0$ and $-\lambda$ is plurisubharmonic in a neighborhood U of $b\Omega$. Then there exists $C > 0$ such that*

$$\begin{aligned}\sum(-\lambda_{z_i\bar z_j}\varphi_i,\varphi_j) + \sum\int_{b\Omega}\lambda r_{z_i\bar z_j}\varphi_i\overline{\varphi}_j dS + \sum\|\lambda^{\frac12}\varphi_{i\bar z_j}\|^2 \\ \leqslant C\left(\|\lambda^{\frac12}\bar\partial\varphi\|^2 + \|\lambda^{\frac12}\bar\partial^*\varphi\|^2 + \left|\left(\sum\lambda_{z_j}\varphi_j,\bar\partial^*\varphi\right)\right|\right)\end{aligned} \tag{1.8}$$

for all $\varphi \in \dot{\mathcal{D}}^1$ with $\operatorname{supp}(\varphi) \subset U \cap \overline{\Omega}$. The same holds when $\lambda \in C^\infty(\Omega)$ is only Hölder continuous on $\overline{\Omega}$.

Near $b\Omega$ we will define the operators Λ^s by means of a partition of unity. We cover a neighborhood of $b\Omega$ by a finite set of coordinate neighborhoods U_ν with coordinates $\{t_1^\nu, t_2^\nu, \dots, t_{2n-1}^\nu, r\}$. For $u \in C_0^\infty(U_\nu)$ we define the partial Fourier transform

$$\mathcal{F}_\nu u(\xi, r) = \int_{U_\nu} e^{-it^\nu\cdot\xi}u(t^\nu, r)dt^\nu, \tag{1.9}$$

where $t^\nu = (t_1^\nu, \dots, t_{2n-1}^\nu), \xi = (\xi_1, \dots, \xi_{2n-1})$, and $dt^\nu = dt_1^\nu \cdots dt_{2n-1}^\nu$. Define $\Lambda_\nu^s u$ by

$$\mathcal{F}_\nu\Lambda_\nu^s(\xi, r) = (1 + |\xi|^2)^{s/2}\mathcal{F}^\nu u(\xi, r)\ . \tag{1.10}$$

Let $0 \leqslant \zeta_\nu \in C_0^\infty(U^\nu)$ be such that $\sum\zeta_\nu = 1$ in a neighborhood of $b\Omega$ and let $0 \leqslant \zeta'_\nu \in C_0^\infty(U^\nu)$ be such that $\zeta'_\nu = 1$ in a neighborhood of $\operatorname{supp}(\zeta_\nu)$. Now we define $\Lambda^s u$ by

$$\Lambda^s u = \sum\zeta'_\nu\Lambda_\nu^s(\zeta_\nu u)\ . \tag{1.11}$$

For forms we define $\widetilde{\Lambda}^s\varphi$ by

$$(\widetilde{\Lambda}^s\varphi)_i = \Lambda^s\varphi_i + r_{\bar z_i}\sum_k[\Lambda^s, r_{z_k}]\varphi_k, \tag{1.12}$$

and since $\sum r_{z_i}r_{\bar z_i} = 1$ on $b\Omega$, we have

$$\sum r_{z_i}(\tilde\Lambda^s\varphi)_i = \Lambda^s(\sum r_{z_i}\varphi_i) \text{ on } b\Omega. \tag{1.13}$$

Hence if $\sum r_{z_i}\varphi_i = 0$ on $b\Omega$, we have $\sum r_{z_i}(\tilde\Lambda^s\varphi)_i = 0$ on $b\Omega$.

Now setting $\lambda = 1$ and substituting $\tilde{\Lambda}^s\varphi$ for φ in (1.8) we get

$$\sum \|(\tilde{\Lambda}^s\varphi)_{i\bar{z}_j}\|^2 \leqslant C\,Q(\tilde{\Lambda}^s\varphi, \tilde{\Lambda}^s\varphi)\,. \tag{1.14}$$

1.15 Definition. Let L_{norm}, $\overline{L}_{\text{norm}}$ denote the normal z-derivative and the normal $\bar{z}$-derivative, defined as follows:

$$L_{\text{norm}} = \sum r_{\bar{z}_i}\frac{\partial}{\partial z_i} \qquad \text{and} \qquad \overline{L}_{\text{norm}} = \sum r_{z_i}\frac{\partial}{\partial \bar{z}_i}\,. \tag{1.16}$$

The operator T is defined by $T = \frac{1}{2}(L_{\text{norm}} - \overline{L}_{\text{norm}})$. We define the operators L_i, $\overline{L}_i$ by

$$L_i = \frac{\partial}{\partial z_i} - r_{z_i}L_{\text{norm}} \qquad \text{and} \qquad \overline{L}_i = \frac{\partial}{\partial \bar{z}_i} - r_{\bar{z}_i}\overline{L}_{\text{norm}}\,. \tag{1.17}$$

Note that, since $\sum |r_{z_i}|^2 = 1$ on $b\Omega$, the operators L_i, $\overline{L}_i$, and T are tangential, that is, $L_i(r) = \overline{L}_i(r) = T(r) = 0$ on $b\Omega$.

In conjunction with (1.14) it is useful to note that any first order partial differential operator can be written as a combination of the $L_i, \overline{L}_i, T$, and the $\frac{\partial}{\partial \bar{z}_k}$. Hence

$$\|D\Lambda^{s-1}u\| \leqslant C\Big(\sum_k \|(\Lambda^{s-1}u)_{\bar{z}_k}\| + \|\Lambda^s u\|\Big)\,. \tag{1.18}$$

In studying global regularity, it is useful to consider the $\bar{\partial}$-Neumann problem with weights (see [K1]). This is based on the weighted inner product $(u,v)_{(t)} = (w_t u, v)$, where $w_t = \exp(-t|z|^2)$ for $t \geqslant 0$. We denote by $\bar{\partial}_t^*$ the adjoint of $\bar{\partial}$ with respect to this inner product. We then have $\mathrm{Dom}(\bar{\partial}_t^*) = \mathrm{Dom}(\bar{\partial}^*)$ and

$$\bar{\partial}_t^* = w_{-t}\bar{\partial}^* w_t\,. \tag{1.19}$$

Let

$$Q_{q,t}(\varphi,\psi) = (\bar{\partial}\varphi, \bar{\partial}\psi)_{(t)} + (\bar{\partial}_t^*\varphi, \bar{\partial}_t^*\psi)_{(t)}$$

for $\varphi, \psi \in \mathcal{D}^q$.

The principal result (see [K1]) is the following.

1.20 Theorem. *If $\Omega \subset \mathbb{C}^n$ is bounded, pseudoconvex and has a smooth boundary, then for each t and q, there exists a unique self adjoint operator $N_{t,q}\colon L_2^q(\Omega) \to \mathcal{D}^q$ with the following properties:*

(i) $Q_{q,t}(N_{t,q}\alpha, \psi) = (\alpha, \psi)_{(t)}$ *for all* $\psi \in \mathcal{D}^q$.

(ii) *There exists a constant c such that if $t \geqslant cs$ then $N_{t,q}(H^s), N_{t,q}\bar\partial(H^s)$, $\bar\partial_t^* N_{t,q}(H^s)$, and $\bar\partial_t^* N_{t,q}\bar\partial(H^s)$ are all contained in H^s.*

Here H^s denotes the subspace of $L_2^q(\Omega)$ with coefficient in the Sobolev s-space.

Boas and Straube (see [BS1] and [BS2]) proved the following identity which we will need in the induction procedure.

$$N_q\bar\partial = B_q w_t N_{t,q}\bar\partial[w_{-t}(I - B_{q-1})] \,. \tag{1.21}$$

Next we will show how the calculations done for $(0,1)$-forms generalize to $(0,q)$-forms. Let $I = (i_1, \dots i_q)$ be a q-tuple $1 \leqslant i_1 < i_2 < \cdots < i_q \leqslant n$. We denote by $d\bar z_I$ the form

$$d\bar z_I = d\bar z_{i_1} \wedge \cdots \wedge d\bar z_{i_q} \,.$$

If $\varphi = \sum \varphi_I d\bar z_I$ then $\bar\partial\varphi = \sum \frac{\partial\varphi_I}{\partial \bar z_i} d\bar z_i \wedge d\bar z_I$.

The condition $\varphi \in \dot{\mathcal{D}}^q$ is characterized by

$$\sum_K \sum_{i=1}^n r_{z_i}\varphi_{iK} = 0 \quad \text{on } b\Omega, \tag{1.22}$$

where K sums over all ordered $(q-1)$-tuples and

$$\varphi_{iK} = \begin{cases} 0 & \text{if } i \in K \\ \operatorname{sgn}\begin{pmatrix} iK \\ \langle iK\rangle \end{pmatrix} & \text{if } i \notin K. \end{cases}$$

Here $\langle iK\rangle$ denotes the ordered q-tuple whose elements are the elements of K and i. Then $\operatorname{sgn}\binom{iK}{\langle iK\rangle}$ denotes the sign of the permutation which takes iK to $\langle iK\rangle$. Then $\bar\partial^*\varphi$ is expressed by

$$\bar\partial^*\varphi = -\sum_K \sum_i \frac{\partial\varphi_{iK}}{\partial x_i} d\bar z_K \,. \tag{1.23}$$

If $-\lambda$ is plurisubharmonic, then for pseudoconvex Ω the generalization of (1.8) to $(0,q)$-forms is:

$$\begin{aligned} &\sum_K \sum_{i,j} (-\lambda_{z_i\bar z_j}\varphi_{iK}, \varphi_{jK}) + \sum_K \sum_{i,j} \int_{b\Omega} \lambda r_{z_i\bar z_j}\varphi_{iK}\bar\varphi_{jK} dS + \sum_I \sum_j \|\lambda^{\frac12}\varphi_{I\bar z_j}\|^2 \\ &\leqslant C\Big(\|\lambda^{\frac12}\bar\partial\varphi\|^2 + \|\lambda^{\frac12}\bar\partial^*\varphi\|^2 + \Big|\Big(\sum_K \sum_j \lambda_{z_j}\varphi_{jK} d\bar z_K, \bar\partial^*\varphi\Big)\Big|\Big) . \end{aligned} \tag{1.24}$$

Finally, $\widetilde{\Lambda}^s\varphi = \sum_I (\widetilde{\Lambda}^s\varphi)_I d\bar{z}_I$ is defined by

$$(\widetilde{\Lambda}^s\varphi)_I = \Lambda^s\varphi_I + r_{\bar{z}_i}\sum_{K\subset I}\sum_j [\Lambda^s, r_{z_j}]\varphi_{jK}, \tag{1.25}$$

where K runs over all order $(q-1)$-tuples that are subsets of I.

2. Estimates involving $|r|^\sigma$

Throughout this section $\Omega \subset \mathbb{C}^n$ will denote a bounded domain with smooth boundary, r the usual defining function, and U a product neighborhood of $b\Omega$ covered by boundary coordinate neighborhoods.

2.1 Lemma. *There exists $C > 0$ such that*

$$\| |r|^\sigma \Lambda^{s+\sigma} u\| \leqslant C\left(\|\Lambda^s u\| + \sum_j \left\| r\Lambda^s \frac{\partial u}{\partial \bar{z} j}\right\|\right) \tag{2.2}$$

for all $\sigma \in [0,1]$ and $u \in C_0^\infty(U \cap \overline{\Omega})$.

Proof. First we prove the estimate for $\sigma = 1$. We have

$$\|r\Lambda^{s+1}u\| = \|\Lambda r \Lambda^s u\| \leqslant C\sum_{i=1}^{2n} \|D_i r\Lambda^s u\| ,$$

where the $D_i = \mathrm{Re}(\frac{\partial}{\partial z_i})$ for $i = 1,\ldots,n$ and $D_i = \mathrm{Im}(\frac{\partial}{\partial z_{i-n}})$ for $i = n+1,\ldots,2n$. Integrating by parts we have

$$\|D_i r\Lambda^s u\| = \left\| \frac{\partial}{\partial \bar{z}_i} r\Lambda^s u\right\| \leqslant C\left(\|\Lambda^s u\| + \left\| r\Lambda^s \frac{\partial u}{\partial \bar{z}_i}\right\| + \left\| r\left[\frac{\partial}{\partial \bar{z}_i}, \Lambda^s\right] u\right\|\right) .$$

Then since $\frac{\partial}{\partial \bar{z}_i}$ is a combination of $\frac{\partial}{\partial r}$ and tangential derviatives, we obtain

$$\left[\frac{\partial}{\partial \bar{z}_i}, \Lambda^s\right] = P_i^{s-1}\frac{\partial}{\partial r} + P_i^s ,$$

where P_i^{s-1} and P_i^s are tangential pseudodifferential operators of order $s-1$ and s, respectively. Since $\frac{\partial}{\partial r}$ is a combination of the $\frac{\partial}{\partial \bar{z}_j}$ and tangential derivatives, we obtain (2.2) in the case $\sigma = 1$.

Next we prove (2.2) for $\sigma = \frac{m}{2^k}$ with $0 \leqslant m \leqslant 2^k$ by induction on k. The above takes care of the case $k = 0$. Assume that the inequality holds for $k-1$ (with C independent of m and k). Then if m is even, we set

$m = 2j$ and $\frac{m}{2^k} = \frac{j}{2^{k-1}}$ hence the inequality holds by induction. If m is odd, then $m = 2j+1$ and we have

$$\left\| |r|^{\frac{2j+1}{2^k}} \Lambda^{s+\frac{2j+1}{2^k}} u \right\|^2 = \left(|r|^{\frac{j+1}{2^{k-1}}} \Lambda^{s+\frac{j+1}{2^{k-1}}} u, |r|^{\frac{j}{2^{k-1}}} \Lambda^{s+\frac{j}{2^{k-1}}} u \right)$$
$$\leqslant C^2 \left(\|\Lambda^s u\| + \sum_i \left\| r\Lambda^s \frac{\partial u}{\partial \bar{z}_i} \right\| \right)^2 .$$

Thus (2.2) holds for $\sigma = \frac{m}{2^k}$ and hence, since $\left\{\frac{m}{2^k}\right\}$ is dense in $[0,1]$, it holds for all $\sigma \in [0,1]$.

With a slight modification of this argument, it follows that for any first order differential operator D, we have

$$\left\| |r|^\sigma D\Lambda^{s+\sigma-1} u \right\| \leqslant C\left(\|\Lambda^s u\| + \sum_j \left\| r\Lambda^s \frac{\partial u}{\partial \bar{z}_j} \right\| \right) \tag{2.3}$$

for $0 \leqslant \sigma \leqslant 1$ and $u \in C_0^\infty(U \cap \overline{\Omega})$.

Now we have

$$\|r\Lambda^s \frac{\partial u}{\partial \bar{z}_j}\|^2 = -\left(r\Lambda^{s-1} \frac{\partial^2 u}{\partial z_j \partial \bar{z}_j}, r\Lambda^{s+1} u \right) + \left(\frac{\partial u}{\partial \bar{z}_j}, \left[r^2\Lambda^{2s}, \frac{\partial}{\partial \bar{z}_j} \right] u \right) .$$

Since $\frac{\partial}{\partial \bar{z}_j}$ is a combination of $\frac{\partial}{\partial r}$ and tangential differential operators, the second term on the right side in the above is bounded by

$$C\|\Lambda^s u\|^2 + \text{small const.} \sum_k \left\| r\Lambda^s \frac{\partial u}{\partial \bar{z}_k} \right\|^2 .$$

Thus we obtain

$$\begin{aligned} \left\| |r|^\sigma \Lambda^{s+\sigma} u \right\| + \left\| |r|^\sigma D\Lambda^{s+\sigma-1} u \right\| &\leqslant C\left(\|\Lambda^s u\| + \sum \left\| r\Lambda^s \frac{\partial u}{\partial \bar{z}_j} \right\| \right) \\ &\leqslant C(\|\Lambda^s u\| + \|r\Lambda^{s-1}\Delta u\|) \end{aligned} \tag{2.4}$$

for $0 \leqslant \sigma \leqslant 1$, $u \in C_0^\infty(U \cap \overline{\Omega})$. Here Δ denotes the Laplacian

$$\Delta = -\frac{1}{4} \sum \frac{\partial^2}{\partial z_j \partial \bar{z}_j} . \tag{2.5}$$

2.6 Lemma. *There exists $C > 0$ such that if $-\frac{1}{2} < \sigma \leqslant 1$*

$$\begin{aligned} \left\| |r|^\sigma \Lambda^{s+\sigma} u \right\| &\leqslant \frac{C}{\sqrt{1+2\sigma}} \left(\|\Lambda^s u\| + \sum_j \left\| r \frac{\partial}{\partial \bar{z}_j} \Lambda^s u \right\| \right) \\ &\leqslant \frac{C}{\sqrt{1+2\sigma}} (\|\Lambda^s u\| + \|r\Lambda^{s-1}\Delta u\|) \end{aligned} \tag{2.7}$$

for $u \in C_0^\infty(U \cap \overline{\Omega})$.

Proof. This has already been proved for $0 \leqslant \sigma \leqslant 1$. If $-\frac{1}{2} < \sigma < 0$, we write $|r|^{2\sigma} = -(-r)^{2\sigma} = -\frac{1}{1+2\sigma}\frac{\partial}{\partial r}((-r)^{1+2\sigma})$ and hence we have

$$\||r|^\sigma \Lambda^{s+\sigma} u\|^2 = -\frac{1}{1+2\sigma}\left(\frac{\partial}{\partial r}(-r)^{1+2\sigma}\Lambda^{s+2\sigma}u, \Lambda^s u\right)$$
$$\leqslant \frac{C}{1+2\sigma}\left(\left\||r|^{1+2\sigma}\frac{\partial}{\partial r}\Lambda^{s+2\sigma}u\right\|^2 + \|\Lambda^s u\|^2\right).$$

Since $1 + 2\sigma > 0$ we apply (2.4) to the above and obtain (2.7).

Next we consider $u = 0$ on $b\Omega$ and we obtain the following.

2.8 Lemma. *There exists* $C > 0$ *such that*

$$\||r|^\sigma \Lambda^{s+\sigma} u\| \leqslant C\sum_j \left\|\frac{\partial}{\partial \bar{z}_j}\Lambda^{s-1}u\right\| \tag{2.9}$$

whenever $-1 \leqslant \sigma \leqslant 0$, $u = 0$ *on* $b\Omega$ *and* $u \in C_0^\infty(U \cap \overline{\Omega})$.

Further

$$\||r|^\sigma \Lambda^{s+\sigma} u\| \leqslant \frac{C}{\sqrt{3+2\sigma}}\left(\sum\left\|\frac{\partial}{\partial \bar{z}_j}\Lambda^{s-1}u\right\| + \|r\triangle\Lambda^{s-1}u\|\right) \tag{2.10}$$

whenever $-\frac{3}{2} < \sigma < -1$, $u = 0$ *on* $b\Omega$ *and* $u \in C_0^\infty(U \cap \overline{\Omega})$.

Proof. First consider $\sigma = 0$. We have $\Lambda^s u = 0$ on $b\Omega$ and hence

$$\|\Lambda^s u\| \leqslant C\sum \|D_j\Lambda^{s-1}u\| \leqslant C'\sum\left\|\frac{\partial}{\partial \bar{z}_j}\Lambda^{s-1}u\right\|.$$

Next, if $\sigma = -1$, we have

$$\||r|^{-1}\Lambda^{s-1}u\|^2 = \left(\frac{1}{r^2}\Lambda^{s-1}u, \Lambda^{s-1}u\right) = -\left(\frac{\partial}{\partial r}\left(\frac{1}{r}\right)\Lambda^{s-1}u, \Lambda^{s-1}u\right)$$
$$\leqslant 2\||r|^{-1}\Lambda^{s-1}u\|\left\|\frac{\partial}{\partial r}\Lambda^{s-1}u\right\|$$
$$\leqslant C\||r|^{-1}\Lambda^{s-1}u\|\sum\left\|\frac{\partial}{\partial \bar{z}_j}\Lambda^{s-1}u\right\|$$

which settles the case $\sigma = 1$. Now we proceed by induction on k with $\sigma = -\frac{m}{2^k}$ with $m = 0, \ldots, 2^k$. Assume (2.9) holds for $k-1$. Then if m

is even, we have $m = 2j$ so $-\frac{m}{2^k} = -\frac{j}{2^{k-1}}$ and (2.9) holds. If m is odd, $m = 2j+1$ and

$$\left\| |r|^{-\frac{2j+1}{2^k}} \Lambda^{s-\frac{2j+1}{2^k}} u \right\|^2 = \left(|r|^{-\frac{j}{2^{k-1}}} \Lambda^{s-\frac{j}{2^{k-1}}} u, |r|^{-\frac{j+1}{2^{k-1}}} \Lambda^{s-\frac{j+1}{2^{k-1}}} u \right)$$

so (2.9) follows for $\sigma = -\frac{m}{2^k}$ with C independent of m and k and hence for all $\sigma \in [-1, 0]$.

Now in case $-\frac{3}{2} < \sigma < -1$ we have

$$\begin{aligned} \left\| |r|^\sigma \Lambda^{s+\sigma} u \right\|^2 &= \left(|r|^{2\sigma} \Lambda^{s+\sigma} u, \Lambda^{s+\sigma} u \right) \\ &= -\frac{1}{1+2\sigma} \left(\frac{\partial}{\partial r} (-r)^{2\sigma+1} \Lambda^{s+\sigma} u, \Lambda^{s+\sigma} u \right) \\ &\quad + \frac{1}{1+2\sigma} \left(|r|^{\sigma+1} \frac{\partial}{\partial r} \Lambda^{s+\sigma} u, |r|^\sigma \Lambda^{s+\sigma} \right). \end{aligned}$$

Then

$$\begin{aligned} \left\| |r|^\sigma \Lambda^{s+\sigma} u \right\| u &\leqslant \left\| |r|^{\sigma+1} \frac{\partial}{\partial r} \Lambda^{s+\sigma} u \right\| \\ &\leqslant \frac{C}{\sqrt{3+2\sigma}} \left(\left\| \frac{\partial}{\partial r} \Lambda^{s-1} u \right\| + \sum \left\| r \frac{\partial}{\partial \bar{z}_j} \frac{\partial}{\partial r} \Lambda^{s-1} u \right\| \right) \\ &\leqslant \frac{C}{\sqrt{3+3\sigma}} \left(\left\| \frac{\partial}{\partial r} \Lambda^{s-1} u \right\| + \sum \left\| r \frac{\partial}{\partial \bar{z}_j} \frac{\partial}{\partial r} \Lambda^{s-1} u \right\| \right). \end{aligned}$$

Let $v = \Lambda^{s-1} u$. Then $v = 0$ on $b\Omega$ and we have

$$\begin{aligned} \sum \|r D_i D_k v\|^2 &\leqslant C \left(\sum \|D_i (r D_k v)\|^2 + \sum \|D_k v\|^2 \right) \\ &\leqslant C \left(\sum \left\| \frac{\partial}{\partial \bar{z}_j} (r D_k v) \right\|^2 + \sum \left\| \frac{\partial v}{\partial \bar{z}_j} \right\|^2 \right) \\ &\leqslant C \left(\sum |(\Delta(r D_k v), r D_k v)| + \sum \left\| \frac{\partial v}{\partial \bar{z}_j} \right\|^2 \right) \\ &\leqslant C \Big(|(\Delta(rv), \sum D_k (r D_k v))| \\ &\qquad + \sum_k |([\Delta \circ r, D_k] v, r D_k v)| + \sum \left\| \frac{\partial v}{\partial \bar{z}_j} \right\|^2 \Big) \\ &\leqslant C \Big(\|r \Delta v\| \left(\sum \|r D_i D_k v\| + \sum \left\| \frac{\partial v}{\partial \bar{z}_j} \right\| \right) \\ &\qquad + \sum \|r D_i D_k v\| \left\| \frac{\partial v}{\partial \bar{z}_j} \right\| + \sum \left\| \frac{\partial v}{\partial \bar{z}_j} \right\|^2 \Big). \end{aligned}$$

Hence

$$\sum \|rD_iD_kv\|^2 \leqslant C\left(\|r\Delta v\|^2 + \sum \left\|\frac{\partial v}{\partial \bar{z}_j}\right\|^2\right).$$

Combining this with the above completes the proof of (2.10).

2.11 Lemma. *There exists $C > 0$ such that*

$$\|\Lambda^s u\| \leqslant C\left(\sum_j \left\||r|^\sigma \frac{\partial}{\partial \bar{z}_j}\Lambda^{s+\sigma-1}u\right\| + \||r|^\sigma \Lambda^{s+\sigma}u\| + \|r\Lambda^{s-1}\Delta u\|\right) \quad (2.12)$$

for $\sigma \in [0,1]$ and $u \in C_0^\infty(U \cap \overline{\Omega})$.

Proof. We have

$$\begin{aligned}
\|\Lambda^s u\|^2 &= \left(\frac{\partial}{\partial r}(r)\Lambda^s u, \Lambda^s u\right) = -2\mathrm{Re}\left(r\frac{\partial}{\partial r}\Lambda^s u, \Lambda^s u\right) \\
&\leqslant 2\left|\left(|r|^\sigma \frac{\partial}{\partial r}\Lambda^{s+\sigma-1}u, |r|^{1-\sigma}\Lambda^{s-\sigma+1}u\right)\right| \\
&\leqslant C\left(\sum \left\||r|^\sigma \frac{\partial}{\partial \bar{z}_j}\Lambda^{s+\sigma-1}u\right\| + \||r|^\sigma \Lambda^{s+\sigma}u\|\right) \\
&\qquad \times \left(\|\Lambda^s u\| + \sum \left\|r\frac{\partial}{\partial \bar{z}_j}\Lambda^s u\right\|\right).
\end{aligned}$$

Then (2.12) follows.

3. Microlocalization in the "bad" direction

In this section we will construct a tangential pseudodifferential operator Γ^+ defined on $C_0^\infty(U \cap \overline{\Omega})$, where U is a neighborhood of $b\Omega$, which has the property that $\sigma(T) < 0$ on $\mathrm{supp}(\sigma(\Gamma^+))$. Here $\sigma(\cdot)$ denotes the principal symbol. Furthermore Γ^+ has the following properties. There exists $\theta \in C_0^\infty(\Omega)$ and $C > 0$ such that

$$\|\Lambda^s(1-\Gamma^+)u\| \leqslant C\left(\sum \left\|\Lambda^{s-1}\frac{\partial u}{\partial \bar{z}_j}\right\| + \|\Lambda^{s-1}u\| + \|\theta u\|_s\right) \quad (3.1)$$

for all $u \in C_0^\infty(U \cap \overline{\Omega})$, where $\|\cdot\|_s$ denotes the Sobolev norm of order s. Also for any first order differential operator D, there exists $C > 0$ such that

$$\|\Lambda^s[D,\Gamma^+]u\| \leqslant C\left(\sum \left\|\Lambda^{s-1}\frac{\partial u}{\partial \bar{z}_j}\right\| + \|\Lambda^{s-1}u\| + \|\theta u\|_s\right). \quad (3.2)$$

To construct Γ^+ we cover $b\Omega$ with a finite number of special coordinate neighborhoods V^ν and set $U = \cup V^\nu$. The V^ν are constructed as follows. For $P \in b\Omega$ we define holomorphic coordinates $z_1^P, \dots, z_n^P$ which are obtained by performing a translation and unitary transformation on the coordinates $z_1, \dots, z_n$ such that $z_i^P(P) = 0$, for $i = 1, \dots, n$ and

$$\frac{\partial r}{\partial z_i^P}(P) = \begin{cases} 0 & \text{if } i < n \\ 1 & \text{if } i = n. \end{cases} \tag{3.3}$$

Let V be a neighborhood of P on which $\frac{\partial r}{\partial z_i^P}$, for $i = 1, \dots, n-1$ and let $\frac{\partial r}{\partial x_n^P} - 1$ be very small (the size will be determined later). Let V^ν be a finite covering of $b\Omega$ with such coordinate neighborhoods with origin $P^\nu \in V^\nu \cap B\Omega$. We will set $z^\nu = z^{P^\nu}$, and on each V^ν, we have the coordinates $\{t_1^\nu, \dots, t_{2n-1}^\nu, r\}$ defined by

$$\begin{aligned} t_{2i-1}^\nu &= \operatorname{Re}(z_i^\nu) \quad \text{for } i = 1, \dots, n-1 \\ t_{2i}^\nu &= \operatorname{Im}(z_i^\nu) \quad \text{for } i = 1, \dots, n-1 \\ t_{2n-1}^\nu &= \operatorname{Im}(z_n^\nu). \end{aligned} \tag{3.4}$$

Then we have

$$\frac{\partial}{\partial z_k^\nu} = \begin{cases} \frac{1}{2}\left(\frac{\partial}{\partial t_{2k-1}^\nu} - \sqrt{-1}\frac{\partial}{\partial t_{2k}^\nu}\right) + h_k^\nu \frac{\partial}{\partial r} & \text{for } k = 1, \dots, n-1 \\ \frac{1}{2}\left(\frac{\partial}{\partial r} - \frac{\sqrt{-1}}{2}\frac{\partial}{\partial t_{2n-1}^\nu}\right) + h_n^\nu \frac{\partial}{\partial r} & \text{for } k = n. \end{cases} \tag{3.5}$$

Here the $h_k^\nu \in C^\infty(V^\nu)$ with $h_k^\nu(P^\nu) = 0$.

On $\mathbb{R}^{2n-1}$, we denote the coordinates by $\{\xi_1, \dots, \xi_{2n-1}\}$. Let $S^{2n-2} = \{\xi \in \mathbb{R}^{2n-1} \big| |\xi| = 1\}$, $C^+ = \{\xi \in S^{2n-2} \big| \xi_{2n-1} > \frac{1}{2}\}$, $C^- = \{\xi \in S^{2n-2} \big| \xi_{2n-1} < -\frac{1}{2}\}$, and $C^0 = \{\xi \in S^{2n-2} \big| -\frac{2}{3} < \xi_{2n-1} < \frac{2}{3}\}$. Let $\gamma^+, \gamma^-, \gamma^0 \in C^\infty(S^{2n-2})$ be non-negative functions with $\operatorname{supp}(\gamma^+) \subset C^+$, $\operatorname{supp}(\gamma^-) \subset C^-$, $\operatorname{supp}(\gamma^0) \subset C^0$ and such that $\gamma^+ + \gamma^- + \gamma^0 = 1$. Now define $\gamma(\xi)$ for $|\xi| \geqslant 1$ by $\gamma(\xi) = \gamma\left(\frac{\xi}{|\xi|}\right)$, thus extending $\gamma^+, \gamma^-, \gamma^0$ to the region $\{\xi \in \mathbb{R}^{2n-1} \big| |\xi| \geqslant 1\}$. Finally we extend each of these functions to the region $\{\xi \in \mathbb{R}^{2n-1} \big| |\xi| < 1\}$ in any way so that $\gamma^+, \gamma^-, \gamma^0 \in C^\infty(\mathbb{R}^{2n-1})$. For $v \in C_0^\infty(V^\nu \cap \overline{\Omega})$, we have

$$\mathcal{F}^\nu v(\xi, r) = \int_{\mathbb{R}^{2n-1}} e^{-it^\nu \cdot \xi} v(t^\nu, r)\, dt^\nu\,. \tag{3.6}$$

Denote by $\mathcal{G}^\nu$ the inverse of $\mathcal{F}^\nu$ so that for a function g on $\mathbb{R}^{2n}$, we have

$$\mathcal{G}^\nu g(t^\nu, r) = \text{const.} \int_{\mathbb{R}^{2n-1}} e^{it^\nu \cdot \xi} g(\xi, r)\, d\xi\,. \tag{3.7}$$

As is customary we will ignore the constant. Now let $0 \leqslant \zeta^\nu, \eta^\nu \in C_0^\infty(U^\nu \cap \overline{\Omega})$ be such that $\sum \zeta^\nu = 1$ in a neighborhood of $b\Omega$ and such that $\eta^\nu = 1$ in a neighborhood of $\text{supp}(\zeta^\nu)$. If $u \in C^\infty(U \cap \overline{\Omega})$, we define Γu by

$$\Gamma u = \sum \eta^\nu \mathcal{G}^\nu \gamma \mathcal{F}^\nu \zeta^\nu u\,. \tag{3.8}$$

Here Γ^+, Γ^-, and Γ^0 are defined by substituting Γ^+, Γ^-, and Γ^0 for Γ and γ^+, γ^-, and γ^0 for γ.

3.9 Proposition. *Let $\theta \in C_0^\infty(\Omega)$ be such that $\sum \zeta^\nu + \theta = 1$. Then there exists $C > 0$ such that for any s*

$$\|\Lambda^s \Gamma^0 u\| + \|\Lambda^s \Gamma^- u\| \leqslant C\Big(\sum \|\Lambda^{s-1} \frac{\partial u}{\partial \bar{z}_j}\| + \|\Lambda^{s-1} u\| + \|\theta u\|_s\Big) \tag{3.10}$$

for all $u \in C_0^\infty(U \cap \overline{\Omega})$.

Proof. On V^ν define the Laplacian Δ^ν by

$$\Delta^\nu = \sum_{i=1}^{2n-1} \left(\frac{\partial}{\partial t_i^\nu}\right)^2 + \frac{\partial^2}{\partial r^2}\,. \tag{3.11}$$

If $v \in C_0^\infty(V^\nu \cap \overline{\Omega})$, we define v_h^ν by

$$v_h^\nu(t^\nu, r) = \int_{\mathbb{R}^{2n-1}} e^{r|\xi|} e^{it^\nu \cdot \xi} \mathcal{F}^\nu v(\xi, 0)\, d\xi\,. \tag{3.12}$$

Then $v_h^\nu(t^\nu, 0) = v(t^\nu, 0)$ and $\Delta^\nu v_h^\nu = 0$. We set $v_0^\nu = v - v_h^\nu$ so that $v_0^\nu = 0$ on $b\Omega$ and $\Delta^\nu v_0^\nu = \Delta^\nu v$. Now we have

$$\zeta^\nu u = (\zeta^\nu u)_h^\nu + (\zeta^\nu u)_0^\nu$$

and

$$\eta^\nu \mathcal{G}^\nu \gamma \mathcal{F}^\nu (\zeta^\nu u)_h^\nu = \eta^\nu \int_{\mathbb{R}^{2n-1}} e^{r|\xi|} e^{it^\nu \cdot \xi} \gamma(\xi) (\mathcal{F}^\nu \zeta^\nu u)(\xi, 0)\, d\xi\,.$$

In terms of the $\{t^\nu, r\}$ coordinates we have

$$[L_{\text{norm}}]_{P^\nu} = r_{\bar{z}_n}^\nu(P^\nu) \frac{\partial}{\partial z_n^\nu} = \frac{\partial}{\partial r} - \sqrt{-1} \frac{\partial}{\partial t_{2n-1}^\nu}\,.$$

So we have

$$\begin{aligned} L_{\text{norm}}&\left(\eta^\nu \mathcal{G}^\nu \gamma \mathcal{F}^\nu (\zeta^\nu u)_h^\nu\right) \\ &= L_{\text{norm}}(\eta^\nu) \mathcal{G}^\nu \gamma \mathcal{F}^\nu (\zeta^\nu u)_h^2 \\ &\quad + \int_{\mathbb{R}^{2n-1}} (|\xi| + \xi_{2n-1}) \gamma(\xi) e^{r|\xi|} e^{it^\nu \cdot \xi} (\mathcal{F}^\nu \zeta^\nu u)(\xi, 0)\, d\xi \\ &\quad + \eta^\nu H D \mathcal{G}^\nu \gamma \mathcal{F}^\nu (\zeta^\nu u)_h^\nu\,, \end{aligned}$$

where $H(P^\nu)=0$ and D is a first order differential operator.

Since $\xi_{2n-1}\gamma^-(\xi) = -|\xi_{2n-1}|\gamma^-(\xi)$ we have

$$(|\xi|+\xi_{2n-1})\gamma^-(\xi) \leqslant \frac{\sum_1^{2n-2}\xi_i^2}{|\xi|}.$$

Setting $\xi' = (\xi_1, \ldots, \xi_{2n-2}, 0)$, we have

$$(|\xi|+\xi_{2n-1})\gamma^-(\xi) \leqslant |\xi'|\,. \tag{3.13}$$

In the $\{z^\nu\}$ coordinates we have, as in (1.17)

$$\begin{aligned} L_i^\nu &= \frac{\partial}{\partial z_i^\nu} - r_{z_i^\nu} L_{\mathrm{norm}} \\ \overline{L}_i^\nu &= \frac{\partial}{\partial \overline{z}_i^\nu} - r_{\overline{z}_i^\nu} \overline{L}_{\mathrm{norm}}\,. \end{aligned} \tag{3.14}$$

Then we have

$$L_i^\nu\Big|_{P^\nu} = \frac{1}{2}\left(\frac{\partial}{\partial t_{2i-1}} - \sqrt{-1}\frac{\partial}{\partial t_{2i}}\right) \qquad \text{for } i = 1, \ldots, n-1$$

and

$$L_n^\nu\Big|_{P^\nu} = 0\,.$$

Hence for $v \in C_0^\infty(V^\nu \cap \overline{\Omega})$ we have

$$\int_{-\infty}^{0}\int_{\mathbb{R}^{2n-1}} (1+|\xi|^2)^{s-1}|\xi'|^2\left|\mathcal{F}^\nu(v)(\xi,r)\right|^2 d\xi dr$$
$$\leq C\left(\sum\left\|L_j\Lambda^{s-1}v\right\|^2 + \sum\left\|\overline{L}_j\Lambda^{s-1}v\right\|^2 + \text{small const.}\|\Lambda^s v\|^2\right). \tag{3.15}$$

Here the size of the small constant depends on the diameter of $V^\nu \cap \overline{\Omega}$. Furthermore, we have

$$\begin{aligned} \|L_j\Lambda^{s-1}v\|^2 &\leqslant \|\overline{L}_j\Lambda^{s-1}v\|^2 + (D\Lambda^{s-1}v, \Lambda^{s-1}v) \\ &\leqslant \|\overline{L}_j\Lambda^{s-1}v\|^2 + C\|\Lambda^{s-1}v\|^2 + \text{small const. } \|\Lambda^s v\|^2. \end{aligned} \tag{3.16}$$

Here D denotes a tangential first order differential operator.

Next observe that in the support of γ^0 we have $|\xi_{2n-1}| \leqslant 2|\xi'|$ and hence

$$(|\xi|+\xi_{2n-1})\gamma^0(\xi) \leqslant 4|\xi'|\,. \tag{3.17}$$

Combining (3.13), (3.15), (3.16), (3.17) with (3.11) we obtain, after some routine calculations,

$$\begin{aligned}\Big\|\Lambda^s \sum \eta^\nu \mathcal{G}^\nu \gamma^- \mathcal{F}^\nu (\zeta^\nu u)_h^\nu\Big\| + \Big\|\Lambda^s \sum \eta^\nu \mathcal{G}^\nu \gamma^0 \mathcal{F}^\nu (\zeta^\nu u)_h^\nu\Big\| \\ \leqslant C\big(\big\|\Lambda^{s-1}\frac{\partial u}{\partial \bar z_j}\big\| + \big\|\Lambda^{s-1}u\big\| + \big\|\theta u\big\|_s\big)\,.\end{aligned} \tag{3.18}$$

To conclude the proof of the proposition, we must show that the inequality (3.18) holds with $(\zeta^\nu u)_h^\nu$ replaced by $(\zeta^\nu u)_0^\nu$. To simplify notation, let $v = \zeta^\nu u$ and let $v_0 = (\zeta^\nu u)_0^\nu$. We then have $\Delta^\nu v_0 = \Delta^\nu v$ and $v_0 = 0$ when $r = 0$.

Now we have

$$\begin{aligned}\big\|\Lambda^s v_0\big\|^2 &\leqslant C\Big(\sum \Big\|\Lambda^{s-1}\frac{\partial v_0}{\partial t_j^\nu}\Big\|^2 + \Big\|\Lambda^{s-1}\frac{\partial v_0}{\partial r}\Big\|^2\Big) \\ &\leqslant C\big(\big|(\Lambda^{s-1}\Lambda^\nu v_0, \Lambda^{s-1}v_0)\big| + \big\|\Lambda^{s-1}v_0\big\|^2\big)\,.\end{aligned} \tag{3.19}$$

Then we have

$$\begin{aligned}\Delta^\nu &= \sum\Big(\frac{\partial}{\partial t_j^\nu}\Big)^2 + \frac{\partial^2}{\partial r^2} \\ &= 4\sum \frac{\partial^2}{\partial z_i \partial \bar z_i} + a^2\frac{\partial^2}{\partial r^2} + \sum a_i \frac{\partial^2}{\partial r \partial t} + \sum a_{ij}\frac{\partial^2}{\partial t_i \partial t_j} \\ &\quad + \sum c_i \frac{\partial}{\partial t_i} + c\frac{\partial}{\partial r}\,,\end{aligned}$$

where $a(P^\nu) = a_i(P^\nu) = a_{ij}(P^\nu) = 0$. Then

$$\begin{aligned}&\big|(\Lambda^{s-1}\Delta^\nu v_0, \Lambda^{s-1}v_0)\big| \\ &\quad = \big|(\Lambda^{s-1}\Delta^\nu v, \Lambda^{s-1}v_0)\big| \\ &\quad \leqslant C\sum \Big\|\Lambda^{s-1}\frac{\partial v}{\partial \bar z_j}\Big\|^2 \\ &\qquad + \text{ small const. }\Big(\big\|\Lambda^s v_0\big\|^2 + \Big\|\Lambda^{s-1}\frac{\partial v_0}{\partial r}\Big\|^2\Big)\,.\end{aligned} \tag{3.20}$$

Combining this with (3.19) we obtain

$$\big\|\Lambda^s v_0\big\| \leqslant C\sum \Big\|\Lambda^{s-1}\frac{\partial v}{\partial \bar z_j}\Big\|\,. \tag{3.21}$$

The same type of calculation that proves (3.18) holds with $(\zeta^\nu u)_h^\nu$ replaced by $(\zeta^\nu u)_0^\nu$. This establishes (3.10), and concludes the proof of Proposition 3.9.

3.22 Proposition. *If D is a first order differential operator, there exists $C > 0$ such that*

$$\|\Lambda^s[D,\Gamma^+]u\| \leqslant C\left(\sum\left\|\Lambda^{s-1}\frac{\partial u}{\partial \bar{z}_i}\right\| + \|\Lambda^{s-1}u\|\right) \tag{3.23}$$

for all $u \in C_0^\infty(U \cap \overline{\Omega})$.

Proof. Evaluating (3.8) at a point $P \in U \cap \overline{\Omega}$ where coordinates are (t^μ, r), we obtain

$$\Gamma^+u(P) = \sum_\nu \eta^\nu(P) \int\limits_{V^\nu\cap\overline{\Omega}\times\mathbb{R}_\xi^{2n-1}} e^{i(t^\nu(P)-s^\nu)\cdot\xi}\gamma^+(\xi)\zeta^\nu(s^\nu,r)u(s^\nu,r)ds^\nu dr d\xi\,. \tag{3.24}$$

Hence

$$\begin{aligned}[D,\Gamma^+]u &= \sum_\nu D\eta^\nu \int\limits_{V^\nu\cap\overline{\Omega}\times\mathbb{R}_\xi^{2n-1}} e^{i(t^\nu-s^\nu)\cdot\xi}\gamma^+(\xi)\zeta^\nu(s^\nu,r)u(s^\nu,r)ds^\nu dr d\xi \\ &+ \sum_\nu \eta^\nu \int\limits_{V^\nu\cap\overline{\Omega}\times\mathbb{R}_\xi^{2n-1}} e^{i(t^\nu-s^\nu)\cdot\xi}\gamma^+(\xi)D\zeta^\nu(s^\nu,r)u(s^\nu,r)ds^\nu dr d\xi \\ &+ \sum_\nu \eta^\nu \int\limits_{V^\nu\cap\overline{\Omega}\times\mathbb{R}_\xi^{2n-1}} e^{i(t^\nu-s^\nu)\cdot\xi}\sum_k a_k(t^\nu,s^\nu)\frac{\partial\gamma^+}{\partial\xi_k}\zeta^\nu(s^\nu,r)u(s^\nu,r)ds^\nu dr d\xi\,. \end{aligned} \tag{3.25}$$

Then the first term of (3.25) is an operator of order $-\infty$, since $\text{supp}(D\eta^\nu) \cap \text{supp}(\zeta^\nu) = \phi$. To estimate the second term, we proceed as follows. Let $I^\mu = \{\nu \mid \text{supp}(\eta^\mu) \cap \text{supp}(\eta^\nu) \neq \phi\}$. Let $\theta^\mu \in C_0^\infty(V^\mu \cap \overline{\Omega})$ such that $\theta^\mu = 1$ on a neighborhood $\text{supp}(\zeta^\mu)$ and such that $\text{supp}(\theta^\mu) \subset \{P \in V^\mu \cap \overline{\Omega} \mid \sum_{\nu\in I_\mu} \zeta^\nu(P) = 1\}$. Since $\sum \zeta^\mu = 1$, the second term can be written as

$$II = \sum_\mu\sum_{\nu\in I^\mu}\int \zeta^\mu e^{i(t^\nu-s^\nu)\cdot\xi}\gamma^+(\xi)\theta^\mu(s^\nu)D\zeta^\nu(s^\nu,r)u(\zeta^\nu,r)ds^\nu dr d\xi + R(u)\,,$$

where R is an operator of order $-\infty$. Changing coordinates and denoting the Jacobian by $\frac{\partial s^\nu}{\partial s^\mu}$, we get

$$\begin{aligned}II = \sum_\mu\sum_{\nu\in I^\mu}\int \zeta^\mu e^{i(t^\mu-s^\mu)\cdot\xi}\gamma^+\left(\frac{\partial s^\nu}{\partial s^\mu}\xi\right)\theta^\mu(s^\mu)D\zeta^\nu(s^\mu,r)u(s^\mu,r)ds^\mu dr d\xi \\ + R^{-1}u\,,\end{aligned}$$

where R^{-1} is an operator of order -1. Since the support of the θ^{μ} is very small, the Jacobian $\frac{\partial s^{\nu}}{\partial s^{\mu}}$ is close to the identity, $\frac{\partial s^{\nu}}{\partial s^{\mu}}(0,\dots,0,1)$ is close to $(0,\dots,0,1)$, and hence $\gamma^{+}(\frac{\partial s^{\nu}}{\partial s^{\mu}}\xi) = 1$ in a conical neighborhood of $(0,\dots,0,1)$. Hence when ξ is in that neighborhood, the integrand has the factor $\sum_{\gamma\in I^{\mu}} D\zeta^{\nu} = 0$. Thus there is a function γ_1 which vanishes in a conical neighborhood of $(0,\dots,0,1)$ such that

$$II = \sum_{\mu}\sum_{\nu\in I^{\mu}} \int \zeta^{\mu} e^{i(t^{\mu}-s^{\mu})\cdot\xi}\gamma^{+}\left(\frac{\partial s^{\nu}}{\partial s^{\mu}}\xi\right)\gamma_1(\xi)\theta^{\mu}(s^{\mu})(D\zeta^{\nu})u\,ds^{\mu}dr\,d\xi + R^{-1}u\,.$$

Then

$$\gamma_1(\xi) \leqslant C(|\xi'| + \gamma^{-}(\xi))$$

and hence

$$\|\Lambda^{s} II\| \leqslant C\left(\sum \left\|\Lambda^{s-1}\frac{\partial u}{\partial \bar{z}_j}\right\| + \|\Lambda^{s-1}u\|\right).$$

Finally we have to estimate the third terms in (3.25). Here we have

$$\left|\frac{\partial \gamma^{+}}{\partial \xi^{k}}\right| \leqslant C\frac{|\xi'| + \gamma^{-}(\xi)}{|\xi|}$$

from which we conclude that $\|\Lambda^{s} III\|$ is also bounded by the right hand side of (3.23), which concludes the proof of Proposition 3.22.

3.26 Definition. For each $s \in \mathbb{R}$ we denote by $T^{(s)}$ an operator whose principal symbol is given by

$$\sigma(T^{(s)}) = \sigma(T)^{s}\sigma(\Gamma^{+})\,. \tag{3.27}$$

Note that, when properly interpreted, this can also be written as $\sigma(T^{(s)}) = \sigma(T)^{s}\gamma^{+}$.

The preceding discussion then implies

Lemma. *There exists $C > 0$ such that*

$$\|\Lambda^{s}u\| \leqslant C\left(\|T^{(s)}u\| + \sum_{j}\left\|\Lambda^{s-1}\frac{\partial u}{\partial \bar{z}_j}\right\| + \|u\|_{s-1}\right) \tag{3.28}$$

for all $u \in C_0^{\infty}(U \cap \overline{\Omega})$.

Let ρ be a smooth defining function for Ω.

We denote by T_{ρ} the differential operator given by

$$T_{\rho} = \frac{1}{\sum |\rho_{z_k}|^2}\left(\sum_{j} \rho_{\bar{z}_j}\frac{\partial}{\partial z_j} - \sum \rho_{z_j}\frac{\partial}{\partial \bar{z}_j}\right). \tag{3.29}$$

Then we have the operators

$$L^\rho_{\text{norm}} = \frac{1}{\sum|\rho_{z_k}|^2}\sum_j \rho_{\bar z_j}\frac{\partial}{\partial z_j}$$
$$L^\rho_i = \frac{\partial}{\partial z_i} - \rho_{z_i}L^\rho_{\text{norm}}\,. \tag{3.30}$$

Hence we have

$$\frac{\partial}{\partial z_j} = \rho_{z_j}T_\rho + L^\rho_j + \rho_{z_j}\overline{L}^\rho_{\text{norm}}\,. \tag{3.31}$$

We define $T^{(s)}_\rho$ by $\sigma(T^{(s)}_\rho) = \sigma(T_\rho)^s\sigma(\Gamma^+)$. Then if $\rho = gr$, we have

$$\|\frac{1}{g^s}\Lambda^s u\| \leqslant C_0\|T^{(s)}_\rho u\| + C\Big(\sum_j\Big\|\Lambda^{s-1}\frac{\partial u}{\partial \bar z_j}\Big\| + \|u\|_{s-1}\Big)\,, \tag{3.32}$$

where $C_0 > 0$ is independent of g.

3.33 Lemma.

$$\Big[\frac{\partial}{\partial \bar z_i}, T_\rho\Big] = -\frac{1}{\sum_k|\rho_{z_k}|^2}\sum_j \rho_{z_j\bar z_i}\rho_{z_j}T_\rho + \sum a_j L^\rho_j + \sum b_j\frac{\partial}{\partial \bar z_j}\,. \tag{3.34}$$

Proof. Differentiating and using (3.31) we get

$$\Big[\frac{\partial}{\partial \bar z_i}, T_\rho\Big] \equiv \sum_j \frac{\partial}{\partial \bar z_i}\Big(\frac{\rho_{\bar z_j}}{\sum|\rho_{z_k}|^2}\Big)\frac{\partial}{\partial z_j} \quad \mathrm{mod}\Big(\frac{\partial}{\partial \bar z_1},\ldots,\frac{\partial}{\partial \bar z_n}\Big)$$
$$\equiv \sum_j\left(\frac{\rho_{\bar z_j\bar z_i}}{\sum|\rho_{z_k}|^2} - \frac{\rho_{\bar z_j}\sum\rho_{z_k\bar z_i}\rho_{\bar z_k}}{\big(\sum|\rho_{z_k}|^2\big)^2} - \frac{\rho_{\bar z_j}\sum\rho_{z_k}\rho_{\bar z_k\bar z_i}}{\big(\sum|\rho_{z_k}|^2\big)^2}\right)\rho_{z_j}T_\rho$$
$$\mathrm{mod}\ \Big(L^\rho_1,\ldots,L^\rho_n,\frac{\partial}{\partial \bar z_1},\ldots,\frac{\partial}{\partial \bar z_n}\Big)$$
$$\equiv -\frac{1}{\sum|\rho_{z_j}|^2}\sum_j \rho_{z_j\bar z_i}\rho_{\bar z_j} \quad \mathrm{mod}\ \Big(L^\rho_1,\ldots,L^\rho_n,\frac{\partial}{\partial \bar z_1},\ldots,\frac{\partial}{\partial \bar z_n}\Big)\,.$$

3.35 Lemma.

$$\Big[\frac{\partial}{\partial \bar z_i}, T^{(2s)}_\rho\Big] = -\frac{2s}{\sum|\rho_{z_k}|^2}\sum_j \rho_{z_i\bar z_i}\rho_{z_j}T^{(2s)}_\rho + \sum P^{2s-1}_j L^\rho_j$$
$$+\sum Q^{2s-1}_j\frac{\partial}{\partial \bar z_j} + P^{2s-1}\,, \tag{3.36}$$

where the P_j^{2s-1}, Q_j^{2s-1} *and* P^{2s-1} *are tangential pseudodifferential operators of order* $2s-1$.

Proof. First, by the calculus of pseudodifferential operators, we have

$$\left[\frac{\partial}{\partial \bar{z}_j}, T_\rho^{(2s)}\right] \equiv 2s\left[\frac{\partial}{\partial \bar{z}_j}, T_\rho\right] T_\rho^{(2s-1)} + \left[\frac{\partial}{\partial \bar{z}_j}, \Gamma^+\right] T_\rho^{(2s)}$$

modulo operators of order $2s-1$. The lemma then follows applying (3.23) and (3.34).

4. A priori estimates

According to a result of Diederich and Fornaess (see [DF2]), given a pseudoconvex domain Ω, with smooth boundary, there exist a defining function $\rho \in C^\infty(\overline{\Omega})$ and a constant $\delta > 0$ such that $-(-\rho)^\delta$ is plurisubharmonic. In [R], Range showed that there exists a $K > 0$ such that if $\rho = re^{-K|z|^2}$, then $-(-\rho)^\delta$ is plurisubharmonic if δ is small enough.

If $-(-\rho)^\delta$ is plurisubharmonic, we have

$$c_{ij} = \frac{\partial^2}{\partial z_i \partial \bar{z}_j}(-(-\rho)^\delta) = \delta|\rho|^{\delta-1}\rho_{z_i\bar{z}_j} + \delta(1-\delta)|\rho|^{\delta-2}\rho_{z_i}\rho_{\bar{z}_j} \tag{4.1}$$

is semidefinite.

4.2 Lemma. *Let* U *be a product neighborhood of* $b\Omega$. *That is,* U *is diffeomorphic to* $b\Omega \times (-a, a)$ *so that there is a map* $\pi\colon U \to b\Omega$ *and the diffeomorphism is given by* $P \mapsto (\pi(P), r(P))$. *Then if* $h \in C^\infty(\overline{\Omega})$ *there exists* $C > 0$ *such that*

$$\|hu\| \leqslant \max_{b\Omega}|h|\|u\| + C\|ru\|$$

for all $u \in C_0^\infty(U \cap \overline{\Omega})$.

Proof. Let $h^\# \in C^\infty(\overline{\Omega})$ be a function such that for $P \in U \cap \overline{\Omega}$, we have $h^\#(P) = h(\pi(P))$. Then there is a function $h_1 \in C^\infty(\overline{\Omega})$ such that $h(P) = h^\#(P) + h_1(P)r(P)$. Hence

$$\|hu\| \leqslant \|h^\# u\| + C\|ru\| \leqslant \max_{b\Omega}|h|\|u\| + C\|ru\|,$$

as required.

4.3 Lemma. *Let* U *be a neighborhood of* $b\Omega$ *which is covered by tangential coordinate systems. There exists a constant* C_0 *such that whenever* ρ *is a* *defining function and $-(-\rho)^\delta$ is plurisubharmonic with $\rho = gr$, then there exists a constant C such that*

$$\begin{aligned}\sum\left(c_{ij}(\widetilde{T}_\rho^{(s+\frac{\delta}{2})}\varphi)_{i'}, (\widetilde{T}_\rho^{(s+\frac{\delta}{2})}\varphi)_j\right) \\ \leqslant C_0\left\{\left\|\frac{1}{g^s}\Lambda^s\overline{\partial}\varphi\right\|^2 + \left\|\frac{1}{g^s}\Lambda^s\overline{\partial}^*\varphi\right\|^2\right. \\ \left. + s^2\max_{b\Omega}\left(1+\frac{|g_{z_i}|}{g}\right)^2\left\|\frac{1}{g^s}\Lambda^s\varphi\right\|^2 + \left\|\frac{r}{g^s}\Lambda^{s-1}\Delta\varphi\right\|^2\right\} \\ + C\left(\left\|\Lambda^{s-1}\overline{\partial}\varphi\right\|^2 + \left\|\Lambda^{s-1}\overline{\partial}^*\varphi\right\|^2\right. \\ \left. + \left\|\Lambda^{s-1}\varphi\right\|^2 + \left\|r\Lambda^{s-2}\Delta\varphi\right\|^2 + \|\varphi\|^2\right)\end{aligned} \tag{4.4}$$

for all $\varphi \in \dot{\mathcal{D}}^q$ with $\operatorname{supp}(\varphi) \subset U \cap \overline{\Omega}$. *$C_0$ is independent of g, δ, and s. Here $\widetilde{T}_\rho^{(s)}$ is defined by*

$$\left(\widetilde{T}_\rho^{(s)}\varphi\right)_i = T_\rho^{(s)}\varphi_i + \frac{\rho_{\overline{z}_i}}{\Sigma|\rho_{z_k}|^2}\sum_j [T_\rho^{(s)}, \rho_{z_j}]\varphi_j\,. \tag{4.5}$$

Proof. We give the proof for the case of $q = 1$. This proof can easily be modified (by writing down the appropriate multi-indices as at the end of Section 1) to establish the result for $q > 1$. Setting $\lambda = (-\rho)^\delta$ and applying Theorem 1.7, with φ replaced by $\widetilde{T}_\rho^{(s+\frac{\delta}{2})}\varphi$, we have

$$\begin{aligned}\sum\left(c_{ij}(\widetilde{T}_\rho^{(s+\frac{\delta}{2})}\varphi)_{i'}, (\widetilde{T}_\rho^{(s+\frac{\delta}{2})}\varphi)_j\right) + \sum\left\||\rho|^{\frac{\delta}{2}}\frac{\partial}{\partial\overline{z}_j}\widetilde{T}_\rho^{(s+\frac{\delta}{2})}\varphi\right\|^2 \\ \leqslant C_0\left\{\left\||\rho|^{\frac{\delta}{2}}\overline{\partial}\widetilde{T}_\rho^{(s+\frac{\delta}{2})}\varphi\right\|^2 + \left\||\rho|^{\frac{\delta}{2}}\overline{\partial}^*\widetilde{T}_\rho^{(s+\frac{\delta}{2})}\varphi\right\|^2\right. \\ \left. + \delta\left|\left(|\rho|^{\delta-1}\sum\rho_{z_j}\left(\widetilde{T}_\rho^{(s+\frac{\delta}{2})}\varphi\right)_j, \overline{\partial}^*\widetilde{T}_\rho^{(s+\frac{\delta}{2})}\varphi\right)\right|\right\}.\end{aligned} \tag{4.6}$$

Now

$$\||\rho|^{\frac{\delta}{2}}\overline{\partial}\widetilde{T}_\rho^{(s+\frac{\delta}{2})}\varphi\| \leqslant \||\rho|^{\frac{\delta}{2}}\widetilde{T}_\rho^{(s+\frac{\delta}{2})}\overline{\partial}\varphi\| + \||\rho|^{\frac{\delta}{2}}(\overline{\partial}\widetilde{T}_\rho^{(s+\frac{\delta}{2})}\varphi) - \widetilde{T}_\rho^{(s+\frac{\delta}{2})}\overline{\partial}\varphi\|\,.$$

Then $\overline{\partial}\widetilde{T}_\rho^{(s+\frac{\delta}{2})}\varphi - \widetilde{T}_\rho^{(s+\frac{\delta}{2})}\overline{\partial}\varphi = (P^{s+\frac{\delta}{2}} + P^{s+\frac{\delta}{2}-1}\frac{\partial}{\partial r})\varphi$, where $P^{s+\frac{\delta}{2}}$ and $P^{s+\frac{\delta}{2}-1}$ are tangential pseudodifferential operators of orders $s+\frac{\delta}{2}$ and $s+\frac{\delta}{2}-1$.

From (2.2) we then obtain

$$\begin{aligned}
\||\rho|^{\frac{\delta}{2}}P^{s+\frac{\delta}{2}}\varphi\| &\leqslant s\max_{b\Omega}\left(1+\frac{|g_{z_i}|}{g}\right)\left\|\frac{1}{g^s}|r|^{\frac{\delta}{2}}\Lambda^{s+\frac{\delta}{2}-1}\varphi\right\| + C\||r|^{\frac{\delta}{2}}\Lambda^{s+\frac{\delta}{2}-1}\varphi\| \\
&\leqslant C_0\, s\max_{b\Omega}\left(1+\frac{|g_{z_i}|}{g}\right)\left\{\|\frac{1}{g^s}\Lambda^s\varphi\| + \|\frac{r}{g^s}\Lambda^{s-1}\Delta\varphi\|\right\} \\
&\quad + C(\|\Lambda^{s-1}\varphi\| + \|\Lambda^{s-2}\Delta\varphi\|)
\end{aligned}$$

and

$$\begin{aligned}
&\||\rho|^{\frac{\delta}{2}}P^{s+\frac{\delta}{2}-1}\frac{\partial}{\partial r}\varphi\| \\
&\quad\leqslant s\max_{b\Omega}\left(1+\frac{|g_{z_i}|}{g}\right)\left\|\frac{1}{g^s}|r|^{\frac{\delta}{2}}\Lambda^{s+\frac{\delta}{2}-1}\frac{\partial}{\partial r}\varphi\right\| \\
&\quad\leqslant C_0\, s\max_{b\Omega}\left(1+\frac{|g_{z_i}|}{g}\right)\left\{\|\frac{1}{g^s}\Lambda^{s-1}\frac{\partial}{\partial r}\varphi\| + \|\frac{r}{g^s}\Lambda^{s-1}\frac{\partial^2}{\partial r^2}\varphi\|\right\}.
\end{aligned}$$

Now we have

$$\begin{aligned}
&\left\|\frac{1}{g^s}\Lambda^{s-1}\frac{\partial}{\partial r}\varphi\right\| \\
&\quad\leqslant C_0\left\|\frac{1}{g^s}\Lambda^s\varphi\right\| + C\left(\sum\left\|\frac{\partial}{\partial_{\bar z_j}}\widetilde{\Lambda}^{(s-1)}\varphi\right\| + \left\|\Lambda^{s-2}\frac{\partial}{\partial r}\varphi\right\|\right) \\
&\quad\leqslant C_0\left\|\frac{1}{g^s}\Lambda^s\varphi\right\| + C(\|\Lambda^{s-1}\bar\partial\varphi\| + \|\Lambda^{s-1}\bar\partial^*\varphi\| + \|\Lambda^{s-1}\varphi\| + \|\varphi\|).
\end{aligned}$$

Note that in tangential coordinates we have

$$\Delta = a\frac{\partial^2}{\partial r^2} + \sum a_i\frac{\partial^2}{\partial t_i\partial r} + \sum a_{ij}\frac{\partial^2}{\partial t_i\partial t_j} + b\frac{\partial}{\partial r} + \sum b_i\frac{\partial}{\partial t_i},$$

hence

$$\begin{aligned}
&\left\|\frac{r}{g^s}\Lambda^{s-1}\frac{\partial^2}{\partial r^2}\varphi\right\| \\
&\quad\leqslant C_0\left(\left\|\frac{r}{g^s}\Lambda^{s-1}\Delta\varphi\right\| + \left\|\frac{r}{g^s}\Lambda^s\frac{\partial}{\partial r}\varphi\right\| + \left\|\frac{r}{g^s}\Lambda^{s+1}\varphi\right\|\right) \\
&\quad\leqslant C_0\left(\left\|\frac{r}{g^s}\Lambda^{s-1}\Delta\varphi\right\| + \sum\left\|\frac{\partial}{\partial\bar z_j}\left(\frac{r}{g^s}\Lambda^s\varphi\right)\right\| + \left\|\frac{1}{g^s}\Lambda^s\varphi\right\|\right) \\
&\qquad + C\|r\Lambda^s\varphi\| \\
&\quad\leqslant C_0\left(\left\|\frac{r}{g^s}\Lambda^{s-1}\Delta\varphi\right\| + \left\|\frac{r}{g^s}\Lambda^s\bar\partial\varphi\right\| + \left\|\frac{r}{g^s}\Lambda^s\bar\partial^*\varphi\right\| + \left\|\frac{1}{g^s}\Lambda^s\varphi\right\|\right) \\
&\qquad + C(\|\Lambda^{s-1}\bar\partial\varphi\| + \|\Lambda^{s-1}\bar\partial^*\varphi\| + \|\Lambda^{s-1}\varphi\| + \|\varphi\|).
\end{aligned}$$

Thus we conclude that the first term on the right of (4.6) is estimated by the right hand side of (4.4). The same argument applies to estimate the second term. To estimate the third term, we note that

$$\sum \rho_{z_j}\left(\widetilde{T}_\rho^{(s+\frac{\delta}{2})}\varphi\right)_j = T_\rho^{(s+\frac{\delta}{2})}\left(\sum \rho_{z_j}\varphi_j\right)$$

vanishes on $b\Omega$, and thus we apply (2.10) to obtain

$$\begin{aligned}
&\left\| |\rho|^{\frac{\delta}{2}-1} T_\rho^{(s+\frac{\delta}{2})}\left(\sum \rho_{z_j}\varphi_j\right)\right\| \\
&\quad\leqslant C_0 \sum \left\| \frac{1}{g^s}\frac{\partial}{\partial \bar{z}_j}\widetilde{\Lambda}^s\varphi\right\| + C\left(\|\Lambda^{s-1}\varphi\| + \sum\left\|\frac{\partial}{\partial \bar{z}_j}\Lambda^{s-1}\varphi\right\|\right) \\
&\quad\leqslant C_0\left(\left\|\frac{1}{g^s}\Lambda^s\bar\partial\varphi\right\| + \left\|\frac{1}{g^s}\Lambda^s\bar\partial^*\varphi\right\| + s\left\|\frac{1}{g^s}\Lambda^s\varphi\right\|\right) \\
&\qquad + C\left(\left\|\Lambda^{s-1}\varphi\right\| + \left\|\frac{\partial}{\partial \bar{z}_j}\Lambda^{s-1}\varphi\right\|\right).
\end{aligned}$$

Combining all these we have proved (4.4), concluding the proof of Lemma 4.3.

If U is a neighborhood of $b\Omega$ as above, let $a > 0$ be such that $\{P \in \overline{\Omega} \,|\, r(P) \geqslant -a\} \subset U$. Let $\zeta \in C_0^\infty(U \cap \overline{\Omega})$ be such that $\zeta(P) = 1$ if $r(P) \geqslant -\frac{a}{2}$ and $\zeta(P) = 0$ if $r(P) \leqslant -\frac{2a}{3}$. Let $\zeta_0 \in C_0^\infty(\Omega)$ be such that $\zeta_0(P) = 1$ if $r(P) \leqslant -\frac{a}{3}$ and $\zeta_0(P) = 0$ if $r(P) \geqslant -\frac{a}{4}$.

4.7 Lemma. *There exists a constant C_0 such that whenever $-(-\rho)^\delta$ is plurisubharmonic with $\rho = gr$, we have*

$$\begin{aligned}
&\left\|\frac{1}{g^s}\Lambda^s\zeta B_q f\right\|^2 \\
&\quad\leqslant C_0\left(\omega + (s+\omega)\sqrt{\frac{1-\delta}{\omega}}\right) s \max_{b\Omega}\left(1 + \frac{|g_{z_i}|}{g}\right) \\
&\qquad \times \left\{\left\|\frac{1}{g^s}\Lambda^s\varphi\right\|^2 + \left\|\frac{1}{g^s}\Lambda^s\zeta B_q f\right\|^2 + \sum\left\|\frac{r}{g^s}\Lambda^s\zeta\frac{\partial}{\partial \bar{z}_j}B_q f\right\|^2\right\} \\
&\qquad + C_0\left(\|\Lambda^s\zeta f\|^2 + \|r\Lambda^{s-1}\zeta\Delta f\|^2 + \sum\left\|r\Lambda^{s-2}\zeta\frac{\partial f}{\partial \bar{z}_j}\right\|^2\right)
\end{aligned}$$

$$
\begin{aligned}
&+C\Big(\|\zeta_0 f\|_s^2+\|f\|^2+\sum_{|\alpha|=0}^{[s]}\|D^\alpha\Lambda^{s-1-|\alpha|}\varphi\|^2\\
&\quad+\sum_{|\alpha|=0}^{[s]}\|D^\alpha\Lambda^{s-1-|\alpha|}\zeta B_q f\|^2+\sum_{|\alpha|=0}^{[s]}\|D^\alpha\Lambda^{s-1-|\alpha|}\zeta f\|^2 \qquad (4.8)\\
&\quad+\|r\Lambda^s\zeta f\|^2+\|r\Lambda^s\zeta B_q f\|^2+\|r\Lambda^s\varphi\|^2\Big)
\end{aligned}
$$

for all $f\in\dot{\mathcal{D}}^q$ with $N_{q+1}\,\bar\partial f\in\dot{\mathcal{D}}^q$ and $\omega\in(0,\frac{\delta}{2})$. Here $\varphi=\zeta N_{q+1}\bar\partial f$.

Here the constant $a>0$ will be taken very small so that the terms after C containing r can be absorbed in the various estimates that follow. Taking a small implies that the derivatives of ζ will be large. However, these derivatives are supported in the interior of Ω so that the terms in which they appear can be estimated (using interior ellipticity) by $C(\|\zeta_0 f\|_s^2+\|f\|^2)$. From now on we will denote the terms appearing after C by "error", these terms are permissible errors which will be estimated by absorption or induction.

Proof. We will prove the estimate for the case $q=0$. The proof for general q is then easily recovered by using multi-indices. We will write $\frac{\partial}{\partial\bar z_j}Bf$ even though these terms vanish, because for $q>0$ the terms $\frac{\partial}{\partial\bar z_j}B_q f$ are not necessarily zero. They will be treated later in an induction argument.

From Lemma 2.11, substituting ω for σ, we obtain

$$\left\|\frac{1}{g^s}\Lambda^s\zeta Bf\right\|\leqslant C_0\left\|\frac{1}{g^s}|r|^\omega\Lambda^{s+\omega}\zeta Bf\right\|+C(\text{error})\,.$$

Since

$$\left\|\frac{1}{g^s}|r|^\omega\Lambda^{s+\omega}\zeta Bf\right\|=\left\||\rho|^\omega T_\rho^{(s+\omega)}\zeta Bf\right\|+C(\text{error})$$

it will suffice to estimate the first term on the right. Setting $\varphi=\zeta N_1\bar\partial f$ we have

$$
\begin{aligned}
\left\||\rho|^\omega T_\rho^{(s+\omega)}\zeta Bf\right\|^2=&\,(|\rho|^\omega T_\rho^{(s+\omega)}\zeta f,|\rho|^\omega T_\rho^{(s+\omega)}\zeta Bf)\\
&-(|\rho|^\omega T_\rho^{(s+\omega)}\bar\partial^*\varphi,|\rho|^\omega T_\rho^{(s+\omega)}\zeta Bf) \qquad (4.9)\\
&+O(\text{error})\,.
\end{aligned}
$$

The first term on the right can be bounded using (2.2), with $\sigma=2\omega$, as follows

$$
\begin{aligned}
&\left|\left(|\rho|^{\omega}T_\rho^{(s+\omega)}\zeta f, |\rho|^{\omega}T_\rho^{(s+\omega)}\zeta Bf\right)\right| \\
&\qquad = \left|\left(T_\rho^{(s)}\zeta f, |\rho|^{2\omega}T_\rho^{(s+2\omega)}\zeta Bf\right)\right| + O(\text{error}) \\
&\qquad \leqslant \text{large const}\left\|\frac{1}{g^s}\Lambda^s\zeta f\right\|^2 \\
&\qquad\quad + \text{small const}\left(\left\|\frac{1}{g^s}\Lambda^s\zeta Bf\right\|^2 + \sum_{j=1}^{n}\left\|\frac{r}{g^s}\Lambda^s\zeta\frac{\partial}{\partial\bar z_j}Bf\right\|^2\right) \\
&\qquad\quad + O(\text{error})\,.
\end{aligned}
$$

The second term on the right in (4.9) is now expressed as follows

$$
\begin{aligned}
&\left|\left(|\rho|^{2\omega}\bar\partial^*\varphi, T_\rho^{(2s+2\omega)}\zeta Bf\right)\right| \\
&\qquad \leqslant 2\omega\left|\sum\left(|\rho|^{2\omega-1}\sum\rho_{z_i}\varphi_i, T_\rho^{(2s+2\omega)}\zeta Bf\right)\right| \\
&\qquad\quad + \left|\sum\left(|\rho|^{2\omega}\varphi_i, \left[\frac{\partial}{\partial\bar z_i}, T_\rho^{(2s+2\omega)}\right]\zeta Bf\right)\right| + O(\text{error}) \qquad (4.10) \\
&\qquad = I_1 + I_2 + O(\text{error})\,.
\end{aligned}
$$

Since $\sum\rho_{z_i}\varphi_i = 0$ on $b\Omega$ we use (2.10) with $\sigma = 2\omega - 1$ and s replaced by $s+1$ to estimate I_1, the first term on the right of (4.10) and we obtain

$$
\begin{aligned}
I_1 &\leqslant C_0\omega\left(\sum_j\left\|\frac{1}{g^s}\frac{\partial}{\partial\bar z_j}\Lambda^s\varphi\right\|^2 + \left\|\frac{1}{g^s}\Lambda^s\zeta Bf\right\|^2\right) + O(\text{error}) \\
&\leqslant C_0\omega\left\{\left\|\frac{1}{g^s}\Lambda^s\bar\partial\varphi\right\|^2 + \left\|\frac{1}{g^s}\Lambda^s\bar\partial^*\varphi\right\|^2 + s^2\left\|\frac{1}{g^s}\Lambda^s\varphi\right\|^2 + \left\|\frac{1}{g^s}\Lambda^s\zeta Bf\right\|^2\right\} \\
&\quad + C(\text{error})\,.
\end{aligned}
$$

Since $\bar\partial N\bar\partial f = 0$ and $\bar\partial^* N\bar\partial f = f - Bf$, we have

$$
I_1 \leqslant C_0\omega\left(s^2\left\|\frac{1}{g^s}\Lambda^s\varphi\right\|^2 + \left\|\frac{1}{g^s}\Lambda^s\zeta Bf\right\|^2 + \left\|\frac{1}{g^s}\Lambda^s\zeta f\right\|^2\right) + C(\text{error})\,.
$$

Next we estimate the second term I_2 on the right of (4.10). We have

$$
I_2 = 2(s+\omega)\left|\sum_{i,j}\left(|\rho|^{2\omega}\varphi_i, \rho_{\bar z_i z_j}\frac{\rho_{z_j}}{\sum|\rho_{z_k}|^2}T_\rho^{(2s+2\omega-1)}\zeta Bf\right)\right| + O(\text{error})\,. \qquad (4.11)
$$

We have

$$c_{ij} = \delta|\rho|^{\delta-1}\rho_{z_i\bar{z}_j} + \delta(1-\delta)|\rho|^{\delta-2}\rho_{z_i}\rho_{\bar{z}_j}$$

so

$$\rho_{z_i\bar{z}_j} = \frac{1}{\delta}|\rho|^{1-\delta}c_{ij} - (1-\delta)|\rho|^{-1}\rho_{z_i}\rho_{\bar{z}_j} .$$

Substituting in (4.11) we obtain

$$\begin{aligned} I_2 \leqslant & 2(s+\omega)C_0\Big\{\frac{1}{\delta}\Big|\sum_{i,j}\Big(c_{ij}(\widetilde{T}_\rho^{(s+\frac{\delta}{2})}\varphi)_{i}, \frac{|\rho|^{2\omega+1-\delta}\rho_{z_j}}{\sum|\rho_{z_k}|^2}T_\rho^{(s+2\omega-\frac{\delta}{2})}\zeta Bf\Big)\Big| \\ & \qquad + (1-\delta)\Big|\Big(|\rho|^{-1}\sum\rho_{z_i}(\widetilde{T}_\rho^{(s)}\varphi)_i, |\rho|^{2\omega}T_\rho^{(s+2\omega)}\zeta Bf\Big)\Big|\Big\} \\ & + C(\text{error}) \\ \leqslant & 2(s+\omega)C_0(I_{21}+I_{22}) + C(\text{error}) . \end{aligned}$$

Since $(c_{ij}) \geqslant 0$ we estimate I_{21}, by using the Schwarz inequality on the integrand and obtain

$$\begin{aligned} I_{21} \leqslant & \frac{1}{\delta}\Big(\sum c_{ij}(\widetilde{T}_\rho^{(s+\frac{\delta}{2})}\varphi)_i, (\widetilde{T}^{(s+\frac{\delta}{2})}\varphi)_j\Big)^{\frac{1}{2}} \times \\ & \Big\{\int_\Omega \frac{|\rho|^{4\omega+2-2\delta}}{(\sum|\rho_{z_k}|^2)^2}\sum c_{ij}\rho_{\bar{z}_i}\rho_{z_j}\Big|T_\rho^{(s+2\omega-\frac{\delta}{2})}\zeta Bf\Big|^2 dV\Big\}^{\frac{1}{2}} + O(\text{error}) . \end{aligned}$$

By Lemma 4.3 the first factor above is estimated by

$$\frac{C_0}{\delta^{\frac{1}{2}}}\Big(\max_{b\Omega}(1+\frac{|g_{z_i}|}{g})\|\frac{1}{g^s}\Lambda^s\varphi\| + \||\rho|^{\frac{\delta}{2}}T_\rho^{(s+\frac{\delta}{2})}\bar{\partial}^*\varphi\|\Big) + O(\text{error}) .$$

To bound the second factor we write c_{ij} in terms of ρ. This factor is then bounded by

$$\begin{aligned} & \delta^{\frac{1}{2}}\Big(\int_\Omega \frac{|\rho|^{4\omega+1-\delta}}{\sum|\rho_{z_i}|^2}\sum\rho_{z_i\bar{z}_j}\rho_{\bar{z}_i}\rho_{z_j}|T_\rho^{(s+2\omega-\frac{\delta}{2})}\zeta Bf|^2 dV\Big)^{\frac{1}{2}} \\ & \qquad + \sqrt{\delta(1-\delta)}\Big(\int_\Omega |\rho|^{4\omega-\delta}|T_\rho^{(s+2\omega-\frac{\delta}{2})}\zeta Bf|^2\Big)^{\frac{1}{2}} = J_1 + J_2 . \end{aligned}$$

Then

$$\begin{aligned} J_1 & \leqslant C\||\rho|^{2\omega+\frac{1}{2}-\frac{\delta}{2}}T_\rho^{(s+2\omega-\frac{\delta}{2})}\zeta Bf\| \\ & \leqslant C\Big(\|\Lambda^{s-\frac{1}{2}}\zeta Bf\| + \sum\|r\Lambda^{s-\frac{1}{2}}\zeta\frac{\partial}{\partial\bar{z}_j}Bf\| + O(\text{error})\Big) . \end{aligned}$$

This then is a lower order term and can be incorporated in the $O(\text{error})$. J_2 is bounded, using (2.7) with $\sigma = 2\omega - \frac{\delta}{2}$, by

$$J_2 \leqslant C_0\sqrt{\delta(1-\delta)}\|\frac{1}{g^s}|r|^{2\omega-\frac{\delta}{2}}\Lambda^{s+2\omega-\frac{\delta}{2}}\zeta Bf\|$$
$$\leqslant C_0\frac{\sqrt{\delta(1-\delta)}}{\sqrt{1-\delta+4\omega}}\left(\|\frac{1}{g^s}\Lambda^s\zeta Bf\| + \sum\|\frac{r}{g^s}\Lambda^s\zeta\frac{\partial}{\partial\overline{z}_j}Bf\|\right) + C(\text{error})\,.$$

Next we estimate $|||\rho|^{\frac{\delta}{2}}T_\rho^{(s+\frac{\delta}{2})}\overline{\partial}^*\varphi\|$ and we have

$$|||\rho|^{\frac{\delta}{2}}T_\rho^{(s+\frac{\delta}{2})}\overline{\partial}^*\varphi\| \leqslant C_0\left(|||\rho|^{\frac{\delta}{2}}T_\rho^{(s+\frac{\delta}{2})}\zeta f\| + |||\rho|^{s+\frac{\delta}{2}}\zeta Bf\|\right) + C(\text{error})$$
$$\leqslant C_0\Big(\|\frac{1}{g^s}\Lambda^s\zeta f\| + s\|\frac{1}{g^s}\Lambda^{s-1}\zeta\Delta f\| + \|\frac{1}{g^s}\Lambda^s\zeta Bf\|$$
$$+ s\sum\|\frac{1}{g^s}\Lambda^s\zeta\frac{\partial}{\partial\overline{z}_j}Bf\|\Big)\,.$$

Finally to conclude the proof we estimate I_{22}. We apply the Schwarz inequality and use (2.9) with $\sigma = -1$ on the first factor and (2.2) with $\sigma = 2\omega$ on the second factor and obtain

$$I_{22} \leqslant (1-\delta)C_0\sum_j\|\frac{\partial}{\partial\overline{z}_j}\widetilde{T}_\rho^{(s)}\varphi\|\,\|\frac{1}{g^s}\Lambda^s\zeta Bf\| + O(\text{error})$$
$$\leqslant (1-\delta)C_0\left(\|\frac{1}{g^s}\Lambda^s\zeta Bf\|^2 + \|\frac{1}{g^s}\Lambda^s\varphi\|^2 + \|\frac{1}{g^s}\Lambda^s\zeta f\|^2\right) + O(\text{error}).$$

Combining all these concludes the proof of Lemma 4.7.

4.12 Proposition. *Under the same hypotheses as above there exist constants η, A, and C with η and A independent of ρ, δ, and s such that if either $s \leqslant \eta$, or $s > \eta$ and*

$$\max_{b\Omega}\left(1 + \frac{|g_{z_i}|}{g}\right)^3 A^s(1-\delta) \leqslant 1$$

then

$$\|Bf\|_s \leqslant C\|f\|_s \tag{4.13}$$

for $f \in C^\infty(\overline{\Omega})$ such that $N\overline{\partial}f \in C^\infty(\overline{\Omega})$.

Proof. When s satisfies the above condition, we will first prove that there exists C_0 independent of ρ, δ, and s such that

$$\|\frac{1}{g^s}\Lambda^s\zeta B_q f\| \leqslant C_0\Big(\|\frac{1}{g^s}\Lambda^s\zeta f\| + \|\frac{r}{g^s}\Lambda^{s-1}\zeta\Delta f\|$$
$$+\sum\|\frac{1}{g^s}\Lambda^{s-1}\zeta\frac{\partial f}{\partial\overline{z}_j}\|\Big) + C(\text{error})\,. \tag{4.14}$$

Setting $q = n - k$, the proof will be by induction on k. For $k = 0$ the operator B_n is the identity so (4.14) holds. Now in (4.9), (4.10) and (4.11) set $\omega = 0$ and we have

$$\begin{aligned}
&\|\frac{1}{g^s}\Lambda^s\zeta B_q f\| \\
&\quad\leqslant C_0\sqrt{s}\Big(\|\frac{1}{g^s}\Lambda^s\varphi\| + \|\frac{1}{g^s}\Lambda^s\zeta B_q f\| + \sum\|\frac{r}{g^s}\Lambda^s\zeta\frac{\partial}{\partial\bar{z}_j}B_q f\|\Big) \\
&\qquad + C_0\left(\|\frac{1}{g^s}\Lambda^s\zeta f\| + \|\frac{r}{g^s}\Lambda^{s-1}\zeta\Delta f\| + \sum\|\frac{1}{g^s}\Lambda^{s-1}\zeta\frac{\partial}{\partial\bar{z}_j}f\|\right) \\
&\qquad + C(\text{error})\,.
\end{aligned} \tag{4.15}$$

From (1.21) we have

$$\|\frac{1}{g^s}\Lambda^s\varphi\| = \|\frac{1}{g^s}\Lambda^s\zeta B_{q+1}(w_t N_{t,q+1}\bar{\partial}[w_{-t}(I - B_q)f])\|\,. \tag{4.16}$$

Assuming that (4.14) holds when q is replaced by $q+1$, we get

$$\begin{aligned}
\|\frac{1}{g^s}\Lambda^s\varphi\| \leqslant C_0&\left(\|\frac{1}{g^s}\Lambda^s\zeta\psi_{q+1}\| + \|\frac{r}{g^s}\Lambda^{s-1}\zeta\Delta\psi_{q+1}\|\right. \\
&\left.+\|\frac{1}{g^s}\Lambda^{s-1}\frac{\partial}{\partial\bar{z}_j}\psi_{q+1}\|\right) + C(\text{error}(\psi_{q+1}))\,,
\end{aligned} \tag{4.17}$$

where

$$\begin{aligned}
\psi_{q+1} &= w_t N_{t,q+1}\bar{\partial}[w_{-t}(I - B_q)f] \\
&= w_t N_{t,q+1}\bar{\partial}\zeta[w_{-t}(I - B_q)f] \\
&\quad + w_t N_{t,q+1}\bar{\partial}(1-\zeta)[w_{-t}(I - B_q)f]\,.
\end{aligned}$$

Since for $t >$ const s the operator $N_{t,q+1}\bar{\partial}$ preserves s-norms and the weighted (t)-norm, we have by error (ψ_{q+1}) = error (f)

$$\|\frac{1}{g^s}\Lambda^s\zeta\psi_{q+1}\| \leqslant C_0^s\left(\|\frac{1}{g^s}\Lambda^s\zeta f\| + \|\frac{1}{g^s}\Lambda^s\zeta B_q f\|\right) + C(\text{error})\,.$$

Now we have $\Delta N_{t,q+1} = I$ plus first order operator, hence

$$\begin{aligned}
\|\frac{r}{g^s}\Lambda^{s-1}\zeta\Delta\psi_{q+1}\| \leqslant C_0\Big(&\|\frac{r}{g^s}\Lambda^{s-1}\zeta\Delta f\| + \|\frac{r}{g^s}\Lambda^s\zeta B_q f\| \\
&+\sum\|\frac{r}{g^s}\Lambda^{s-1}\zeta\frac{\partial}{\partial\bar{z}_j}B_q f\| + \|\frac{1}{g^s}\Lambda^s\zeta f\|\Big) \\
&+ C(\text{error})\,.
\end{aligned}$$

Note that the same estimate holds for the third term on the right of (4.17). So we get

$$\|\frac{1}{g^s}\Lambda^s\varphi\| \leqslant C_0\Big(\|\frac{1}{g^s}\Lambda^s\zeta f\| + \|\frac{1}{g^s}\Lambda^s\zeta B_q f\| + \|\frac{r}{g^s}\Lambda^{s-1}\zeta\Delta f\| + \sum\|\frac{r}{g^s}\Lambda^{s-1}\zeta\frac{\partial}{\partial\bar{z}_j}B_q f\|\Big) + C(\text{error})\,. \tag{4.18}$$

We also have

$$\sum\|\frac{r}{g^s}\Lambda^s\zeta\frac{\partial}{\partial\bar{z}_j}B_q f\| \leqslant C_0(\|\frac{r}{g^s}\Lambda^{s-1}\zeta\Delta f\| + \|\frac{1}{g^s}\Lambda^s\zeta B_q f\|) + C(\text{error})\,. \tag{4.19}$$

Substituting (4.18) and (4.19) in (4.15) we obtain

$$\begin{aligned}\|\frac{1}{g^s}\Lambda^s\zeta B_q f\| \leqslant &C_0^s\sqrt{s}(\|\frac{1}{g^s}\Lambda^s\zeta B_q f\| + \|\frac{1}{g^s}\Lambda^s\zeta f\| + \|\frac{r}{g^s}\Lambda^{s-1}\zeta\Lambda f\|)\\ &+ C_0\Big(\|\frac{1}{g^s}\Lambda^s\zeta f\| + \|\frac{r}{g^s}\Lambda^{s-1}\zeta\Delta f\| + \sum\|\frac{1}{g^s}\Lambda^{s-1}\zeta\frac{\partial f}{\partial\bar{z}_j}\|\Big)\\ &+ C(\text{error})\,.\end{aligned} \tag{4.20}$$

Hence if we choose η such that $C_0^s\sqrt{s} \leqslant \frac{1}{2}$ when $s \leqslant \eta$, then we obtain (4.14).

If $s \geqslant \eta$, we substitute (4.17), (4.18), and (4.19) in (4.8), and we get the same inequality as (4.20) with $C_0^s\sqrt{s}$ replaced by

$$\frac{C_0^{(1+\frac{2}{\eta})s}}{\eta}\left(\omega + \sqrt{\frac{1-\delta}{\omega}}\right)\max_{b\Omega}\left(1 + \frac{|g_{z_j}|}{g}\right).$$

Setting $\omega = (1-\delta)^{\frac{1}{3}}$ the above becomes

$$\frac{2}{\eta}C_0^{(1+\frac{2}{\eta})s}(1-\delta)^{\frac{1}{3}}\max_{b\Omega}\left(1 + \frac{|g_{z_i}|}{g}\right). \tag{4.21}$$

Since $s \geqslant \eta$ there exists A independent of s such that

$$\frac{2}{\eta}C_0^{(1+\frac{2}{\eta})s} \leqslant \frac{1}{2}A^{\frac{s}{3}}\,.$$

Hence when $s \geqslant \eta$ and

$$A^s\max_{b\Omega}\left(1 + \frac{|g_{z_i}|}{g}\right)^3 \leqslant 1$$

the coefficient given by (4.21) is less than or equal to $\frac{1}{2}$, and hence (4.14) holds.

To prove (4.13) we use the operator $P_h\colon L_2(\Omega) \to \mathcal{H}arm(\Omega)$, where P_h denotes the orthogonal projection on the space of complex valued harmonic functions denoted by $\mathcal{H}arm(\Omega)$. Since $\mathcal{H}(\Omega) \subset \mathcal{H}arm(\Omega)$ we have $BP_h = B$. We also have (see [Be]) $P_h(H^s(\Omega)) \subset H^s(\Omega)$ and

$$\|P_h f\|_s \leqslant C\|f\|_s. \tag{4.22}$$

Hence substituting $P_h f$ for f in (4.14) (with $q = 0$), we obtain (4.13), concluding the proof of Proposition 4.12.

5. Conclusion

Main Theorem. *Given $\Omega \subset \mathbb{C}^n$ pseudoconvex, bounded with a smooth boundary, let $\rho = gr$ be a C^∞ defining function for Ω such that $-(-\rho)^\delta$ is plurisubharmonic. Then there exist constants η and A independent of ρ and δ such that whenever either $s \leqslant \eta$, or $s > \eta$ and*

$$A^s(1-\delta)\max_{b\Omega}\left(1+\frac{|g_{z_i}|}{g}\right)^3 \leqslant 1$$

then $B(H^s(\Omega)) \subset H^s(\Omega)$ and there exists C such that $\|Bf\|_s \leqslant C\|f\|_s$ for all $f \in H^s(\Omega)$.

Proof. Let $\rho^{(\varepsilon)} = \rho+\varepsilon$, and let $\Omega_\varepsilon = \{P \in \Omega | \rho_\varepsilon(P) \leqslant 0\}$ when ε is small enough so that the gradient of $\rho^{(\varepsilon)}$ is non-zero. Then Ω_ε is pseudoconvex and, furthermore, if $\tilde{\rho}^{(\varepsilon)} = -(-\rho)^\delta + \varepsilon^\delta$ then, when $\varepsilon > 0$, $\tilde{\rho}^{(\varepsilon)}$ is a smooth plurisubharmonic defining function for Ω_ε. Let $B^{(\varepsilon)}\colon L_2(\Omega_\varepsilon) \to \mathcal{H}(\Omega_\varepsilon)$ denote the Bergman projection operator on Ω_ε. We have

$$-\sum((-\rho^{(\varepsilon)})^\delta)_{z_i\bar{z}_j}\zeta_i\bar{\zeta}_j \geqslant -\sum((-\rho)^\delta)_{z_i\bar{z}_j}\zeta_i\bar{\zeta}_j \geqslant 0$$

so $-(-\rho^{(\varepsilon)})^\delta$ is plurisubharmonic. Then by a theorem of Boas and Straube (see [BS1]) we have $B^{(\varepsilon)}(C^\infty(\overline{\Omega}^{(\varepsilon)})) \subset C^\infty(\overline{\Omega}^{(\varepsilon)})$. Furthermore, if $\rho^{(\varepsilon)} = g^{(\varepsilon)}r^{(\varepsilon)}$, where $r^{(\varepsilon)}$ is a normalized defining function for Ω_ε, then $\lim_{\varepsilon\to 0} g^{(\varepsilon)} = g$ and $\lim_{\varepsilon\to 0} g^{(\varepsilon)}_{z_i} = g_{z_i}$ so that $\max_{b\Omega_\varepsilon}\left(1+\frac{|g^{(\varepsilon)}_{z_i}|}{g^{(\varepsilon)}}\right)$ is equivalent to $\max_{b\Omega}\left(1+\frac{|g_{z_i}|}{g}\right)$. Hence $\rho^{(\varepsilon)}$ satisfies the hypotheses of Proposition 4.12, and thus there exists C independent of ε and A and η independent of $s, \rho^{(\varepsilon)}, \delta$, and ε so that if either $s \leqslant \eta$, or $s > \eta$ and

$$\max_{b\Omega}\left(1+\frac{|g_{z_i}|}{g}\right)^3 A^s(1-\delta) \leqslant 1$$

then

$$\|B^{(\epsilon)}f\|_{s,\Omega_\epsilon} \leqslant C\|f\|_{s,\Omega_\epsilon} .$$

Hence we conclude that $Bf \in H^s(\Omega)$ and satisfies $\|Bf\|_s \leqslant C\|f\|_s$ whenever $f \in C^\infty(\overline{\Omega})$. Finally, since $C^\infty(\overline{\Omega})$ is dense in $H^s(\Omega)$, we have $B(H^s(\Omega)) \subset H^s(\Omega)$ and the theorem is proved.

5.1 Remark. The hypothesis $-(-\rho)^\delta$ is plurisubharmonic can be replaced by the hypothesis

$$(-(-\rho)^\delta)_{z_i\overline{z}_j} \geqslant -\delta C|\rho|^\delta \delta_{ij} , \tag{5.2}$$

with C independent of δ. When (5.2) is satisfied we set

$$c_{ij} = (-(-\rho)^\delta)_{z_i\overline{z}_j} + \zeta C|\rho|^\delta \delta_{ij}$$

and the proofs of Lemmas 4.3 and 4.7 go through with minor modifications and hence the main theorem follows in this case.

We observe that if ρ is a defining function which is plurisubharmonic on $b\Omega$, then (5.2) is satisfied for every $\delta \in (0,1]$. To see this, let $\pi\colon U \to b\Omega$ be a smooth mapping of a neighborhood U of $b\Omega$ onto $b\Omega$ with $\pi(P) = P$ for $P \in b\Omega$. Then for $P \in U$, we have

$$\rho_{z_i\overline{z}_j}(P) = \rho_{z_i\overline{z}_j}(\pi(P)) + O(|\rho(P)|),$$

hence in $U \cap \overline{\Omega}$

$$\begin{aligned}
-\sum_{i,j}((-\rho)^\delta)_{z_i\overline{z}_j}\zeta_i\overline{\zeta}_j &= \delta|\rho|^{\delta-1}\sum_{i,j}\rho_{z_i\overline{z}_j}\zeta_i\overline{\zeta}_j \\
&\quad + \delta(1-\delta)|\rho|^{\delta-2}\left|\sum_i \rho_{z_i}\zeta_i\right|^2 \\
&= \delta|\rho|^{\delta-1}\left\{\sum \rho_{z_i\overline{z}_j}\circ\pi\zeta_i\overline{\zeta}_j + O(|\rho|\,|\zeta|^2)\right\} \\
&\quad + \delta(1-\delta)|\rho|^{\delta-2}\left|\sum_i \rho_{z_i}\zeta_i\right|^2 \\
&\geqslant -\delta C|\rho|^\delta|\zeta|^2 .
\end{aligned}$$

5.3 Remark. Let $\mathcal{W} \subset b\Omega$ be the set defined as follows. For each $P \in \mathcal{W}$, there exists a neighborhood U of P with the following property. If $\zeta, \zeta' \in C_0^\infty(U \cap \overline{\Omega})$ with $\zeta' = 1$ on a neighborhood of the support of ζ, then for every s there exists C_s such that

$$\|\zeta N_{q+1}\overline{\partial}\psi\|_s + \|\zeta\overline{\partial}^* N_{q+1}\overline{\partial}\psi\|_s \leqslant C_s(\|\zeta'\psi\|_s + \|\psi\|) \tag{5.4}$$

for all $(0,q)$-forms in $\dot{\mathcal{D}}^q$. Let $\mathcal{Z} = b\Omega - \mathcal{W}$. Now suppose that there exists a defining function ρ which satisfies (5.2) in $V \cap \overline{\Omega}$, where V is a neighborhood of $\mathcal{Z}$. Then by a straightforward modification of the proofs, the main theorem holds with $\max\limits_{b\Omega}$ replaced by $\max\limits_{\mathcal{Z}}$.

The estimate (5.4) holds whenever P is of finite D'Angelo type (see [Ca1] and [D'A]) and also under a variety of weaker hypoellipticity conditions (see [Chr2] and [K2]).

References

[Ba1] Barrett, David, E., *Behavior of the Bergman projection on the Diederich-Fornaess Worm,* Acta Math. **168** (1992), 1–10.

[Ba2] Barrett, David, E., *Irregularity of the Bergman projection on a smooth bounded domain in* $\mathbb{C}^2$, Ann. of Math. **119** (1984), 431–436.

[Be] Bell, Steven, R., *A duality theorem for harmonic functions,* Mich. J. of Math. **29** (1982), 123–128.

[BS1] Boas, Harold, P. and Straube, Emil, J., *Sobolev estimates for the $\overline{\partial}$-Neumann operator on domains in* $\mathbb{C}^n$ *admitting a defining function that is plurisubharmonic on the boundary,* Math Z. **206** (1991), 81–88.

[BS2] Boas, Harold, P. and Straube, Emil, J., *Equivalence of regularity for the Bergman projection and the $\overline{\partial}$–Neumann operator,* Manuscr. Math. **67** (1990), 25–33.

[BS3] Boas, Harold, P. and Straube, Emil, J., *De Rham cohomology of manifolds containing the points of infinite type, and Sobolev estimates for the $\overline{\partial}$-Neumann problem,* J. Geom. Anal. **3** (1993), 225–235.

[BC] Bonami, A. and Charpentier, P., *Une estimation Sobolev $\frac{1}{2}$ pour le projecteur de Bergman,* C. R. Acad. Sci. Paris **307** (1988), 173–176.

[Ca1] Catlin, D., *Subelliptic estimates for the $\overline{\partial}$-Neumann problem on pseuodoconvex domains,* Ann. of Math. **126** (1987), 131–191; *Necessary conditions for the subellipticity of the $\overline{\partial}$–Neumann problem,* Ann. of Math. **117** (1983), 147–171.

[Ca2] Catlin, D., *Global regularity for the $\overline{\partial}$-Neumann problem,* Proc. Symp. Pure Math. A.M.S. **41** (1982), 39–49.

[Che] Chen, S.–C., *Global regularity of the $\overline{\partial}$-Neumann problem in dimension two,* Proc. Symp. Pure Math. A.M.S. **52**, Part 3 (1991), 55–61.

[Chr1] Christ, M., *Global C^∞ irregularity of the $\overline{\partial}$–Neumann problem for worm domains,* J. of A.M.S. **9** (1996), 1171–1185.

[Chr2] Christ, M., *Superlogarithmic gain implies hypoellipticity,* preprint 1997.

[D'A] D'Angelo, J. P., *Real hypersurfaces, orders of contact and applications,* Ann. of Math. **115** (1982), 615–637.

[DF1] Diederich, K. and Fornaess, J. E., *Pseudoconvex domains: An example with nontrivial Nebenhülle,* Math. Ann. **225** (1978), 275–292.

[DF2] Diederich, K. and Fornaess, J. E., *Pseudoconvex domains: Bounded strictly plurisubharmonic exhaustion functions,* Invent. Math. **39** (1977), 129–147.

[FK] Folland, G. B. and Kohn, J. J., The Neumann problem for the Cauchy-Riemann complex, Annals of Math. Studies **75**, Princeton U. Press, 1972.

[Ki] Kiselman, C. O., *A study of the Bergman projection on certain Hartogs domains,* Proc. Symp. Pure Math. **52**, Part 3 (1991), 219–231.

[K1] Kohn, J. J., *Global regularity for $\overline{\partial}$ on weakly pseudoconvex manifolds,* Trans. A.M.S. **181** (1973), 273–292.

[K2] Kohn, J. J., *Hypoellipticity of some degenerate subelliptic operators,* J. of Func. Anal. **159** (1998), 203–216.

[R] Range, M., *A remark on bounded strictly plurisubharmonic exhaustion functions,* Proc. A.M.S. **81**, No. 2 (1981), 220–222.

[S] Siu, Y.–T., *Non-Hölder property of Bergman projection of worm domain,* preprint 1997.

Department of Mathematics
Princeton University
Princeton, NJ 08544-1000

CHAPTER VII

Pdes Associated to the CR Embedding Theorem

Masatake Kuranishi †

The purpose of the present article is to explain a construction of a complex tubular neighborhood of a strongly pseudoconvex CR manifold, say M, using the CR geometry of M. The outline of the results obtained prior to 1996 are published in [11]–[14]. For the sake of the convenience of the readers, we will give the description from the beginning. The construction may be used to give a new proof of the CR embedding theorem. We also discuss and compare this attempt with the known proof of the CR embedding theorems (cf. [1], [3], [10], [17], [23]).

The CR embedding theorem was first proved using the $\overline{\partial}_b$-operator of M. Then Catlin gave a new proof by constructing a suitable complex structure on a differentiable tubular neighborhood of M. He is thus led to use pdes closely related to the $\overline{\partial}$-operator. Since $\overline{\partial}$-operators have better a priori estimates than $\overline{\partial}_b$-operators, he was able to improve the embedding theorem. When we use our method, we end up with a pde defined on $M \times \mathbf{C}^n$, $\dim M = 2n - 1$, which has the feature of $\overline{\partial}_b$-operators of M as well as of the $\overline{\partial}$-operators of $\mathbf{C}^n$. This situation hopefully helps us to improve the embedding theorem further.

Since the construction directly connects the ambient complex structure to the CR geometry of M, it is expected to have other applications. When the construction of the asymptote of the singularity of the Bergman kernel is rewritten in this language, it will give the expression of the asymptote in terms of the CR curvature. Since the construction of the CR geometry can also be carried out for the deformations of Shilov boundary of the bounded symmetric domains, it seems to suggest a way to construct the theory of CR manifold of higher codimension (under some conditions).

†Work partially supported by the National Science Foundation under grant DMS-9203974

1. CR structure and embedding theorem

Let M be a hypersurface in $\mathbf{C}^n$ of codimension 1. The set of type $(0,1)$ complex tangent vectors of $\mathbf{C}^n$ located at points in M and tangential to M forms a subbundle L of the complex valued tangent vector bundle $\mathbf{C}TM$. Such L satisfies the following conditions:

1) The fiber dimension is $n-1$,
2) $L \cap \overline{L} = \{0\}$,
3) the set of smooth sections of L is closed under bracket.

Let M be a manifold of dimension $2n-1$. A subbundle L of $\mathbf{C}TM$ satisfying the above three conditions is called a *CR structure*. When a CR structure L is associated with M, we call M a CR manifold. Let M be a CR manifold and $p \in M$ a reference point. We say that (M,p) (or (L,p)) is embeddable when, for an open neighborhood U of p in M, we can find a smooth embedding $U \to \mathbf{C}^n$ such that the image of $L|U$ is induced as above by the type $(0,1)$ tangent vectors of $\mathbf{C}^n$. An embedding as above will be called a CR embedding. In this paper we always discuss such local properties. Hence we usually shrink M if necessary and also omit mentioning reference points, unless it is necessary.

We discuss the local embeddability of a CR manifold under suitable assumptions. We use the Levi-form to state such assumptions. By conditions 1) and 2), we see that we can pick $F \in TM$ such that we have a direct sum decomposition: $\mathbf{C}TM = \mathbf{C}F + L + \overline{L}$. Denote by ρ (resp. $\overline{\rho}$) the projection to the factor L (resp. $\overline{L}$). For smooth sections X, Y of $\mathbf{C}TM$ set

$$[X, \overline{Y}] = i\,\mathcal{L}(X,Y)F + \rho[X,\overline{Y}] + \overline{\rho}[X,\overline{Y}]. \tag{1}$$

By taking the bar of the above formula we find that $\mathcal{L}(X,Y)$ defines on each fiber of M a Hermitian form, which is called the Levi form of L. $\mathcal{L}$ depends on the choice of F. But its conformal class of $\mathcal{L}$ is uniquely determined by L. Therefore we may say that L is called strongly pseudoconvex (or nondegenerate) when $\mathcal{L}$ is definite (or nondegenerate).

When $\dim M \geq 7$ and L is strongly pseudoconvex, L is locally embeddable. Strong pseudoconvexity in the above can be replaced by the property: $\mathcal{L}$ has at least 3 positive eigen-values or $n-1$ negative eigen-values. (cf. [3]). When $\dim M = 3$, there are examples of nonembeddable strongly pseudoconvex CR structures. (cf. [7], [20]). The question of embeddability in case of $\dim M = 5$ is an open problem.

We answer the embedding question by solving a system of partial differential equations. Consider a differential equation Σ on an unknown complex valued function f given by

$$Xf = 0 \quad \text{for all} \quad X \in L. \tag{2}$$

Assume that we can find solutions $f_1, \ldots, f_n$ of Σ such that

$$M \ni p \to (f_1(p), \ldots, f_n(p)) \in \mathbf{C}^n \tag{3}$$

is an embedding. This is a CR embedding for L. We solve Σ in the above sense by considering the $\overline{\partial}_b$-complex. Namely, by a k-form of type L we mean a smooth assignment of a skew-symmetric multilinear function in k variables in each fiber of L. Denote by $\Lambda^k(L)$ the vector space of k-forms of type L. By imitating the construction of the de Rham complex, we define a complex of differential operators

$$\overline{\partial}_b \colon \Lambda^k(L) \to \Lambda^{k+1}(L), \quad \text{where}$$

$$\begin{aligned}(\overline{\partial}_b u)(X_0, \ldots, X_k) &= \sum_j [X_j, u(X_0, \ldots, \hat{X}_j, \ldots, X_k)] \\ &\quad + \sum_{i<j} (-1)^{i+j} u([X_i, X_j], X_0, \ldots, \hat{X}_i, \ldots, \hat{X}_j, \ldots, X_k)\end{aligned} \tag{4}$$

for $u \in \Lambda^k(L)$, where the terms with hats are omitted. This is the $\overline{\partial}_b$-complex.

Before proceeding further, note the following well-known proposition.

(5) Proposition. *Any CR structure L is formally embeddable.*

Proof. There is clearly a chart (z', x), $z' = (\ldots, z^k, \ldots) \in \mathbf{C}^{n-1}$ and $x \in \mathbf{R}$, with center a reference point p_0 , so that L at p_0 has a base $(\partial/\partial \overline{z^k})_{p_0}$. It follows that L has generators L^k of the form:

$$L^k = \frac{\partial}{\partial \overline{z^k}} + A^l_{\overline{k}} \frac{\partial}{\partial z^l} + B^k \frac{\partial}{\partial x}. \tag{6.1}$$

We then find by the integrability condition that

$$[L^k, L^l] = 0. \tag{6.2}$$

Set $z^n = x$. By our choice of the chart

$$L^k z^l = A^l_{\overline{k}} = O(1), \quad L^k z^n = B^k = O(1). \tag{6.3}$$

We now consider a general chart (z, x) with center p_0 and a system of generators L^k which has the expression (6.1). We prove the following statement $(*)_\nu$ for any natural number ν by induction on ν: We can find a chart (z', x) as above so that, when we set $z^n = x + H$ with a suitable $H = O(2)$,

$$L^k z^t = O(\nu) \quad \text{for} \quad t = 1, \ldots, n. \tag{$(*)_\nu$}$$

Our original chart with $H = 0$ is the case $\nu = 1$. Assume that $(*)_\nu$ holds. We now construct a new chart (w, x) and new H' of the form:

$$w^k = z^k + h^k, \quad H' = H + h^n, \quad \text{where } h^t = O(\nu + 1) \tag{6.4}$$

such that $(*)_{\nu+1}$ holds under the new chart with the old L^k. We then see that $(*)_{\nu+1}$ holds also with a new L^k of the form (6.1) with respect to the chart (w, x). Clearly we can carry out this inductive construction, provided we have the following lemma: Let f be a complex valued function such that $L^k f = O(\nu)$. Then we can find that

$$h = O(\nu + 1) \text{ so that } L^k(f + h) = O(\nu + 1). \tag{6.5}$$

To see the above set

$$L^k f = \sum_p f_p^k(z', x; \overline{z'}) + O(\nu + 1), \tag{6.6}$$

where f_p^k is a homogeneous polynomial of degree ν and homogeneous of degree p in $\overline{z'}$. We find by (6.1)–(6.3) that

$$\frac{\partial}{\partial \overline{z^l}} f_p^k = -\frac{\partial}{\partial \overline{z^k}} f_p^l. \tag{6.7}$$

Then

$$h = \sum_p \frac{1}{p+1} \overline{z^k} f_p^k \tag{6.8}$$

works, q.e.d.

Using the smooth functions of which Taylor series are formal solutions, we can embed M in $\mathbf{C}^n$ so that the given CR structure L is very close to the CR structure induced by the ambient complex structure.

A pde that solves a certain problem may not be unique. For example Newton's equation can be also solved by considering Hamilton's equation. This equation has various advantages. The proof of the embedding theorem of Riemann geometry was simplified considerably by setting up a clever pde. In the case of the CR embedding theorem, there has been no such success so far. But a number of new pdes, which have some advantages compared to the above, were tried. In this paper we discuss two of these pdes besides the equation (2). These are the methods of Catlin, which will be referred to as II, and the one now being developed by the present author, which will be referred to as III. The method using the equation (2) will be referred as I.

2. The complex structure on the tubular neighborhood

In method II we construct a complex structure on a differential tubular neighborhood $\widetilde{M}$ of M such that the CR structure it induces on the boundary M is the given L. Therefore in this approach the unknown is a complex structure, say T'', on $\widetilde{M}$. To write down the pde for T'', note that as mentioned above, we have a complex structure, say $\widetilde{T}''$, on $\widetilde{M}$ which induces a CR structure on M very close to L. Hence we may regard our unknown complex structure T'' as a small deformation of $\widetilde{T}''$. Therefore we can use standard deformation theory machinery. Namely, denote by $\overline{\partial}_{\widetilde{T}''}$ the $\overline{\partial}$ exterior derivative with respect to the complex structure $\widetilde{T}''$. Let $\widetilde{T}'$ be the bar of $\widetilde{T}''$. We write

$$T'' = \{\overline{Z} + \omega(\overline{Z}) \ : \ \overline{Z} \in \widetilde{T}''\}, \tag{7}$$

where ω is an unknown $\widetilde{T}'$-valued 1-form of type $(0,1)$ with respect to the complex structure $\widetilde{T}''$. Then our equation is given by

$$\overline{\partial}_{\widetilde{T}''}\omega + \frac{1}{2}[\omega, \omega] = 0 \tag{8}$$

with the boundary condition: The CR structure on M induced by the complex structure determined by ω is the given L.

We may simplify the boundary condition by considering, instead of $\widetilde{T}''$, an almost complex structure, say $L^\sharp$, which induces L on M. Clearly we can write down the integrability condition of deformations of almost complex structure $L^\sharp$ imitating the case of complex structure. Namely, we define k-form of type $L^\sharp$ as a smooth assignment of a skew-symmetric multilinear form in k variables on each fiber of $L^\sharp$. Denote by $\Lambda^k(L^\sharp)$ the vector space of k forms of type $L^\sharp$. Let ρ be the projection to $L^\sharp$ with respect to the decomposition $\mathbf{C}TM = L^\sharp + \overline{L^\sharp}$. We define the $L^\sharp$ exterior derivative

$$\overline{\partial}_{L^\sharp} : \Lambda^k(L^\sharp) \to \Lambda^{k+1}(L^\sharp) \tag{9}$$

using the formula (4) where X_j are in $L^\sharp$ and the "[" is replaced by "ρ[". Then almost complex deformations of $L^\sharp$ are parameterized by an $\overline{L^\sharp}$ valued 1-form of type $L^\sharp$, say ω, by

$$T''_\omega = \{X + \omega(X) : \ X \in L^\sharp\}. \tag{10}$$

Denote by $\Lambda^k(L^\sharp, \overline{L^\sharp})$ the vector space of $\overline{L^\sharp}$ valued k-forms of type $L^\sharp$. Then we still have the exterior derivative

$$\overline{\partial}_{L^\sharp} : \Lambda^k(L^\sharp, \overline{L^\sharp}) \to \Lambda^{k+1}(L^\sharp, \overline{L^\sharp}) \tag{11}$$

as in (4) by inserting suitable projections where needed. We see clearly that T''_ω is integrable if and only if

$$\omega(\rho[X + \omega(X), Y + \omega(Y)]) = (I - \rho)[X + \omega(X), Y + \omega(Y)].$$

This condition can be rewritten as

$$\overline{\partial}_{L^\sharp}\omega + R(\omega) = 0, \tag{12}$$

where

$$\begin{aligned} R(\omega)(X,Y) =& (I-\rho)([\omega(X), \omega(Y)] + [X,Y]) \\ & - \omega(\rho([X, \omega(Y)] - [Y, \omega(X)] - [\omega(X), \omega(Y)])). \end{aligned}$$

Therefore the pde for the CR embedding theorem by method II is given by

$$\begin{gathered} \overline{\partial}_{L^\sharp}\omega + R(\omega) = 0, \\ \text{with the boundary condition } \omega(X) = 0 \text{ for all } X \in L. \end{gathered} \tag{13}$$

In the above we simplified the boundary condition in (7) but we ended up with a more complicated pde. Note also that $\overline{\partial}_{L^\sharp} \circ \overline{\partial}_{L^\sharp}$ may not be zero. However, from the standpoint of the general elliptic pde there is not much difference between the two. Catlin used equation (13), which we also use.

3. The CR geometry

In method III we use the CR frame bundle of the CR structure L. Since all the local geometric information on L is stored in the CR frame bundle and the associated Cartan connection, the aim of this method is to make more explicit the role played by the geometry of the CR structure, in particular its curvature. The method is based on the following idea of E. Cartan: When we consider a type of mathematical structure, we first pick a model case, which is a homogeneous structure. We study the model case in detail and the general case is discussed as a deformation of the model case. More specifically, we write the model case $N = G/H$ for a Lie group G and its closed subgroup H. We regard G as a principal bundle:

$$G \to N, \quad \text{with the structure group } H. \tag{14}$$

Denote by $\mathbf{g}$, $\mathbf{h}$ the Lie algebra of G, H. We have the Maurer-Cartan forms

$$\omega_G : TG \to \mathbf{g}, \quad \omega_H : TH \to \mathbf{h}. \tag{15}$$

The 1-form ω_G is a special case of Cartan connection forms. Namely, for any principal bundle (P, N, H) with the structure group H over a manifold N we define a Cartan connection form as a 1-form:

$$\omega : TP \to \mathbf{g} \tag{16}$$

with the conditions:

1)° At each $\mathbf{f} \in P$, ω restricted to the tangent vector space $T_{\mathbf{f}}P$ at $\mathbf{f}$ is an isomorphism.

2)° $(R_h)^*\omega = \mathrm{Ad}(h^{-1})\omega$ for any $h \in H$, where R_h is the action of h on the fibers of P.

3)° Under any injection of H compatible with the principal bundle structure, ω is pulled back to the Maurer-Cartan form ω_H of H.

For the model case $N = G/H$, its frame bundle is by definition the principal bundle (14). The associated Cartan connection is, by definition, ω_G in (15). This pair constitutes the geometry of N.

We try to find a construction by which we associate, for each deformation of the model N, a principal bundle and a Cartan connection ω imitating the case of the model case. When we succeed, the curvature of the structure is given by

$$k = d\omega + \frac{1}{2}[\omega, \omega]. \tag{17}$$

In the model case the curvature is zero. In the non-model case, it seems, there is no unique way of such construction. There could be many geometries associated to deformations of N. We simply choose one which seems most natural. The chosen one is usually called normal. Even in the Riemannian metric case, we may consider a geometry different from Levi-Civita geometry by imposing conditions on the torsion. In the strongly pseudoconvex CR case, the geometries in [4] and [22] are isomorphic (cf. Yang Liu [18]). In [12] a geometry depending on a parameter is constructed, and the previous cases correspond to the special points in the parameter space.

In our case of the strongly pseudoconvex CR structure, the model is the boundary CR structure on the boundary sphere S of the unit ball B in $\mathbf{C}^n$. Denote by G the connected component of the holomorphic automorphism group of B. Pick reference points p_S, p_B of S, B, respectively. Since the connected component of the CR automorphism group of S is equal to the restriction to S of G, we may set

$$S = G/H_S, \quad B = G/H_B, \tag{18}$$

where H_S, H_B are the isotropy groups at p_S, p_B, respectively. We thus have the principal bundles (G, B, H_B) and (P_S, S, H_S), $P_S = G$. As mentioned above the CR frame bundle of S is (P_S, S, H_S). The Maurer-Cartan form ω_G of G regarded as a 1-form on P_S is the Cartan connection form associated to the CR structure S, which we denote by ω_S. We examine (P_S, S, H_S, ω_S) closely in order to construct the Cartan geometry of strongly pseudoconvex CR structures.

The CR structure subbundle $L_S \subset \mathbf{C}TS$ of S is defined by equations of the form:

$$\omega^t = 0, \quad t = 1, \ldots, n, \tag{19.1}$$

where ω^k, $(k = 1, \ldots, n-1)$, are complex valued 1-forms and ω^n is a real valued 1-form. By the strong pseudoconvexity we find that we may choose the above so that

$$d\omega^n \equiv \omega^k \wedge \overline{\omega^k} \quad (\text{mod. } \omega^n). \tag{19.2}$$

The totality of such choices of $\ldots, \omega^\alpha, \ldots$, considered pointwise, forms a bundle P'_S over S. Denote by Ω^α the tautology form on P'_S associated to the choices of ω^α. Introduce a $\mathbf{C}^{n-1} \times \mathbf{R}$-valued form $\Omega'_S = (\Omega^1, \ldots, \Omega^n)$. We thus have a bundle over S with a 1-form

$$(P'_S, S, \Omega'_S). \tag{20}$$

Note that this construction can be carried out for any strongly pseudoconvex CR structure.

We next observe that the principal bundle $P_S \to S$ is a prolongation of $P'_S \to S$. To see this, we have to write down more explicitly the action of G. For calculations it is more convenient to use, instead of B, its Cayley transform. This is a domain Q^+ in $\mathbf{C}^n$ given by

$$Q^+ : \Im z^n > \frac{1}{2}|z'|^2, \quad z' = (z^1, \ldots, z^{n-1}). \tag{21}$$

Therefore instead of S we use the boundary CR structure Q on the boundary of Q^+ defined by

$$Q : \Im z^n = \frac{1}{2}|z'|^2. \tag{22}$$

Q^+ is biholomorphic to B and Q is CR isomorphic to S minus a point. As a reference point we use $p_{Q^+} = (0, i)$ and $p_Q = 0$. Let H_{Q^+}, H_Q be the isotropy group of p_{Q^+}, p_Q, respectively. Denote by $\mathbf{PC}^n$ the projective space with homogeneous coordinate $[\xi^0, \ldots, \xi^n]$. We embed

$$Q^+ \ni z \mapsto [1, z] \in \mathbf{PC}^n. \tag{23}$$

Introduce a quadratic form:

$$\Phi(\xi, \xi) = 2\Re\, i\, \xi^0 \overline{\xi^n} - |\xi'|^2, \quad \xi = (\xi^0, \ldots, \xi^n),\ \xi' = (\xi^1, \ldots, \xi^n), \tag{24}$$

and denote by S_Φ (resp. S_Φ^+) the real hypersurface (resp. the domain) in $\mathbf{C}P^n$ defined by the equation:

$$\Phi = 0, \quad (\text{resp.} \quad \Phi > 0). \tag{25}$$

Then by the embedding (23) we may regard Q^+ and Q as open submanifolds

$$Q^+ \subset S_\Phi^+, \quad Q \subset S_\Phi. \tag{26}$$

We can now represent G as the connected component of the group of projective transformations induced by the group $\widetilde{G}$ of special linear transformations of $\mathbf{C}^{n+1}$ which leave Φ invariant. Therefore we now represent $g \in G$ by an $(n+1) \times (n+1)$ matrix mod. the center of $\widetilde{G}$, which is the finite group of scalar matrices of determinant 1. We are thus able to calculate explicitly the elements of G.

For a column vector $\zeta = (\zeta', \zeta^n) \in \mathbf{C}^n$ set

$$T(\zeta) = \begin{pmatrix} 1 & 0 & 0 \\ \zeta' & I & 0 \\ \zeta^n & i(\zeta')^* & 1 \end{pmatrix}. \tag{27}$$

The set of $T(\zeta)$ forms a matrix group $\widetilde{\mathcal{N}}$ which acts holomorphically and simply transitively on $\mathbf{C}^n$. Its operation is given by

$$T(\zeta)z = (z' + \zeta', z^n + \zeta^n + i\langle z', \zeta' \rangle). \tag{28}$$

Set

$$\mathcal{N} = \{T(z) : z \in Q\}, \quad \mathcal{N}^+ = \{T(\zeta) : \zeta \in Q^+\}. \tag{29}$$

$\mathcal{N} \subset G$ is the Heisenberg group which acts simply transitively on Q. Therefore

$$\mathcal{N} \cap H_Q = \{I\}, \quad \text{and } \mathcal{N} H_Q \text{ is an open subset of } G. \tag{30}$$

$\mathcal{N}^+$ is closed under the composition and its operation preserves Q^+. We find by explicit calculation that

$$\begin{aligned} H_Q = \{h(a, u, \beta, b) : a, b \in \mathbf{C}, u^*u = I,\ (a/\overline{a}) \det u = 1, \\ \Im(b/a) + (1/2)|\beta|^2 = 0\}, \end{aligned} \tag{31}$$

where

$$h(a,u,\beta,b)=\begin{pmatrix} a & \nu^* & b \\ 0 & u & \beta \\ 0 & 0 & 1/\overline{a} \end{pmatrix}, \quad \nu = i\,\overline{a}\,u^*\beta.$$

We set

$$H_{\mathbf{C}} = \{h_{\mathbf{C}}(r,\beta,b) : r>0, \beta\in\mathbf{C}^n, b\in\mathbf{C}, \Im b = -(r^2/2)|\beta|^2\}, \tag{32}$$

where

$$h_{\mathbf{C}}(r,\beta,b)=\begin{pmatrix} 1/r & -i\beta^* & (1/r)b \\ 0 & I & r\beta \\ 0 & 0 & r \end{pmatrix},$$

and

$$H_{\mathbf{su}} = \{h_{\mathbf{su}}(\theta,u) : \theta\in\mathbf{R},\ u^*u=I, e^{2i\theta}\det u = 1\}, \tag{33}$$

where

$$h_{\mathbf{su}}(\theta,u)=\begin{pmatrix} e^{i\theta} & 0 & 0 \\ 0 & u & 0 \\ 0 & 0 & e^{i\theta} \end{pmatrix}.$$

Then we find by calculation that the above are subgroups of H_Q and

$$H_Q = H_{\mathbf{C}}H_{\mathbf{su}}, \quad H_{\mathbf{C}}\cap H_{\mathbf{su}} = \{I\}. \tag{34}$$

Moreover the underlying manifold of $H_{\mathbf{C}}$ is diffeomorphic to the manifold Q^+ by the mapping

$$H_{\mathbf{C}} \ni h_{\mathbf{C}}(r,\beta,b) \mapsto \zeta \in Q^+, \quad \text{where } \zeta' = \frac{ir^2}{1+ib}\beta,\ \zeta^n = \frac{ir^2}{1+ib}. \tag{35}$$

We thus find that the isotropy group of Q is carrying a copy of the interior domain.

The Lie algebra of $\mathcal{N}$ is represented by the matrix Lie algebra

$$\mathbf{n} = \{n(\dot z', \dot x) : \dot z' \in \mathbf{C}^{n-1}, \dot x \in \mathbf{R}\}, \tag{36}$$

where

$$n(\dot z',\dot x)=\begin{pmatrix} 0 & 0 & 0 \\ \dot z' & 0 & 0 \\ \dot x & i(\dot z')^* & 0 \end{pmatrix}.$$

The Lie algebra of H_Q is represented by the matrix Lie algebra

$$\mathbf{h}_Q = \{h(\dot a, \dot u, \dot\beta, \dot b) : \dot a\in\mathbf{C},\ \dot u \in \mathbf{su}(n-1),\ \dot\beta\in\mathbf{C}^{n-1},\ \dot b\in\mathbf{C}\}, \tag{37}$$

where

$$h(\dot{a}, \dot{u}, \dot{\beta}, \dot{b}) = \begin{pmatrix} \dot{a} & -\dot{\beta}^* & \dot{b} \\ 0 & \dot{u} & \dot{\beta} \\ 0 & 0 & -\overline{\dot{a}} \end{pmatrix}.$$

Denote by $\rho_{\mathbf{n}}$ (resp. by $\rho_{\mathbf{h}_Q}$) the projection of $\mathbf{g}$ to $\mathbf{n}$ (resp. to $\mathbf{h}_Q$). Then we have a complex valued invariant 1-form Ω_Q^k and a real-valued invariant 1-form Ω_Q^n so that

$$\rho_{\mathbf{n}}\omega_Q = n(\Omega_Q). \tag{38.1}$$

Similarly, we have complex valued invariant 1-form Φ_Q^k and real-valued invariant 1-forms Π'_Q, Ψ'_Q such that with a suitable $\dot{u}$

$$\rho_{\mathbf{h}_Q}\omega_Q = h(\Pi'_Q, \dot{u}, \Phi'_Q, \Psi'_Q). \tag{38.2}$$

We now go back to the study of the CR Cartan connection form on Q. Since $\mathcal{N}$ acts simply transitively on Q, we see by (30) that the CR frame bundle of Q (which is the open subbundle over Q of $P_S \to G/H_Q$) is

$$P_Q = \mathcal{N}H_Q \to \mathcal{N} = Q. \tag{39}$$

We can now calculate explicitly the bundle P'_Q which we obtain for Q as we obtained P'_S for S in (20). Note first that in view of the decomposition (30) we have by (36) an isomorphism

$$\mathbf{g}/\mathbf{h} = \mathbf{n} \to \mathbf{C}^{n-1} \times \mathbf{R}. \tag{40}$$

Then the adjoint action $\mathrm{Ad}(h^{-1})$ on $\mathbf{g}/\mathbf{h}$ with $h \in H_Q$ is transplanted to the action of H_Q on $\mathbf{C}^{n-1} \times \mathbf{R}$ which is given in terms of the representation (31) by

$$\mathrm{Ad}(h^{-1})(\dot{z}, \dot{x}) = (au^*\dot{z} - |a|^2\dot{x}u^*\beta, |a|^2\dot{x}). \tag{41}$$

Denote by H' the quotient group of H_Q which makes the above action effective. Then we find by calculation that P'_Q is a principal bundle over Q with structure group H'. Moreover the isomorphism (40) allows us to regard Ω'_S in (20) as valued in $\mathbf{g}/\mathbf{h}$. Therefore we have a principal bundle with $\mathbf{g}/\mathbf{h}$-valued 1-form Ω'_Q

$$(P'_Q, Q, H', \Omega'_Q). \tag{42}$$

Further we find that Ω'_Q satisfies a condition similar to 2)° in the definition of Cartan connection, where R_h should be replaced by the action of H' on P'_Q. Therefore we conclude that

$$\begin{aligned} &(P_Q, Q, H_Q, \Omega_Q) \text{ may be regarded as a prolongation} \\ &\text{of } (P'_Q, Q, H', \Omega'_Q). \end{aligned} \tag{43}$$

The above observation opens the way to construct CR geometry of an arbitrary strongly pseudoconvex CR manifold M. Namely, as we noted earlier the construction of Ω^α through (19)–(20) can be applied to M. We thus have (P'_M, M, H', Ω'_M). Following Cartan's program we consider an unknown principal bundle with a Cartan connection (P_M, M, H', ω_M) which is a prolongation of (P'_M, M, H', Ω'_M). We see that this is the exact analogy of the construction of Levi-Civita connection (using the orthonormal frame bundle) in the case of Riemann geometry. In that case $P'_M = P_M$ and an extension is uniquely determined by imposing the torsion-free condition. In our case we find that, when we impose a suitable torsion-free condition, the principal bundle P_M is uniquely determined and Ω_M is also determined uniquely up to a family of Cartan connections depending on one arbitrary function. We make the choice unique by imposing a suitable trace zero condition. We call the Cartan connection thus obtained the CR Cartan connection. We now complete the construction of the CR geometry (cf. [4], [12], [22]).

4. The CR geometry and the embedding theorem

We now discuss method III of the construction of CR embeddings. Following Cartan's plan, we first examine the case of S. In this case the problem is to construct B using only the CR geometry of S. We interpret this to mean that we construct the projection $\rho_B\colon G \to B = G/H_B$ and the complex structure on B using the projection $P_S = G \to S = G/H_S$ and the Maurer-Cartan form of G. However it is more convenient to use Q^+. We thus try to construct the projection

$$\rho_{Q^+}\colon P_Q = \mathcal{N} H_Q \to Q^+ = G/H_{Q^+}, \tag{44}$$

using only the projection $\rho_Q\colon \mathcal{N} H_Q \to Q = \mathcal{N}$ and its CR Cartan connection $\omega_G = \omega_Q$. Denote by I the identity element.

We first make the following observation: The complex structure of Q^+ is induced by the complex structure of $\mathbf{C}^n$ (cf. (21)) and invariant under the operation of G. Therefore, the pull back of the bundle of type (1,0) tangent vectors of Q^+ by ρ_{Q^+} is determined by $(\Xi^t)_I = ((\rho_{Q^+})^* dz^t)_I$, $t = 1, \ldots, n$, where $(z^1, \ldots, z^n)$ is the standard chart of $\mathbf{C}^n$ and ρ_{Q^+} is the projection in (44). Note that there is a Maurer-Cartan form, say Ξ^t, on G which agrees with the above form at I. Denote by Σ_Q the subbundle of $\mathbf{C}T^*G$ generated by $\Xi^1, \ldots, \Xi^n$. By the observation made above we see that Σ_Q has the following two properties:

1)$^\sharp$ The projection ρ_{Q^+} is determined by the subbundle $\Sigma_Q + \overline{\Sigma_Q}$, i.e., the fibers are the maximal integral submanifolds of Pfaff's equation

$\Sigma_Q = \overline{\Sigma_Q} = 0.$

2)$^\sharp$ A function defined on an open subset of Q^+ is holomorphic if and only if $d(h \circ \rho_{Q^+})$ is a section Σ_Q.

Therefore our problem in the model case is solved when we have Σ_Q. Since we know explicitly how the elements in G act, we can easily write down Σ_Q explicitly. Even though we used the projection ρ_{Q^+} to define Ξ^α initially, we are now able to write down Σ_Q using only P_Q and ω_Q. In fact,

$$\Xi^k = \Omega_Q^k + i\Phi_Q^k, \quad \Xi^n = \Omega_Q^n - 2i\,\Pi_Q' + \Psi_Q'. \tag{45}$$

We may rewrite the above more conveniently as follows: There is a linear map $\widetilde{\pi}\colon \mathbf{g} \to \mathbf{C}^n$ such that

$$\Sigma_Q \text{ is generated by components of } \widetilde{\pi}\omega_Q. \tag{46}$$

The properties 1)$^\sharp$, 2)$^\sharp$, and (45) show how we should proceed for the general case. This is based on the generalized Newlander-Nirenberg Theorem (cf. [6], [21]):

(47) Theorem. *Let N be a manifold and Σ a subbundle of $\mathbf{C}T^*N$ of rank n satisfying the following conditions:*

i) $\Sigma \cap \overline{\Sigma} = \{0\}$,

ii) *Σ is integrable, i.e., the exterior derivatives of the sections of Σ are in the ideal generated by the sections of Σ.*

Then there is (locally) a projection, say ρ_Σ, of N to a complex manifold, say N', such that Σ is generated by differentials of the pull back of holomorphic functions of N'.

Proof. By the classical Frobenius Theorem applied to $\Sigma + \overline{\Sigma}$, we find the projection ρ_Σ to a manifold N'. Therefore we may identify N with $N' \times W$ for a manifold W so that ρ_Σ is the projection to the first factor. Pick a point $w_0 \in W$ and consider the injection $j\colon N' \ni x \mapsto (x, w_0) \in N' \times W$. By applying the Newlander-Nirenberg Theorem to $j^*\Sigma$, we find a complex chart $(z^1, \dots, z^n)$ of N'. By the definition of ρ_Σ we see that Σ is generated by $dz^t, \overline{dz^t}$, where we wrote z^t instead of $z^t \circ \rho_\Sigma$ for simplicity. Therefore there is a generator of Σ of the form: $dz^t + a^{ts}\overline{dz^s}$. By the integrability condition of Σ we find immediately that a^{ts} is a constant. Since it is zero at $N' \times w_0$, we conclude that dz^t generate Σ. q.e.d.

The above complex manifold N' will be referred to as the complex manifold obtained by the integrals of Σ. When the rank of Σ is n, we call

Σ satisfying the above conditions a complex structure on N of rank n. Also we may say almost complex structure of rank n when the condition i) is satisfied but the integrability condition may not be satisfied.

By the above discussion we may now conclude that the CR embedding theorem is reduced to the following problem:

(48) Problem. *Find a complex structure Σ on P_M of rank n satisfying the following condition: We can attach M as the boundary to a complex structure $\widetilde{M}$ obtained by the integrals of Σ so that the CR structure it induces is the given one.*

Using the CR Cartan connection form ω_M, we may try $\Sigma_1 = \widetilde{\pi} \circ \omega_M$ (cf. (46)), which is generated by

$$\Xi_M^t = \widetilde{\pi}^t \omega_M. \tag{49}$$

However, Σ_1 is not integrable. The curvature of M is the obstruction to the integrability. Thus we have to deform Σ_1 to a complex structure. It turns out this is rather difficult to handle. The space P_M is too big. We first reduce the problem to a problem on a quotient bundle of P_M. To carry out this reduction we have to examine the projection $\rho_{Q^+} : \mathcal{N}H_Q \to Q^+$ more closely.

In terms of the decomposition (34), we have the projection

$$\rho_1 : \mathcal{N}H_Q \to \mathcal{N}H_{\mathbf{C}}. \tag{50.1}$$

In view of (29) and (35) we may use a chart

$$(z, \zeta) \in Q \times Q^+ \tag{50.2}$$

of $\mathcal{N}H_{\mathbf{C}}$. For convenience denote by $z^\sharp$ the general element in $Q^+ \subset \mathbf{C}^n$. We then have a map

$$\rho^\sharp : \mathcal{N}H_{\mathbf{C}} \to Q^+ \text{ given by } (z, \zeta) \mapsto z^\sharp = T(z)\zeta. \tag{51}$$

(cf. (28)). By explicit calculation we find that

$$\rho_{Q^+} = \rho^\sharp \circ \rho_1, \tag{52}$$

and that Ξ_Q^α in (45) is the pull back of a 1-form on $\widetilde{P}_Q = \mathcal{N}H_{\mathbf{C}}$, i.e.

$$\Sigma_Q \text{ can be reduced to a subbundle } \widetilde{\Sigma}_Q \subset \mathbf{C}T^*\widetilde{P}_Q. \tag{53}$$

In view of the decomposition (34) we may identify

$$\widetilde{P}_Q = \mathcal{N}H_{\mathbf{C}} = (\mathcal{N}H_Q)/H_{\mathbf{su}} = P_Q/H_{\mathbf{su}}. \tag{54}$$

The formula suggests we consider

$$\widetilde{P}_M = P_M / H_{\mathbf{su}}. \tag{55.1}$$

We then find by explicit calculation (cf. (49)) that

$$\Sigma_1 \text{ can be reduced to a subbundle } \widetilde{\Sigma}_1 \subset \mathbf{C}T^*\widetilde{P}_M. \tag{55.2}$$

Therefore we try to deform $\widetilde{\Sigma}_1$ to an integrable subbundle of $\mathbf{C}T^*\widetilde{P}_M$. We now forget P_M and work on $\widetilde{P}_M$. For simplicity we use the same symbol Ξ_M^α to denote the 1-form in (49) reduced to $\widetilde{P}_M$. Therefore our problem is to find 1-forms, say A^t, on $\widetilde{P}_M$ satisfying the following two conditions:

(56.1) The subbundle $\widetilde{\Sigma}_M$ generated by $\Xi_M^t - A^t$ is integrable.

(56.2) We can attach M to the complex manifold obtained by integrals of the above subbundle in such a way that it induces the CR structure of M.

To write down the pde for an n-vector valued 1-form $A = (\dots, A^t, \dots)$ to satisfy (56.1), note that any CR structure M can be considered as a small deformation of the induced CR structure of a hypersurface, say M_1. In the case of a strongly pseudoconvex CR structure, by a change of variables we can make the equation of the hypersurface M_1 very close to the equation of Q as was done in the early stage of the construction of Chern-Moser normal form (cf. [4]). Therefore by a diffeomorphism we may identify the manifold M as an open submanifold of Q so that the given CR structure $L_M \subset \mathbf{C}TQ$ is very close to the CR structure L_Q of Q. Then we may identify $\widetilde{P}_Q$ with $\widetilde{P}_M$ by a diffeomorphism. Thus we are in a situation:

$$M = Q, \quad \widetilde{P}_M = \widetilde{P}_Q = Q \times Q^+, \tag{57}$$

where Q is, as mentioned in the beginning, now replaced by a small open substructure of the old Q. What distinguishes between the CR structures Q and M is the Cartan connections ω_Q and ω_M on $Q \times Q^+$. We use its standard chart (z, ζ) or (z', x, ζ), $z^n = x + i|z'|^2/2$.

We say that a form θ on $Q \times Q^+$ is basic when θ can be written using only $dz^k, \overline{dz^k}, dx$ with functions of (z, x, ζ) as coefficients. Set $\Xi_M = (\dots, \Xi_M^t, \dots)$. By explicit calculation we find that

$$\Xi_M = d\zeta + \Xi_0, \quad \text{where } \Xi_0 \text{ is basic and holomorphic in } \zeta. \tag{58.1}$$

Moreover Ξ is actually defined on $Q \times \mathbf{C}^n$ and we have the following property:

$$\Xi_0 \equiv \omega_M \pmod{\zeta^t} \tag{58.2}$$

where $\omega_M = (\dots, \omega_M^t, \dots)$ is as in (19.1)–(19.2) (where ω is replaced by ω_M), defining the CR structure of M. (58.1) implies that we may only consider A of the form

$$A = A_M + A_{[\overline{\alpha}]} d\overline{\zeta^\alpha}, \text{ where } A_M \text{ is basic, and} \tag{59}$$

$$d\Xi_M = U \wedge \Xi_M + B, \quad B \text{ is basic} \tag{60}$$

for a 1-form U and a 2-form B which depend on M. Denote by d_M the exterior derivative with respect to the variable (z', x) of the standard chart (z', x, ζ) of $\widetilde{P}_M$. Then we find easily that A satisfies (56.1) if and only if

$$d_M A - U \wedge A - (\Xi_0^t - A^t) \wedge \frac{\partial A}{\partial \zeta^t} + d\overline{\zeta^t} \wedge \frac{\partial A}{\partial \overline{\zeta^t}} - B = 0. \tag{61}$$

Therefore in the approach III the unknown is an n-vector valued 1-form A on $Q \times Q^+$ of the form (59) and the associated pde is (61).

As for the condition (56.2), we note that the equation (61) is defined on $Q \times \mathbf{C}^n$. Hence we will try to find a small solution A on a small open neighborhood of 0×0. Now in the case where the CR structure M is equal to Q, $A = 0$ is a solution, and the induced projection is $\rho^\sharp$ in (51). We then see by (28) that $\rho^\sharp$ induces the identity map of $Q \times 0 =$ (the submanifold $\zeta = 0$ of $Q \times Q^+$) to the boundary Q of Q^+. Since M is very close to Q and A is small, the projection induced by the subbundle $\widetilde{\Sigma}_M$ will also induce a diffeomorphism of $Q \times 0 = M \times 0$ to a submanifold of the complex manifold defined by $\widetilde{\Sigma}_M$. We regard this as the attaching map of M. We then see by (58.2) that condition (56.2) is satisfied when the following condition is satisfied:

$$A = 0 \text{ on the submanifold } \zeta = 0 \text{ of } Q \times Q^+. \tag{62}$$

We thus find that for method III we impose the above condition instead of a boundary condition.

5. Construction of the solution

We now discuss how to solve the equations for methods I, II, III. (cf. (2)–(3), (13), (61)–(62)). Let us denote the unknown by u. Then our equations are of the form

$$Du + R(u) = 0, \tag{63}$$

where D is a linear differential operator. u is a some type of differential form of degree p, where $p = 0$ in case I and $p = 1$ in case II, III. So far all the methods follow a similar line: We first find a solution. Then denoting by $u = u_0$ the approximate solution of which the Taylor series is equal to the formal solution, the error

$$Du_0 + R(u_0) = b_0 \tag{64.1}$$

is very small. We then try to find u_1 so that its error is smaller than b_0. We use the Newton's method. Namely, we consider the differential of the map $u \mapsto Du + R(u)$ at u_0, obtaining a linear operator D_{u_0}. We then try to solve the correction equation

$$D_{u_0} v = b_0. \tag{64.2}$$

It turns out that we can not solve the equation. We find that we can solve it with an error with magnitude something like the square of the magnitude of b_0. Let $v = v_0$ be such an approximate solution. Then $u_1 = u_0 - v_0$ is the next approximation to a solution of (63). We repeat the process, hoping that the sequence u_ν converges to a solution. Actually this does not work. Following the Nash-Moser procedure (cf. [18]) we use not v_0 but the smoothed version of v_0. This approach works for cases I and II. Case III is now being tried but not yet finished.

The existence of a formal solution of I is already done in (6). We can similarly find formal solutions for II and III. For equation III it is quite involved (cf. [14]) and we already see that the curvature of the CR geometry comes in the construction.

We are thus forced to consider the differentials of $Du + R(u)$ at, say u_0. These differentials in our cases can be made parts of sequences of differential operators of the following form: We are given

i) linearly independent complex valued 1-forms ξ^λ on a smooth manifold N, where λ runs in a set of indices, say Π,

ii) a first order (matrix coefficients) partial differential operators X_λ for $\lambda \in \Pi$ such that its principal parts are diagonal operators, and

iii) a set of matrix-valued functions $c^{\sigma_1 \cdots \sigma_p}_{\lambda_0 \cdots \lambda_p}$ (where the indices are considered as skew-symmetric) for each p.

Denote by $\mathcal{X}$ the system of such chosen $\xi^\lambda, X_\lambda, c's$. Let $\Lambda^p(\mathcal{X})$ be the vector space of differential p-forms u of the form

$$u = \frac{1}{p!} \sum u_{\lambda_1 \cdots \lambda_p} \xi^{\lambda_1} \wedge \cdots \wedge \xi^{\lambda_p} \tag{65}$$

where $u_{\lambda_1\cdots\lambda_p}$ is skew-symmetric in $\lambda_1,\ldots,\lambda_p$. Define

$$D_{\mathcal{X}}\colon \Lambda^p(\mathcal{X}) \to \Lambda^{p+1}(\mathcal{X})$$

by

$$\begin{aligned} D_{\mathcal{X}}u = \frac{1}{p!}\sum \big(&X_\lambda u_{\lambda_1\cdots\lambda_p}\xi^\lambda \wedge \xi^{\lambda_1}\wedge\cdots\wedge\xi^{\lambda_p} \\ &+ c^{\sigma_1\cdots\sigma_p}_{\lambda_0\cdots\lambda_p}u_{\sigma_1\cdots\sigma_p}\xi^{\lambda_0}\wedge\cdots\wedge\xi^{\lambda_p}\big). \end{aligned} \tag{66}$$

Since there are no commutativity relations among X_λ, $D_{\mathcal{X}}\circ D_{\mathcal{X}}$ may not be zero.

In Case I we consider the case $N = M$. $\{X_\lambda\}$ is an arbitrary base of the CR structure L. ξ^λ is the dual base extended to the 1-form taking value zero on the supplementary bundle $\overline{L} + F$. The sequence $D_{\mathcal{X}}$ is independent of the choice of u_0 and equal to the $\overline{\partial}_b$-complex. In the case of III at $u_0 = A$, for example, $N = Q \times Q^+$ and the set of indices Π consists of

$$t = 1,\ldots,n;\ \overline{k},\ k = 1,\ldots,n-1;\ [\overline{t}],\ t = 1,\ldots,n. \tag{67.1}$$

Thus Π consists of $3n - 1$ elements.

$$\xi^t = \omega^t_M, \text{ as in (58.2)},\quad \xi^{\overline{k}} = \overline{\xi^k},\quad \xi^{[\overline{t}]} = d\overline{\zeta^t}. \tag{67.2}$$

In terms of the standard chart (z', x, ζ) of $N = Q \times Q^+$

$$\begin{aligned} X_k &= \frac{\partial}{\partial z^k} + \frac{i}{2}\overline{z^k}\frac{\partial}{\partial x} - \frac{\partial}{\partial \zeta^k} + R_k, \quad X_n = \frac{\partial}{\partial x} - \frac{\partial}{\partial \zeta^0} + R_n, \\ X_{\overline{k}} &= \frac{\partial}{\partial \overline{z^k}} - \frac{i}{2}z^k\frac{\partial}{\partial \xi} - i\zeta^k\frac{\partial}{\partial \zeta^n} + R_{\overline{k}}, \quad X_{[\overline{t}]} = \frac{\partial}{\partial \overline{\zeta^t}} + R_{[\overline{t}]}, \end{aligned} \tag{67.3}$$

where R_λ is a small term depending on the CR structure of M and A. The curvature appears in the expression of R_λ. In this case we have

$$d\xi^\lambda = \frac{1}{2}c^\lambda_{\nu\sigma}\xi^\nu \wedge \xi^\sigma, \tag{68.1}$$

and we choose c's so that

$$c^{\sigma_1\cdots\sigma_p}_{\lambda_0\cdots\lambda_p}\xi^{\lambda_0}\wedge\cdots\wedge\xi^{\lambda_p} = d(\xi^{\sigma_1}\wedge\cdots\wedge\xi^{\sigma_p}). \tag{68.2}$$

Going back to our problem of solving the correction equation, it so happens that in each of our cases I to III, there is associated a chosen $\mathcal{X}$ and our unknown u can be regarded as being in $\Lambda^p(\mathcal{X})$. Moreover, the differential of the map $u \mapsto Du + R(u)$ at an approximating solution

u_0 is equal to $D_{\mathcal{X}}$. Therefore we now want to solve approximately the correction equation in an unknown u:

$$D_{\mathcal{X}} u = b \tag{69}$$

for the error $b = Du_0 + R(u_0)$, which is in $\Lambda^q(\mathcal{X})$, where $q = p+1$. We may use the a priori estimate method or explicit construction in terms of an integral kernel to solve (69). We discuss here the method of the a priori estimate.

Fix a L_2-norm on vector valued functions. We introduce an L_2-norm in $\Lambda^q(\mathcal{X})$ by

$$\|w\|^2 = \frac{1}{q!} \sum \|w_{\lambda_1 \cdots \lambda_q}\|^2, \tag{70}$$

and consider the Laplacian

$$\Delta_{\mathcal{X}} = D_{\mathcal{X}}(D_{\mathcal{X}})^* + (D_{\mathcal{X}})^* D_{\mathcal{X}}. \tag{71}$$

We need an a priori estimate: For a suitable norm $|||w|||$

$$Q(w) = \|D_{\mathcal{X}} w\|^2 + \|(D_{\mathcal{X}})^* w\|^2 \geq |||w|||^2, \tag{72}$$

for $w \in \Lambda^q(\mathcal{X})$, for which $Q(w)$ is well defined and which satisfies a suitable boundary condition. We then show that there is a unique solution $w = w_0$ of the equation

$$\Delta_{\mathcal{X}}\, w = b \tag{73}$$

with the above boundary condition. It turns out that $v_0 = (D_{\mathcal{X}})^* w_0$ is the approximate solution of the correction equation.

The standard method originating in J.J. Kohn's work [8] on $\overline{\partial}$-operators can be carried out in our general setting. We assume that $q \geq 1$ and define a $(q-1)$-form w_λ by

$$w_\lambda = \frac{1}{(q-1)!} \sum w_{\lambda\lambda_1 \cdots \lambda_{q-1}} \xi^{\lambda_1} \wedge \cdots \wedge \xi^{\lambda_{q-1}}. \tag{74}$$

Let χ be a smooth function with compact support inside N. Then we find the following (cf. [13]).

(75)Proposition. *Let w be a section of $\Lambda^p(\mathcal{X})$ which is smooth on the support of χ. Then*

$$\|\chi D_{\mathcal{X}} w\|^2 + \|\chi (D_{\mathcal{X}})^* w\|^2 = P_1^\chi + P_2^\chi + P_3^\chi + P_4^\chi + P_5^\chi + P_6^\chi,$$

where

$$P_1^\chi = \frac{1}{p!} \sum_{\lambda,\lambda_1,\ldots,\lambda_p} \|\chi X_\lambda w_{\lambda_1\cdots\lambda_p}\|^2,$$

$$P_2^\chi = \frac{1}{(p+1)!} \sum_{\lambda_0,\ldots,\lambda_p} \|\sum \chi c_{\lambda_0\cdots\lambda_p}^{\sigma_1\cdots\sigma_p} w_{\sigma_1\cdots\sigma_p}\|^2$$

$$+ \frac{1}{p!p} \sum_{\lambda_1,\ldots,\lambda_{p-1}} \|\chi \sum \overline{c_{\sigma_1\cdots\sigma_p}^{\lambda_1\cdots\lambda_{p-1}}} w_{\sigma_1\cdots\sigma_p}\|^2$$

$$+ \frac{2}{p!} \sum \Re\langle \chi^2 w_{\lambda\sigma_1\cdots\sigma_{p-1}}, [X_\lambda, \overline{c_{\lambda_1\cdots\lambda_p}^{\sigma_1\cdots\sigma_{p-1}}}] w_{\lambda_1\cdots\lambda_p}\rangle,$$

$$P_3^\chi = \frac{2}{p!} \sum \Re\langle \chi^2 X_\lambda w_{\lambda_1\cdots\lambda_p}, c_{\lambda\lambda_1\cdots\lambda_p}^{\sigma_1\cdots\sigma_p} w_{\sigma_1\cdots\sigma_p} + c_{\lambda_1\cdots\lambda_p}^{\sigma_1\cdots\sigma_{p-1}} w_{\lambda\sigma_1\cdots\sigma_{p-1}}\rangle,$$

$$P_4^\chi = \sum_{\sigma\lambda} \Re\langle \chi^2 [X_\sigma, X_\lambda^*] w_\lambda, w_\sigma\rangle,$$

$$P_5^\chi = \sum_{\sigma\mu} \Re\langle [X_\sigma, [X_\mu^*, \chi^2]] w_\mu, w_\sigma\rangle,$$

$$P_6^\chi = -2\Re\langle D^* w, [D^*, \chi^2] w\rangle.$$

We expect that $|||w||| \geq C\|w\|$, where C may be as large as we want by choosing sufficiently small N. Hence P_2^χ is a junk term. P_3^χ may be regarded as a junk term, in view of P_1^χ. When we impose the Spencer boundary condition, P_6^χ converges to zero when χ converges to the characteristic functions of N. Therefore we have the following

(76) Proposition. *Under the Spencer boundary condition,*

$$\|\chi D_\mathcal{X} w\|^2 + \|\chi (D_\mathcal{X})^* w\|^2 = P_1^\chi + P_4^\chi + P_5^\chi + \text{junk terms}.$$

Let us consider case I on a strongly pseudoconvex M. In this case $N = M$. Fix a smooth real-valued function $R(z,x)$ such that its Taylor series at the origin starts with a positive definite quadratic term in (z,x). Assume that M is defined by

$$r = r_0 - R(z,x) > 0. \tag{77}$$

Then we find that

$$\lim_\chi P_5^\chi \geq c(\|w\|^\flat)^2 + \text{junk terms}, \tag{78.1}$$

where $\|w\|^b$ is the L_2-norm of the restriction to the boundary. It is very good so far. However, the trouble is the term P_4^χ. Namely, in view of (1)

$$\begin{aligned} P_4^\chi = &- \Re\langle i\chi^2 \mathcal{L}(X_\sigma, X_\lambda)Fw_\lambda, w_\sigma\rangle \\ &+ \Re\langle [\overline{\rho}[X_\sigma, \overline{X_\lambda}], \chi^2]w_\lambda, w_\sigma\rangle \\ &- \Re\langle \chi^2(\rho[X_\sigma, \overline{X_\lambda}] + (\overline{\rho}[X_\sigma, \overline{X_\lambda}])^*)w_\lambda, w_\sigma\rangle \\ &+ \text{junk term}. \end{aligned} \tag{78.2}$$

Since we do not have any information on Fw, all we may hope for is to eliminate the term containing Fw using nonnegative P_1^χ and P_5^χ. The way to do this is to introduce the CR Hessian of a function f by

$$H^f(X, Y) = X\overline{Y}f - (\overline{\rho}[X, \overline{Y}])f. \tag{79}$$

for $X, Y \in L$. We then check easily that $H^f(X, Y)$ is a quadratic form on L. We then find by (78.2) that

$$\begin{aligned} P_4^\chi + P_5^\chi = &- \Re\langle (i\chi^2 \mathcal{L}(X_\sigma, X_\lambda)F + H^{\chi^2}(X_\sigma, X_\lambda))w_\lambda, w_\sigma\rangle \\ &+ \text{junk terms}, \end{aligned} \tag{80}$$

We now apply integration by parts to P_1^χ. We consider the tangential part of X_λ. Namely, we set

$$\gamma_\lambda = X_\lambda r, \quad |\gamma|^2 = \sum |\gamma_\lambda|^2, \quad Q_{\lambda\overline{\sigma}} = \delta_\lambda^\sigma - \frac{1}{|\gamma|^2}\gamma_\lambda\overline{\gamma_\sigma}, \tag{81.1}$$

$$Y = \frac{1}{|\gamma|}\overline{\gamma_\lambda}X_\lambda, \quad W_\lambda = Q_{\lambda\overline{\sigma}}X_\sigma. \tag{81.2}$$

Then we have

$$X_\lambda = \gamma_\lambda Y + W_\lambda, \quad \overline{\gamma_\lambda}W_\lambda = 0, \quad W_\lambda r = 0. \tag{81.3}$$

We then find after some calculation that

$$\sum \|\chi W_\lambda f\|^2 = \sum \|\chi \overline{W_\lambda} f\|^2 + A(f) + B(f) + \text{junk terms}, \tag{82.1}$$

where

$$\begin{aligned} A(f) &= \Re\langle (\chi^2 i\mathcal{L}(W_\lambda, W_\lambda)F + H^{\chi^2}(W_\lambda, W_\lambda))f, f\rangle, \\ B(f) &= -\Re\langle \chi^2([X_\lambda, [\overline{X_\sigma}, Q_{\sigma\overline{\lambda}}]] + \overline{\rho}[X_\lambda, \overline{X_\sigma}]Q_{\sigma\overline{\lambda}})f, f\rangle. \end{aligned}$$

We try to eliminate $P_4^\chi + P_5^\chi$ as expressed in (80) by using $\sum A(u_{\lambda_1 \cdots \lambda_p})$. However, two quadratic forms, $\mathcal{L}$ and H, are too many to handle. Hence we assume that R in (77) is so chosen so that

$$h\mathcal{L} = H^r \tag{82.2}$$

for a scalar valued function h. We choose our χ as a function of r in (77). Then because of the Spencer boundary value problem

$$\begin{aligned}\langle H^{\chi^2}(X_\sigma, X_\lambda)w_\lambda, w_\sigma\rangle =&\langle(\chi^2)' H^r(X_\sigma, X_\lambda)w_\lambda, w_\sigma\rangle \\ &+ \text{junk terms.}\end{aligned} \tag{82.3}$$

Since the generator X_λ of sections of L can be chosen arbitrarily, we may consider the case where they form an orthonormal base with respect to $\mathcal{L}$. Then we find by (80) and (82) that modulo junk terms

$$P_4^\chi + P_5^\chi = -\Re\langle(\chi^2 F + (\chi^2)'h)w_\lambda, w_\lambda\rangle \tag{83.1}$$

and

$$A(f) = (n-2)\Re\langle(\chi^2 F + (\chi^2)'h)f, f\rangle. \tag{83.2}$$

Since they are of the opposite sign when $n > 2$, we can manipulate the formulae to cancel out $P_4^\chi + P_5^\chi$. We end up with the estimate

$$Q(w) \geq c(n-2-q)\left(\sum \|X_\lambda w\|^2 + \sum \|\overline{W_\lambda} w\|^2 + \| |\gamma|^{-1} w\|^2\right). \tag{84}$$

The term $\| |\gamma|^{-1}w\|^2$ comes from $B(f)$ in (82.1). The factor $n-2-q$ comes up because of the presence of $1/q!$ in (65) and w_λ is of degree $q-1$. Therefore to make our idea work, we need condition (82.2). Unfortunately, we could not find M satisfying the condition unless we assumed that our CR manifold is embeddable. However, if we assume the embeddability, there are a lot of them. For example, $M \subset \mathbf{C}^n$ and

$$R = \Re(\text{a holomorphic function}). \tag{85}$$

Note that, since we had to eliminate $P_5^\chi + P_4^\chi$, we no longer have (78.1).

Since we do not have the necessary estimate to solve the correction equation (69) for general M, we consider a submanifold $M_0 \subset M$ with an associated CR structure given by the approximating embedded CR manifold associated to u_0 and satisfying condition (85). We now use $\overline{\partial}_{M_0}$ instead of $\overline{\partial}_M$ in the correction equation so that we can find its approximating solution with estimate. We can now repeat this process. Since, in our case, we obtain better approximating embedding at each stage, we use the latest approximation in the correction equation. We thus construct an approximating embedding u_ν defined on $M_\nu \subset M_{\nu-1}$. We have to take care so that $\cap M_\nu$ contains an open subset. In this way we find an embedding. In our case $q = 1$. Since we need $n-2-q > 0$ in (84), we need $n \geq 4$.

In method II, D_χ is a small deformation of $\overline{\partial}$. Hence we should obtain a better estimate. In fact, P_4^χ is not a problem in this case, because the principal parts of $X_\lambda, \overline{X_\lambda}$ generate the whole tangent vector bundle and we have the boundary estimate (78.1) where the constant c can be chosen independent of r_0 in (77). In handling P_4^χ we have to introduce a boundary integral, but its effect can be made arbitrarily small by choosing a sufficiently small r_0. However, the difficulty in method II comes from the fact that the boundary of the smooth tubular neighborhood $M \times I$, $I = [-1, 0]$ has singularities. Also it is not clear what boundary condition should be put on the boundary outside of M. Assume that M is an open subset of a manifold $M^\sharp$ defined by an inequality $r > 0$ for a function r defined on $M^\sharp$ with the condition $dr \neq 0$. Catlin considers the domain

$$\widetilde{M} = \{(x,t) \in M \times I : \; -\epsilon\sigma^3 r(x)^4 < t < 0\}, \tag{86}$$

where ϵ, σ are positive parameters we choose later to be very small. Thus $\widetilde{M}$ is a lens-like space with a very sharp edge. The regular part of the boundary has two components: $M' = \{t = 0\}$ and $M'' = \{t < 0\}$. It turns out that the boundary condition which works is the standard Spencer boundary condition on M' and the dual Spencer boundary condition: (the symbol of $\overline{\partial}_{L^:})v = 0$ on M''. He also has to use weight in L^2-norm to control the disturbance coming from the edge in the boundary.

As for the weight we use the system developed by Hörmander (cf. [5]), which fits very well with the estimate we need. Namely, when we have a sequence of first order differential operators $D \colon C^\infty(E^p) \to C^\infty(E^{p+1})$ for a sequence of vector bundles E^p, $p = 0, 1, \ldots$, we pick real-valued functions ψ, ϕ and use the weight

$$e^{\phi + p\psi} \quad \text{on } C^\infty(E^p). \tag{87.1}$$

For a weight e^ϕ we set

$$\|v\|_\phi = \|e^{\frac{1}{2}\phi} v\|. \tag{87.2}$$

Going back to case II, Catlin had to use the case:

$$e^\phi = \epsilon^{-(n+2)} r^{-4(n+2)}, \quad \psi = 0 \tag{88}$$

to make the estimate work. He uses a more sophisticated micro-local analysis which allows him to consider the case of a more general Levi-form. Still he needs at least 3 positive eigenvalues in the Levi-form, which we would expect from the condition $n \geq 4$.

For case III we fix a positive definite quadratic form $R(z, x, \zeta)$ and consider the domain

$$M : \; r = r_0 - R(z, x, \zeta) > 0. \tag{89.1}$$

Therefore we work on the manifold $N = M \times Q^+$. In view of (67.3) we see that our case sort of includes $\overline{\partial}$ as well as $\overline{\partial}_b$. We should get a better estimate than in cases I and II. However, the main difficulty in this case comes from the requirement (62). To accommodate this we use the weight as in (87) where

$$e^{\phi} = \left(\frac{1}{|\zeta|}\right)^{m_0}, \quad e^{\psi} = \left(\frac{1}{|\zeta|}\right)^{m_1} \tag{89.2}$$

for a suitable choice of m_1, m_0. The intention is to prove, with the help of the weight,

$$\text{the continuity of } |\zeta|^{-\alpha} v \text{ for an } \alpha > 0 \tag{90}$$

for the solution v of (69) so that $v = 0$ when $\zeta = 0$.

Since our weight has singularity at $\zeta = 0$, we always consider measurable w smooth where $\zeta \neq 0$. We call such w admissible. If w is admissible and X is an admissible differential operator, Xw is defined as an admissible function which coincides with the standard one on the subspace $\zeta \neq 0$. Let w be admissible and

$$\|w\|_{\phi_q}, \ \|D_{\mathcal{X}} w\|_{\phi_{q+1}}, \ \|(D_{\mathcal{X}})^* w\|_{\phi_{q-1}} < \infty, \tag{91.1}$$

where $(D_{\mathcal{X}})^*$ denotes the (naive) adjoint of $D_{\mathcal{X}}\colon \Lambda^{q-1}(\mathcal{X}) \to \Lambda^q(\mathcal{X})$ with the weighted metrics on $\Lambda(\mathcal{X})$. We then set

$$Q(w) = \|D_{\mathcal{X}} w\|^2_{\phi_{q+1}} + \|(D_{\mathcal{X}})^* w\|^2_{\phi_{q-1}}. \tag{91.2}$$

We look for an estimate

$$Q(w) \geq c\|w\|_{\phi_q} \tag{91.3}$$

for w satisfying the Spencer boundary condition. We consider a smooth real-valued function χ with support inside $N \cap \{\zeta \neq 0\}$. We estimate

$$Q_{\chi}(v) = \|\chi D_{\mathcal{X}} v\|^2_{\phi_{q+1}} + \|\chi (D_{\mathcal{X}})^* v\|^2_{\phi_{q-1}}. \tag{91.4}$$

and let χ converge to the characteristic function of N. We can handle the proof of our estimate as in case I. However, because of the presence of the weight, we have to make some modifications in (75): For P_1^{χ} to P_5^{χ} we have to replace "$\rangle$" by "$\rangle_{\phi_{q+1}}$". P_6^{χ} is now

$$\begin{aligned} \widetilde{P}_6^{\chi} = &- 2\Re\langle e^{-\psi} D^* u, \chi^2 [D^*, \psi] u + [D^*, \chi^2] u\rangle_{\phi_{a+1}} \\ &- \Re\langle [D^*, \psi] u, [D^*, \chi^2] u\rangle_{\phi_{a+1}}. \end{aligned} \tag{91.5}$$

We also have a new term

$$\begin{aligned} \widetilde{P}_7^{\chi} = &- \Re\langle \chi^2 [X_\sigma, [\overline{X_\mu}, \phi_{a+1}]] u_\mu, u_\sigma\rangle_{\phi_{a+1}} \\ &+ (\|\chi[D^*, \phi_a] u\|_{\phi_{a+1}})^2 - (\|\chi[D^*, \phi_{a+1}] u\|_{\phi_{a+1}})^2 \\ &+ 2\Re\langle \chi^2 [D^*, \phi_a] u, [D^*, \psi] u\rangle_{\phi_{a+1}}. \end{aligned} \tag{91.6}$$

In view of (69) we consider the case $a = q$. Since $*$ is now with respect to the weighted L^2-norm, we find

$$P_4^X + P_5^X \geq (\|w\|^\flat_{\phi_{q+1}})^2 - (m_1 + 1)(\text{non negative}) + \text{junk term}. \tag{91.7}$$

We are thus forced to assume $m_1 + 1 \leq 0$. Actually we consider the case

$$m_1 = -1, \tag{91.8}$$

which seems to fit in well in our set up. We find also that

$$\begin{aligned} P_7^X &\geq c\widetilde{m}(\sum \|w_t\|^2_{\phi_q} + \sum \|w_{[\bar{t}]}\|^2_{\phi_q}) + \text{junk term}, \\ \widetilde{m} &= m_0 + (q+1)m_1. \end{aligned} \tag{91.9}$$

Unfortunately we do not have the estimate of $\|w_{\bar{k}}\|^2_{\phi_q}$ in the above. Hence we have to go through the calculation for (82.1) with weight and use $B(f)$. Since we need $|\zeta|^{-1}$ instead of $|\gamma|^{-1}$, we also have to modify (81) to accommodate the requirement. When we carry out the calculation along the lines indicated above, we end up with

$$\begin{aligned} Q(w) \geq & c(\sum \|X_\lambda w\|^2_{\phi_{q+1}} + \sum \|\overline{W_\lambda} w\|^2_{\phi_{q+1}} + \sum(\|w\|^\flat_{\phi_{q+1}})^2 \\ & + (n - \widetilde{m} - \delta)\||\zeta|^{-1} w\|_{\phi_{q+1}}). \end{aligned} \tag{92.1}$$

We thus have our estimate and the solution of the correction equation under the assumption $\widetilde{m} \geq 0$, $n - \widetilde{m} > 0$, i.e.

$$n + (q+1) > m_0 \geq q + 1. \tag{92.2}$$

To realize (90) we use m_0 as positive as we are allowed.

6. Construction of the embeddings

We now have to see that when we use the Moser-Nash procedure the approximating solutions converge to a solution. For this we need a good regularity theorem. We discuss here the method of using the a priori estimate. We first recall how the solution of (73) is constructed. Denote by H the completion with respect to the norm defined by $Q(w)$ in (91.2) of the space of admissible w satisfying (91.1) and the Spencer boundary condition. Clearly $D_\mathcal{X} w$, $(D_\mathcal{X})^* w$ are defined for w in H. For simplicity of notation we set

$$Pw = (D_\mathcal{X} w, e^{-\psi}(D_\mathcal{X})^* w) \tag{93.1}$$

so that for $v, w \in H$

$$\begin{aligned} Q(v, w) &= \langle D_\mathcal{X} v, D_\mathcal{X} w\rangle_{\phi_{q+1}} + \langle (D_\mathcal{X})^* v, (D_\mathcal{X})^* w\rangle_{\phi_{q-1}} \\ &= \langle Pv, Pw\rangle_{\phi_{q+1}}. \end{aligned} \tag{93.2}$$

For given b in (73) with finite $\|b\|_{\phi_q}$ the solution $w = v$ of the equation (73) is the unique $v \in H$ such that

$$Q(v,w) = \langle b, w\rangle_{\phi_q} \quad \text{for all } w \in H \tag{94.1}$$

(cf. [9]). Then the existence and uniqueness of the solution of the equation (73) follows when we have the following estimate: For a constant $c > 0$

$$Q(w) \geq c\|w\|^2_{\phi_q} \tag{94.2}$$

for all $w \in H$. And for the solution v we have the estimate

$$\|v\|_{\phi_q} \leq c^{-1}\|b\|_{\phi_q}. \tag{94.3}$$

Now to obtain the regularity of the solution v we apply the following equality: Let $w \in H$ be smooth on $\{\zeta \neq 0\}$. Then for a tangential admissible differential operator V of order 1 (with $V^* = V + R_V$)

$$\begin{aligned} \Re\langle Pw, &PV^*Vw\rangle_{\phi_{q+1}} \\ &= \|PVw\|^2_{\phi_{q+1}} - \|[P,V^*]w\|^2_{\phi_{q+1}} + S(w), \\ S(w) &= \Re\langle Pw, [[P,V^*],V]w\rangle_{\phi_{q+1}} \\ &+ \Re\langle PR_Vw, [P,V^*]w\rangle_{\phi_{q+1}} \\ &- \Re\langle PVw, [P,R_V]w\rangle_{\phi_{q+1}} \\ &+ \Re\langle [V,\phi_{q+1}]Pw, PVw - [P,V^*]w\rangle_{\phi_{q+1}}, \end{aligned} \tag{95.1}$$

provided the terms in the above are well-defined. This we check by calculation. In order that the terms in the above are in a form we can handle, we have to use V which is 0 mod. $|\zeta|$. Apply the above when $w = u$ is the solution of the problem (94.1). It then follows that, when V^*Vu is in H,

$$\Re\langle Pu, PV^*Vu\rangle_{\phi_{a+1}} = \langle Vb, Vu\rangle_{\phi_a} + \langle [V,\phi_a]b, Vu\rangle_{\phi_a}. \tag{95.2}$$

Therefore

$$\begin{aligned} C_\delta\|&(V+[V,\phi_a])b\|^2_{\phi_a} \\ &\geq \|PVu\|^2_{\phi_{a+1}} - \delta\|Vu\|^2_{\phi_a} - \|[P,V^*]u\|_{\phi} + |S(u)|. \end{aligned} \tag{95.3}$$

When $Vu \in H$, we can use our a priori estimate to $\|PVu\|^2_{\phi_{a+1}}$. We see $(c-\delta)\|Vu\|_{\phi_a}$ on the right hand side. Taking $\delta < c$, we get the estimate of $\|Vu\|_{\phi_a}$, provided $S(u)$ and $\|[P,V^*]u\|^2_{\phi_a+1}$ do not cause trouble. $|S(u)|$ is no problem. Therefore this approach works only when $[P,V^*]$ is very small mod. vector fields for which we already have the estimate.

Note that we have the inside estimate of $|\zeta|\overline{X_\lambda}u$ by integration by parts. Note also that we have the estimate of $\|X_\lambda v\|_{\phi_{q+1}}$ and $\|\overline{W_\lambda}v\|_{\phi_{q+1}}$ (cf. (92.1)). Therefore, for V with compact support in N, $[P, V]u$ may be regarded as very small in the above sense. Therefore our problem is to find V as above such that the principal parts of $|\zeta|^{-1}V$, X_λ, $\overline{W_\lambda}$ generate the complex tangent vector bundle.

We first write down vector fields which are generated by the principal part of $X_\lambda, \overline{W_\lambda}$. Note that they, together with the principal part of $\overline{Y}$, generate the tangent bundle (cf. (81.2)). We find by calculation and (67.3) that (cf. (81.1)) the following holds.

(96) Proposition. *$\kappa_\alpha \overline{Y}$ is a linear combination of $X_\lambda, \overline{W_\lambda}$ modulo 0-th order terms and very small terms, $(\alpha = k, n, \overline{k}\ ;\ k = 1, \ldots, n-1)$, where*

$$\kappa_k = \gamma_k + i\overline{\zeta^k}\gamma_{[\overline{0}]},\ \kappa_n = \gamma_n - \gamma_{[\overline{n}]},\ \kappa_{\overline{k}} = \gamma_{\overline{k}} - \gamma_{[\overline{k}]}. \tag{96.1}$$

This is a phenomena which is not present in methods I and II. This means in effect that we may restrict our attention to a small neighborhood of the subset $\{\kappa_\alpha = 0\}$ for the regularity.

We now find that it is enough to find a vector field V (with $\overline{V} = -V$) outside of the span of the principal parts of $|\zeta|X_\lambda, |\zeta|\overline{W_\lambda}$ so that $[P, V]$ is very small (mod κ, X_λ, $\overline{W_\lambda}$). After a long calculation we succeed when our domain is such that $R(z', x, \zeta)$ in (89.1) is very special. Namely, we assume now that

$$R(z', x, \zeta) = ax^2 + |z'|^2 + bx(\zeta^n + \overline{\zeta^n}) + b|\zeta^n|^2 + |\zeta'|^2, \quad a > b > 0 \tag{97}$$

with a very large. In this case the vector field V we need is

$$\begin{aligned} V =& (|\zeta^n|^2 + 2|\zeta'|^2 + g_\alpha \kappa_{\overline{\alpha}}) \frac{1}{|\gamma|}(\overline{L} - L) \\ &+ \overline{p_\lambda} W_\lambda - p_\lambda \overline{W_\lambda}. \end{aligned} \tag{98}$$

with suitable g_α and p_λ.

In order to apply the above V in the way mentioned in (95), we have to apply a mollifier to make it smooth up to the boundary. We also have to modify it by changing the 0-th order term to satisfy the Spencer boundary condition. In this way we obtain the estimate of the first order derivatives of u.

The estimate of the higher order derivatives can be obtained by the similar method and induction. Note that we started with V of homogeneity ≥ 1 in ζ. When we differentiate it we lose one degree of homogeneity.

This suggests we consider

$$N(l,w)^2 = \sum_{k\le l} \left|\left||\zeta|^k Y_{\tau_1}\cdots Y_{\tau_k} w\right|\right|_{\phi_q}. \tag{99.1}$$

We then prove that for the solution v of (94.1)

$$N(l,v) \le cN(l,b). \tag{99.2}$$

It is remarkable that in all of our approach we have to use special domains which are capable of providing the necessary estimates.

References

[1] T. Akahori, *A new approach to the local embedding theorem of CR-structures for* $n \ge 4$, Mem. Amer. Math. Soc. **366**, Amer. Math. Soc. Providence R.I. (1987).

[2] D. Catlin, *Extension of CR structures,* Pro. Symp. Pure Math. **52**, Part 3 (1991) 27–34.

[3] D. Catlin, *Sufficient conditions for the extension of CR structures*, J. Geom. Anal. **4** (1994) 467–538.

[4] S.S. Chern and J. Moser, *Real hypersurfaces in complex manifolds*, Acta Math. **133** (1974) 219–271.

[5] L. Hörmander, *L^2-estimate and extension theorems for $\overline{\partial}$ operator*, Acta Math. **113** (1965) 89–152.

[6] L. Hörmander, *Frobenius Nirenberg theorem*, Ark. for Math. **5** (1964) 425–432.

[7] H. Jacobowitz and F. Treves, *Aberrant CR structures*, Hokkaido Math. Jour. **12** (1983) 276–292.

[8] J.J. Kohn, *Harmonic integrals on strongly pseudo-convex manifolds I and II*, Ann. of Math. **78** (1963) 112–148; **79** (1964) 450–472.

[9] J.J. Kohn and L. Nirenberg, *Non-coercive boundary value problem*, Comm. Pure Appl.Math. **18** (1965) 443–492.

[10] M. Kuranishi, *Strongly pseudoconvex CR structures over small balls I–III*, Ann. of Math. **115** (1982) 451–500; **116** (1982) 1–64, 249–330.

[11] M. Kuranishi, *The frame bundles of CR structures and the Bergman kernel*, Pro. Symp. Pure Math. **52**, Part 3. (1991) 233–237.

[12] M. Kuranishi, *CR geometry and Cartan geometry*, Forum Math **7** (1995) 147–205.

[13] M. Kuranishi, *On a priori estimate on a manifold with singularity on the boundary*, Cont. Math. **205** (1997) 155–168.

[14] M. Kuranishi, *CR structure and tubular complex neighborhoods*, Geometric Complex Analysis. Hayama, Japan. World Scientific (1995) 371–368.

[15] M. Kuranishi, *The complex tubular neighborhoods of CR structures*, §1–§7, Preprint.

[16] Y. Liu, *Cartan geometry, CR geometry and complex geometry,* Thesis., Columbia University (1998).

[17] L. Ma and J. Michel, *Regularity of local embeddings of strictly pseudoconvex CR structures*, J. Reine Angew. Math. **447** (1994) 147–164.

[18] J. Moser, *A new technique for the construction of solutions of nonlinear differential equations*, Proc. Nat. Acad. Sci. **47** (1961) 1824–1831.

[19] A. Newlander and L. Nirenberg, *Complex coordinates in almost-complex manifolds,* Ann. of Math. **65** (1957) 391–404.

[20] L. Nirenberg, *On a problem of Hans Lewy*, Uspeki Math. Nauk. **292** (1974) 241–251.

[21] L. Nirenberg, *A complex Frobenius theorem, Seminars on analytic functions,* Institute for Advanced Study, Princeton (1957) 172–189.

[22] N. Tanaka, *On the pseudoconformal geometry of hypersurfaces of the space of n complex variables,* J. Math. Soc. Japan **14** (1962) 397–420.

[23] S. Webster, *On the proof of Kuranishi's embedding theorem,* Ann. Inst. Henri Poincaré **9** (1989) 183–207.

Department of Mathematics
Columbia University
New York, NY 10027

CHAPTER VIII

$\overline{\partial}$ and $\overline{\partial}_b$ Problems on Nonsmooth Domains

*Joachim Michel and Mei-Chi Shaw**

Introduction

In this survey article we want to describe our method for constructing barriers on weakly pseudoconvex domains with smooth boundaries and give some applications to nonsmooth pseudoconvex domains. We also consider annuli of a very general type which are difference sets of a large piecewise smooth pseudoconvex domain with a union of piecewise smooth pseudoconvex domains in the interior. We do not impose any further conditions on the Levi form. For more details on the proofs, see the papers [24], [27], [28], [29].

In the following we shall use some notions concerning the integral formula method and in particular the Koppelman formula calculus. By a generating form or a barrier we understand a differential form of bidegree $(1,0)$ which will be plugged into integral kernels in order to obtain integral representation formulae. These barriers are very often obtained from the decomposition of a given function $\Phi(z,\zeta)$ of the kind

$$\Phi(z,\zeta)=\sum_{i=1}^{n}P_i(z,\zeta)(z_i-\zeta_i)$$

by

$$w(z,\zeta)=\sum_{i=1}^{n}\frac{P_i(z,\zeta)}{\Phi(z,\zeta)}d\zeta_i.$$

This makes sense for a given domain Ω if one has (for example in the pseudoconvex case) that $\Phi(z,\zeta)\neq 0$ if $\zeta\in b\Omega$ and $z\in\Omega$. The function Φ and its generalizations will be called support functions and the set

$$\{z\in U(\overline{\Omega})|\Phi(z,\zeta)=0\}$$

*Partially supported by NSF grant DMS 98-01091.

a supporting hypersurface in the boundary point ζ. Here U denotes a neighborhood of the domain. In the present article we need to strongly generalize these notions.

There have always been attempts to generalize the Cauchy integral formula in one complex variable to higher dimensions. The generating form of the Cauchy integral formula is

$$w = \frac{d\zeta}{\zeta - z}.$$

Here ζ and z vary in $\mathbb{C}$ with $\zeta \neq z$. w depends holomorphically on both variables. Martinelli and Bochner have given independently a generalization for functions in $\mathbb{C}^n$, where the generating form is

$$w_0(z,\zeta) = \frac{(\overline{\zeta} - \overline{z})d\zeta}{\|\zeta - z\|^2}.$$

This is the first example of a barrier in $\mathbb{C}^n$. Later this was generalized to integral formula representations for $(0,q)$ differential forms in $\mathbb{C}^n$. Obviously w_0 has lost its holomorphic properties when $n \geq 2$. Nevertheless the corresponding integral formula has some applications to the $\overline{\partial}$-problem, especially when one wants to solve the $\overline{\partial}$-equation for compactly supported differential forms.

In the case of a strictly pseudoconvex domain, a breakthrough was given by Henkin and Ramirez when they constructed new barriers to which the Koppelman formulae were applied. The corresponding integral representations were studied independently by Henkin and Lieb in order to prove uniform and Hölder estimates for the $\overline{\partial}$-equation.

Let $\Omega \subset\subset \mathbb{C}^n$ be a strictly pseudoconvex domain, $U = U(b\Omega)$ a neighborhood of the boundary and $r \in C^\infty(U)$ a real-valued function such that $\Omega \cap U = \{z \in U | r(z) < 0\}$ and $dr \neq 0$ if $z \in b\Omega$. In the following we shall mainly consider the regularity class C^∞. Henkin and Ramirez have constructed a function $\Phi(z,\zeta)$ of class C^∞, where ζ varies in a neighborhood $U' \subset U$ of $b\Omega$ and z in $U' \cup \Omega$ with the following properties.

(i) $\Phi(\cdot,\zeta)$ is holomorphic with respect to z;

(ii) there exists a positive constant c such that

$$|\Phi(z,\zeta)| \geq |\mathrm{Im}\Phi(z,\zeta)| + c|\zeta - z|^2;$$

(iii) there exist C^∞-smooth functions $P_1, P_2, \ldots, P_n$, which are holomorphic with respect to the z-variable, such that

$$\Phi(z,\zeta) = \sum_{i=1}^{n} P_i(z,\zeta)(\zeta_i - z_i);$$

(iv) locally near each boundary point z, $r(\zeta)$ and $\mathrm{Im}\Phi(z,\zeta)$ can serve as a part of a coordinate system. The same is true if the role of z and ζ is interchanged.

In particular these properties show that for $\zeta_0 \in b\Omega$

$$\{z \in \Omega \cup U' | \mathrm{Re}\Phi(z,\zeta_0) = 0\}$$

is a hypersurface with pluriharmonic defining function, which touches the boundary in ζ_0 of second order from the exterior (here U' denotes a sufficiently small neighborbood of ζ_0).

Because of the good estimates from below and (iv), one can show that Φ gives rise to a continuous linear integral operator T_q solving the $\bar\partial$-equation which maps the space of bounded $(0,q)$ forms into the space of $\frac{1}{2}$-Hölder continuous $(0,q-1)$ forms.

The essential barrier to be inserted into the integral kernels of T_q is

$$w(z,\zeta) = \frac{P(z,\zeta)d\zeta}{\Phi(z,\zeta)}.$$

One can find more details of these classical constructions in [31]. One can easily find a support function for convex domains. Let r be a convex defining function. Set

$$\Phi(z,\zeta) = \sum_{i=1}^{n} \frac{\partial r}{\partial z_i}(\zeta)(\zeta_i - z_i).$$

Then $\{z \in U \cup \Omega | \mathrm{Re}\Phi(z,\zeta_0) = 0\} \cap \Omega = \emptyset$ if $\zeta_0 \in b\Omega$. It is easy to construct analogous solution operators T_q. But in general T_q will not allow good estimates and the solutions will not be regular up to the boundary. It is still open if on such domains there exist solution operators with uniform estimates. For convex domains of finite type, there is a method to overcome the problems recently found by Cumenge [5].

The above given barrier for convex domains, with $w_i = \frac{\partial r}{\partial z_i}(\zeta)/\Phi$, allows the estimate

$$|w(z,\zeta)| \leq C\delta(\zeta)^{-1},$$

where δ is the boundary distance function. The constant C is independent of $z \in \Omega$ and $\zeta \in U \setminus \overline{\Omega}$. Similar estimates with higher powers of δ^{-1} hold for the derivatives of w. The next generalization would be the construction of support functions on weakly pseudoconvex domains. But in [18] it was shown that there exist weakly pseudoconvex domains with real-analytic boundary, which do not admit supporting complex hypersurfaces. A method to overcome this problem was hidden in papers of

Lieb/Range [19] and Aizenberg/Dautov [1] on strictly pseudoconvex domains. The above authors gave modified solution operators $\tilde{T}_q$, where the terms containing the Henkin and Ramirez barrier w are of the form

$$\int_S \overline{\partial} Ef(\zeta) \wedge w_0 \wedge w \wedge (\overline{\partial}_\zeta w_0)^i \wedge (\overline{\partial}_\zeta w)^j \wedge (\overline{\partial}_z w_0)^{q-1}.$$

Here $S = \Omega_0 \setminus \overline{\Omega}$, with a larger domain $\overline{\Omega} \subset \Omega_0$. E denotes a linear extension operator which respects C^k-norms, with $\operatorname{supp} Ef \subset \Omega_0$.

With these integral operators it was possible to show C^k-estimates for $\tilde{T}_q$. Later on this was generalized in [23] and [25] to piecewise strictly pseudoconvex domains. In all these cases the main advantage of $\tilde{T}_q$ is given by the presence of the term $\overline{\partial} Ef$. Let f be a C^k-smooth $\overline{\partial}$-closed $(0,q)$ form on $\overline{\Omega}$. Then $\overline{\partial} Ef$ vanishes of order $k-1$ on Ω and fulfills the inequality

$$|\overline{\partial} Ef| \leq C\delta(\zeta)^{k-1}.$$

So if we could construct similar barriers in more general situations and if k were very large, then w and its derivatives could be allowed to be bounded by high powers of $\delta(\zeta)^{-1}$. This would give rise to solution operators with C^k-estimates but with a considerable loss of regularity. Chaumat/Chollet [3], [4] have used these ideas in order to obtain regularity results for so-called H-s-convex domains. Here the loss of regularity is still connected with geometrical properties of the domains.

Now let Ω be a weakly pseudoconvex domain with smooth boundary. The first author has constructed a barrier [24]

$$w(z,\zeta) = \sum_{i=1}^{n} w_i(z,\zeta) d\zeta_i,$$

which is holomorphic in the z-variable with $1 = \sum_{i=1}^n w_i(z,\zeta)(\zeta_i - z_i)$. The coefficients w_i are obtained by solving the decomposition problem for 1, which is indicated by the last formula. For any fixed $\zeta \in \mathbb{C} \setminus \overline{\Omega}$, $w_i(\cdot,\zeta)$ is smooth up to the boundary with respect to $z \in \overline{\Omega}$. The blowing-up, when ζ approaches the boundary, is controlled by powers of $\delta(\zeta)^{-1}$.

Therefore this barrier gives rise to solution operators which map $C^\infty_{(0,q)}(\overline{\Omega}) \cap \ker(\overline{\partial})$ into $C^\infty_{(0,q-1)}(\overline{\Omega})$. Once the barriers are constructed in the smooth case one can easily give integral formula representations and solution operators on real transversal intersections of such domains. This problem is hardly accessible by using the weighted Neumann problem as in [15]. However, we have to stress here that the construction of the barrier was done by merging the ideas of Skoda [40] and Kohn [15] and by introducing the so-called $\mathcal{L}$-complex. The $\mathcal{L}$-complex generalizes the Dolbeault complex to vector valued forms. Here we replace the complex Laplacian by a self-adjoint operator

$$\Lambda = \mathcal{L}_0\mathcal{L}_0^* + \mathcal{L}_1^*\mathcal{L}_1.$$

The definition of the $\mathcal{L}$-complex

$$\mathcal{L}_q \colon W^q \longrightarrow W^{q+1}$$

for $0 \leq q \leq n$, is given in sections 1 and 2. Here $\mathcal{L}_q$ denotes a vector valued first order differential operator where the principal part is $\bar\partial$. W^q are Hilbert spaces. Because of $\mathcal{L}_{q+1} \circ \mathcal{L}_q = 0$ we can use the analogy with the Dolbeault complex. By treating $\mathcal{N} = \Lambda^{-1}$ in an analogous way to the $\bar\partial$-Neumann operator we can look for "canonical" solutions $\mathcal{L}_0^*\mathcal{N}\alpha$ of the equation

$$\mathcal{L}_0\beta = \alpha,$$

if $\mathcal{L}_1\alpha = 0$. The space W^1 is constructed in such a way that the $n+1$-tuple $\alpha = (0, \ldots, 0, 1)$ is an element. Then $\beta = \mathcal{L}_0^*\mathcal{N}\alpha = (\beta_1, \ldots, \beta_n)$ gives a solution of the decomposition problem

$$1 = \sum_{i=1}^{n} \beta_i(z, \zeta)(\zeta_i - z_i)$$

where the coefficients are holomorphic with respect to z. In this approach much advantage has been taken from the analogy between the complex Laplacian and Λ. Also in the $\mathcal{L}$-complex there are built-in weight functions coming from the papers of Kohn and Skoda.

Now let Ω be a strictly **pseudoconcave** domain. That means that

$$\Omega = \Omega_1 \setminus \overline{\Omega}_2,$$

where $\Omega_2 \subset\subset \Omega_1$ are strictly pseudoconvex domains. In this situation Ω_2 has to be treated as the exterior of the annulus Ω. For $b\Omega_1$ we can take the support function Φ_1 of Henkin and Ramirez. But for $b\Omega_2$ we have to exchange the roles played by z and ζ, and $\Phi^*(z, \zeta) = \Phi(\zeta, z)$ is now an appropriate function for a barrier.

This is no longer the case if Ω_2 is merely a weakly pseudoconvex domain. Let w_+ be the barrier of Ω_2 as it was constructed in [24]. The corresponding barrier w_+^* with z and ζ interchanged is C^∞-smooth with respect to $\zeta \in \overline{\Omega}_2$. On the other hand it will blow up if $z \in \Omega$ approaches $b\Omega_2$. In the corresponding solution operator the barrier w_+^* and its derivatives are multiplied by $\bar\partial Ef(\zeta)$ and the domain of integration with respect to ζ is Ω_2. So regularity will meet regularity and there is no need for compensation. On the other hand, the resulting solutions will blow up at $b\Omega_2$.

Therefore one has to solve a completely different division problem in the weakly pseudoconcave case. From the above considerations it follows that one needs a barrier $w(z,\zeta)$, which is, before interchanging z and ζ, holomorphic with respect to $z \in \Omega_2$, smooth for $\zeta \in \overline{\Omega}_1 \setminus \Omega_2$ and which blows up in a controlled way if z approaches $b\Omega_2$. The latter assertion means that w and its derivatives should be bounded from above by powers of $\delta_2(z)^{-1}$. We shall denote this latter barrier, after having interchanged ζ and z, by w_-.

Thus for a weakly pseudoconvex smooth boundary we should deal with two kinds of barriers: w_+ and w_-. w_- is holomorphic with respect to ζ, which is varying in the pseudoconvex domain. It is regular up to the boundary with respect to z, which is varying in the exterior of the pseudoconvex domain, and blows up if ζ approaches the boundary.

The construction of w_- is explained in more detail in section 2 and completely in [27]. It is again based on the $\mathcal{L}$-complex, where an additional plurisubharmonic weight function is added. This function comes from a special bounded plurisubharmonic exhaustion function and was given by Diederich–Fornaess in [7].

Once the two barriers of a weakly pseudoconvex domain are at our disposal, we can solve numerous problems by applying the integral formula method. In particular we can give smooth solutions for the $\overline{\partial}$-equation on annuli

$$\Omega = \Omega_1 \setminus \overline{\Omega}_2,$$

with $\Omega_2 \subset\subset \Omega_1$, where Ω_1 is the intersection of finitely many pseudoconvex domains and Ω_2 is a finite union of intersections of pseudoconvex domains. Details are given in [29].

Rosay [33] was the first to construct smooth solutions for the $\overline{\partial}_b$-equation on the boundary of a weakly pseudoconvex domain. Henkin considered the local problem on strictly pseudoconvex boundaries in [12]. Let Ω be a strictly pseudoconvex domain and h a holomorphic function on an open set U, which intersects the boundary of Ω. Set $M = \{z \in b\Omega | \text{Re}h(z) \geq 0\}$. If M is a smooth and compact manifold, then one can solve

$$\overline{\partial}_b u = f$$

for a $\overline{\partial}_b$-closed $(0,q)$ form f on M, if $1 \leq q \leq n-3$. For $q = n-2$, Henkin needs an additional condition on polynomial convexity, which is always satisfied if M is sufficiently small.

We know already how to solve the $\overline{\partial}$-equation on a weakly pseudoconcave annulus. The global problem for $\overline{\partial}_b$ on $M = b\Omega$ could be looked at as a limit case for $\overline{\partial}$, when the two boundary pieces of the annulus converge to M. This indicates why in the $\overline{\partial}_b$ problem for strictly pseudoconvex boundaries the support function of Henkin and Ramirez $\Phi(z, \zeta)$ appears together with $\Phi(\zeta, z)$, as in the annulus case. Analytically the above described shrinking process is represented by the Bochner–Martinelli jump formula.

In [28] we generalize this to the case where $b\Omega$ is a smooth weakly pseudoconvex boundary and M is given as above by the real part of a holomorphic function. By extending the $\overline{\partial}_b$-closed $(0, q)$ form f, $1 \leq q \leq n-2$, in an appropriate way to the ambient space, we can show the solvability of the $\overline{\partial}_b$-equation with a solution which is C^∞-smooth up to the boundary.

We mention here that we could get rid of the condition on polynomial convexity for $q = n-2$. This is new even in the strictly pseudoconvex case. The proof relies on the construction of an Oka map and on the existence of solution operators on intersections of pseudoconvex domains.

The solution operators on M are given by the barriers w_+ and w_-. The integration of critical kernels containing w_+ and w_- takes place on certain $2n$-dimensional wedges. The compensation of the blowing-up of the kernels is due to the presence of the factor $\overline{\partial} Ef$, which vanishes of high order. Here Ef denotes an extension of f to the ambient space.

One even can show homotopy formulae for nonclosed $(0, q)$ forms f, $q \leq n-3$,

$$f = \overline{\partial}_b T_q(f) + T_{q+1}(\overline{\partial}_b f).$$

More details are given in section 3. Rosay has shown that there are no such formulas possible for $q = n-2$. By using homotopy formulae, Webster has given in [44], [45] a considerable simplification and improvement of the proof of the Kuranishi embedding theorem for abstract strictly pseudoconvex Cauchy–Riemann structures. See also [20] and [21] in this context where his results have been improved. One of our motivation for studying more general situations is the hope of obtaining embedding results for weakly pseudoconvex structures in the future.

Up to now a direct solution operator for $\overline{\partial}_b$ does not exist if one replaces the hypersurface $\{\mathrm{Re} h = 0\}$ by a general Levi-flat one. We have treated this problem in [29] in all details and in section 4 with some indications by solving several $\overline{\partial}$-equations on auxiliary pseudoconvex-concave domains. Here we show the solvability of the $\overline{\partial}_b$-equation in the C^∞ class up to the boundary of an open submanifold M of a weakly pseudoconvex boundary. But now the boundary of M is given by k Levi flat hypersurfaces and f

is a $(0,q)$ form, with $1 \leq q \leq n-k-2$.

A counterexample shows that in general one cannot solve for $q \geq n-1-k$. So even for $k=1$ the situation, where M is given by $\{\mathrm{Re} h = 0\}$ with a holomorphic function h, is much more special than the general Levi-flat case.

1. The first barrier

The content of this section has been published in [24]. We describe it here for the convenience of the reader. Let Ω be a bounded pseudoconvex domain in $\mathbb{C}^n$ with C^∞-smooth boundary and φ a plurisubharmonic function on Ω. We denote by $L^2_{p,q}(\Omega,\varphi)$ the weighted L^2-space in the sense of Hörmander.

Let $\zeta \notin \overline{\Omega}$ be a fixed point and $t>0$ be a fixed real parameter. Set for $z \in \Omega$

$$\varphi_1(z,\zeta,t) = (1+t)(n-1)\log\|\zeta - z\|^2,$$

$$\varphi_2(z,\zeta,t) = \varphi_1(z,\zeta,t) + \log\|\zeta - z\|^2.$$

Let

$$\overline{\partial}\colon L^2_{p,q}(\Omega,\varphi_i) \to L^2_{p,q+1}(\Omega,\varphi_i),$$

for $i=1,2$, be the densely defined $\overline{\partial}$-operator and

$$\overline{\partial}^*\colon L^2_{p,q+1}(\Omega,\varphi_i) \to L^2_{p,q}(\Omega,\varphi_i)$$

its adjoint. The domains of definition do not depend on ζ, z, t.

We set

$$H_1^q = L^2_{0,q}(\Omega,\varphi_1)^n,$$

$$H_2^q = L^2_{0,q}(\Omega,\varphi_2),$$

$$W^q = H_1^q \times H_2^{q-1}.$$

$$T\colon \begin{cases} H_1^q \to H_2^q \\ (a_1,\dots,a_n) \mapsto \sqrt{1+t}\sum_{i=1}^n(\zeta_i - z_i)a_i, \end{cases}$$

$$\mathcal{L}_q\colon \begin{cases} W^q \to W^{q+1} \\ (a,b) \mapsto (\overline{\partial}a, Ta - \overline{\partial}b), \end{cases}$$

where $a=(a_1,\dots,a_n) \in H_1^q$, $b \in H_2^{q-1}$ and $\overline{\partial}a = (\overline{\partial}a_1,\dots,\overline{\partial}a_n)$. $\mathcal{L}_q$ is a densely defined closed operator. These operators define what we shall call the $\mathcal{L}$-complex. We have

$$\mathcal{L}_{q+1} \circ \mathcal{L}_q = 0.$$

The Hilbert space adjoint of $\mathcal{L}_q$ is given by

$$\mathcal{L}_q^*: \begin{cases} W^{q+1} \to W^q \\ (a,b) \mapsto (\overline{\partial}^* a + T^* b, \overline{\partial}^* b). \end{cases}$$

The domains of definition are again independent of the parameters. Therefore the $\mathcal{L}$-complex has similar properties to the weighted $\overline{\partial}$-complex of Kohn [15]. Finally we set

$$\Lambda = \mathcal{L}_0 \mathcal{L}_0^* + \mathcal{L}_1^* \mathcal{L}_1,$$

with

$$\operatorname{dom} \Lambda = \{\Psi \in \operatorname{dom}(\mathcal{L}_1) \cap \operatorname{dom}(\mathcal{L}_0^*) | \mathcal{L}_1 \Psi \in \operatorname{dom}(\mathcal{L}_1^*), \mathcal{L}_0^* \Psi \in \operatorname{dom}(\mathcal{L}_0)\}.$$

The weight functions are chosen in order to give basic estimates for Λ. In particular it follows from them that Λ is bijective from its domain of definition onto W^1. We denote the inverse operator by $\mathcal{N}_{\zeta,t} = \mathcal{N}$ in analogy to the Neumann operator. $\mathcal{N}$ will be bounded and self-adjoint.

The main idea of the above construction is the following. Let $\alpha = (0, 0, \ldots, 0, \sqrt{1+t}) \in W^1$. Obviously $\mathcal{L}_1 \alpha = 0$. Then $\mathcal{L}_0^* \mathcal{N}_{\zeta,t} \alpha = (w_1, w_2, \ldots, w_n, 0) = \beta \in W^0$ solves

$$\mathcal{L}_0 \beta = \alpha.$$

This means that $\overline{\partial}_z (w_1, w_2, \ldots, w_n) = 0$ and

$$\sum_{i=1}^{n} w_i (\zeta_i - z_i) = 1.$$

Proceeding like in [15] we can give Sobolev estimates with respect to z. The problem is to control the dependence on ζ, because in the integral formulas at least first derivatives with respect to ζ occur.

Now we shall roughly describe how this has to be done. Firstly, by combining the estimates of Skoda [40] and Kohn [15] one obtains the basic estimate

$$\mathcal{Q}(\Phi, \Phi) \geq \|\Phi\|^2,$$

for $\Phi \in \operatorname{dom} \mathcal{L}_1 \cap \operatorname{dom} \mathcal{L}_0^*$. Here $\mathcal{Q}$ denotes the quadratic form

$$\mathcal{Q}(\Phi, \Psi) = (\mathcal{L}_0^* \Phi, \mathcal{L}_0^* \Psi) + (\mathcal{L}_1 \Phi, \mathcal{L}_1 \Psi),$$

for $\Phi, \Psi \in \operatorname{dom} \mathcal{L}_1 \cap \operatorname{dom} \mathcal{L}_0^*$. If we denote s-Sobolev norms with respect to z by $\| \,.\, \|_s$ we obtain, by imitating Kohn's proof, the following theorem.

Here U denotes a fixed neighborhood of $\overline{\Omega}$ and δ the boundary distance function.

Theorem 1. *There exists an increasing sequence $0 < K_0 < K_1 < \cdots$ of real numbers, such that for every smooth mapping $\alpha\colon U \setminus \overline{\Omega} \to W^1$ and every nonnegative integer s we have the estimates*

$$\|\mathcal{N}_{\zeta,t}\alpha(\cdot,\zeta)\|_s \le \frac{c(t,s)}{\delta(\zeta)^{2s}}\|\alpha(\cdot,\zeta)\|_s.$$

For $\mathcal{L}_0^*\mathcal{N}_{\zeta,z}\alpha$ similar estimates hold.

In order to control the dependence on ζ one has to make use of the fact that our solution of the decomposition problem is given by the canonical solution of a Neumann-like problem. Thus by varying ζ one can eventually control the necessary derivatives with respect to ζ. We do not give the details here.

From these considerations one obtains the following.

Theorem 2. *For each positive integer r there exists an increasing sequence $0 < t_0 < t_1 < \cdots$ of real numbers and a C^∞-smooth map $w\colon \overline{\Omega} \times (U \setminus \overline{\Omega}) \to \mathbb{C}^n$, such that*

$$\overline{\partial}_z w(z,\zeta) = 0, \qquad \sum_{i=1}^{n} w_i(z,\zeta_i)(\zeta_i - z_i) = 1.$$

For any differentiation D_ζ^a of order $a \le r$ with respect to ζ and for all s there exists a constant $c(s,r)$ with

$$\|D_\zeta^a w(\cdot,\zeta)\| \le \frac{c(s,r)}{\delta(\zeta)^{t_s}}.$$

To give an idea how the integral formula method works we will shortly describe the construction of the homotopy formula for a smooth weakly pseudoconvex domain. Let

$$w_0 = \frac{\partial_\zeta \|\zeta - z\|^2}{\|\zeta - z\|^2}$$

be the Martinelli–Bochner barrier and let $E\colon C^0(\overline{\Omega}) \to C_c^0(\overline{\Omega})$ be a Seeley extension operator (cf. [19], [35]). Let Δ_{01} be the standard simplex in $\mathbb{R}^2$ and Δ_0, Δ_1 its corners. Set $R = U \setminus \overline{\Omega}$. We define a form depending on parameters $(\lambda_0, \lambda_1) \in \Delta_{01}$ by

$$\eta = \lambda_0 w_0 + \lambda_1 w.$$

Here we identified w with the form $\sum w_i d\zeta_i$. For $q = 0, 1, \ldots, n$ set, with a certain constant $c_{n,q}$,

$$\mathcal{D}_{n,q}(z, \zeta, \lambda) = c_{n,q}\eta \wedge ((\overline{\partial}_\zeta + d_\lambda)\eta)^{n-q-1} \wedge (\overline{\partial}_z \eta)^q.$$

Let $f \in C^\infty_{0,q}(\overline{\Omega})$. We set

$$R_q(f) = \int_{R\times\Delta_1} (E\overline{\partial} f - \overline{\partial} E f) \wedge \mathcal{D}_{n,q}(z, \zeta, \lambda),$$

$$T_q(f) = \int_{R\times\Delta_{01}} (E\overline{\partial} f - \overline{\partial} E f) \wedge \mathcal{D}_{n,q-1}(z, \zeta, \lambda) - \int_{U\times\Delta_0} Ef \wedge \mathcal{D}_{n,q-1}(z, \zeta, \lambda).$$

The last term is the Martinelli–Bochner integral. $E\overline{\partial} f - \overline{\partial} E$ vanishes of infinitely high order on Ω, so that all integrals are well defined. The holomorphy of w in z implies the vanishing of R_q if $q > 0$.

An easy calculation by using the well-known Koppelman formula

$$(\overline{\partial}_\zeta + d_\lambda)\mathcal{D}_{n,q} = (-1)^q \overline{\partial}_z \mathcal{D}_{n,q-1}$$

gives the following.

Theorem 3. *Let Ω be a pseudoconvex domain with smooth boundary. Then there exist for $q = 0, 1, \ldots, n$ linear integral operators R_q, T_q, with $R_q = 0$ if $q > 0$, $R_0\colon C^\infty(\overline{\Omega}) \to C^\infty(\overline{\Omega}) \cap \mathcal{O}(\Omega)$, $T_q\colon C^\infty_{0,q}(\overline{\Omega}) \to C^\infty_{0,q-1}(\overline{\Omega})$, such that for $f \in C^\infty_{0,q}(\overline{\Omega})$ the following homotopy formula holds:*

$$f = R_q(f) + \overline{\partial} T_q(f) + T_{q+1}(\overline{\partial} f).$$

An analogous formula can be shown for the transversal intersection of finitely many such domains. If f is $\overline{\partial}$-closed, then $T_q(f)$ yields a smooth solution of the $\overline{\partial}$-equation. If f depends smoothly on some parameters, then $T_q(f)$ does the same.

2. The second barrier

The detailed proofs of this section are described in [27]. Let $\Omega \subset \mathbb{C}^n$ be a bounded pseudoconvex domain with defining function $\tilde{\rho}$ of class C^K, $2 \leq K \leq \infty$. Let $\nu \geq 1$ be a real number such that $\rho = -(-\tilde{\rho})^{1/\nu}$ is a strictly plurisubharmonic exhaustion function on Ω (cf. [7]). For a bounded open set $\Omega^0 \supset \overline{\Omega}$, we obtain for $S = \overline{\Omega}^0 \setminus \Omega$ the following theorem.

Theorem 4. *Let k be a nonnegative integer and $t_k = [2\nu \max(4 + 3k, \frac{n-1}{5}) + 1]$. Then there exists a C^k map $w^k = (w^k_1, \ldots, w^k_n)\colon \Omega \times S \to \mathbb{C}^n$ with the following properties:*

(i) *$w^k(\cdot,\zeta)$ is holomorphic for all $\zeta \in S$ and for $(z,\zeta) \in \Omega \times S$ one has*

$$sum_{i=1}^n w_i^k(z,\zeta)(\zeta_i - z_i) = 1;$$

(ii) *there exists a constant $C(k)$ such that for all I, with $|I| \leq k$, and for all $\zeta \in S$ and every $z \in \Omega$*

$$|D_\zeta^I w^k(z,\zeta)| \leq C(k)\delta(z)^{-[(5t_k^2+3|I|+1)/\nu+2n]}.$$

($[a]$ denotes the largest integer $j \leq a$ and D_ζ^I a derivative with respect to ζ of order $|I|$.)

Remark. Almost in every step of the proof of Theorem 4 one could slightly improve the regularity of the barrier. But it seems to us that in our approach in order to have a C^k-smooth barrier function w^k, one always has to compensate with a growth of $w(\cdot,\zeta)$ of asymptotic order $k^2\nu$. But compare the preprint of Thilliez [42] where a different method is given. In [27] we applied w^k to the $\overline{\partial}$ problem on a weakly pseudoconcave annulus. More applications will be given in section 4.

In order to prove Theorem 4 we use the $\mathcal{L}$-complex defined in [24] for a C^∞-smooth boundary, but now the weight functions have to be modified. In [24] we constructed a barrier function $w(z,\zeta)$ where $w(\cdot,\zeta)$ is holomorphic and smooth up to the boundary for $\zeta \notin \overline{\Omega}$ and where C^k norms of $w(\cdot,\zeta)$ on $\overline{\Omega}$ blow up to some order if ζ approaches the boundary. For solving interesting $\overline{\partial}$-problems on annuli or for $\overline{\partial}_b$, one needs a barrier function which is smooth up to the boundary in the ζ-variable and where one has to compensate for this with a certain growth with respect to the variable $z \in \Omega$. The proof in [24] was modeled in some aspects after Kohn's original proof in [15]. In the present section, however, we were forced to apply a completely different method. One step in the proof was inspired by a paper of Berndtsson [2].

Let us sketch some details of the proof of Theorem 4. Let $\alpha = (\alpha_1,\ldots,\alpha_n) \in \mathbb{C}^n$. Then we set $\langle\partial\rho,\alpha\rangle = \sum_{i=1}^n \dfrac{\partial\rho}{\partial z_i}\cdot\alpha_i$. For a function φ we denote by $\mathrm{Lev}(\varphi)$ its complex Hessian, i.e., its Levi form. We set $\psi := -\log(-\rho)$. Since

$$\mathrm{Lev}(\psi)[\alpha] = \frac{1}{-\rho}\,\mathrm{Lev}(\rho)[\alpha] + \frac{|\langle\partial\rho,\alpha\rangle|^2}{\rho^2}$$

ψ is strictly plurisubharmonic on Ω. Let $t > 0$, $\tau \geq 0$ and $\zeta \in \mathbb{C}^n \setminus \Omega$ be parameters. We define for $z \in \Omega$

$$\varphi_1(z,\zeta,t,\tau) := (1+t)(n-1)\log|\zeta - z|^2 + t|z|^2 + (5t^2+\tau)\psi(z),$$
$$\varphi_2(z,\zeta,t,\tau) := \varphi_1(z,\zeta,t,\tau) + \log|\zeta - z|^2.$$

In order to simplify the notations, we shall often drop z, ζ and t. However, if we drop τ we always mean the corresponding term for $\tau = 0$. For $\epsilon > 0$ we set $\Omega_\epsilon = \{z \in \Omega |\ \tilde{\rho}(z) < -\epsilon\}$. So for ϵ sufficiently small, Ω_ϵ is a strictly pseudoconvex domain of class C^K.

For $i = 1, 2$ and $\epsilon > 0$ let $L^2(\Omega_\epsilon, \varphi_i)$ denote the Hilbert space of functions on Ω_ϵ, which are square integrable with respect to the measure $e^{-\varphi_i} dV$ (dV denotes the Lebesque measure on $\mathbb{C}^n$). We denote the spaces of differential forms of bidegree (p, q) by $L^2_{p,q}(\Omega_\epsilon, \varphi_i)$, $p, q \in \mathbb{N}_0$. The respective scalar products are denoted by $(f, g)_{i,\epsilon,t,\tau}$.

Now we insert these weight functions into the $\mathcal{L}$-complex. For the first barrier we supposed that $\zeta \notin \overline{\Omega}$. This is no longer possible for the second barrier. Here we have to study the function spaces over Ω_ϵ for ϵ positive.

$$H^{q,\epsilon}_{1,\tau} = [L^2_{0,q}(\Omega_\epsilon, \varphi_1)]^n, \quad H^{q,\epsilon}_{2,\tau} = L^2_{0,q}(\Omega_\epsilon, \varphi_2), \quad W^{q,\epsilon}_\tau = H^{q,\epsilon}_{1,\tau} \times H^{q-1,\epsilon}_{2,\tau},$$

with $H^{-1,\epsilon}_{i,\tau} = \{0\}$, $q \geq 0$, $\zeta \in \mathbb{C}^n \setminus \Omega$, $t > 0$, $\tau \geq 0$, $\epsilon > 0$.

Let $\overline{\partial}\colon L^2_{0,q}(\Omega_\epsilon, \varphi_i) \to L^2_{0,q+1}(\Omega_\epsilon, \varphi_i)$ be the maximal weak extension of the $\overline{\partial}$ operator. Note that $\operatorname{dom}(\overline{\partial})$, the domain of definition of $\overline{\partial}$, does not depend on i, ζ, t and τ. Moreover we set

$$T\colon \begin{cases} H^{q,\epsilon}_{1,\tau} \to H^{q,\epsilon}_{2,\tau} \\ (a_1, \ldots, a_n) \mapsto \sqrt{1+t}\sum_{i=1}^n (\zeta_i - z_i) a_i, \end{cases}$$

$$\mathcal{L}^\epsilon_{q,\tau}\colon \begin{cases} W^{q,\epsilon}_\tau \to W^{q+1,\epsilon}_\tau \\ (a, b) \mapsto (\overline{\partial} a, Ta - \overline{\partial} b), \end{cases}$$

with $\overline{\partial} a = (\overline{\partial} a_1, \ldots, \overline{\partial} a_n)$, $a \in H^{q,\epsilon}_{1,\tau} \cap [\operatorname{dom}(\overline{\partial})]^n$, $b \in H^{q-1,\epsilon}_{2,\tau} \cap \operatorname{dom}(\overline{\partial})$.

Denote by $\overline{\partial}^{*,\epsilon}_\tau$ the adjoint of $\overline{\partial}$ in $L^2_{0,q}(\Omega_\epsilon, \varphi_i)$ and by $\mathcal{L}^{*,\epsilon}_{q,\tau}$ the adjoint of $\mathcal{L}^\epsilon_{q,\tau}$. For $\tau = 0$ we shall drop τ.

$$\mathcal{L}^{*,\epsilon}_{q,\tau}\colon \begin{cases} W^{q+1,\epsilon}_\tau \to W^{q,\epsilon}_\tau \\ (a, b) \mapsto (\overline{\partial}^*_\tau a + T^* b, -\overline{\partial}^{*,\epsilon}_\tau b). \end{cases}$$

Let $\Phi = (a, b)$, $\Psi = (a', b')$ be elements of $W^{q,\epsilon}_\tau$. Then we denote by

$$(\Phi, \Psi)_{\epsilon,t,\tau} = \sum_{i=1}^n (a_i, a'_i)_{1,\epsilon,t,\tau} + (b, b')_{2,\epsilon,t,\tau}$$

the scalar product and by $\|\Phi\|_{\epsilon,t,\tau}$ the norm of Φ. For $\Phi, \Psi \in \operatorname{dom}(\mathcal{L}^\epsilon_{1,\tau}) \cap \operatorname{dom}(\mathcal{L}^{*,\epsilon}_{0,\tau})$ we set

$$\mathcal{Q}^\epsilon_\tau(\Phi, \Psi) = (\mathcal{L}^{*,\epsilon}_{0,\tau}\Phi, \mathcal{L}^{*,\epsilon}_{0,\tau}\Psi)_{\epsilon,t,\tau} + (\mathcal{L}^\epsilon_{1,\tau}\Phi, \mathcal{L}^\epsilon_{1,\tau}\Psi)_{\epsilon,t,\tau},$$

and for $\Phi \in \{\Psi \in \mathrm{dom}(\mathcal{L}^{\epsilon}_{1,\tau}) \cap \mathrm{dom}(\mathcal{L}^{*,\epsilon}_{0,\tau}) | \ \mathcal{L}^{\epsilon}_{1,\tau}\Psi \in \mathrm{dom}(\mathcal{L}^{*,\epsilon}_{1,\tau}), \mathcal{L}^{*,\epsilon}_{0,\tau}\Psi \in \mathrm{dom}(\mathcal{L}^{\epsilon}_{0,\tau})\}$

$$\Lambda^{\epsilon}_{\tau}\Phi = \mathcal{L}^{\epsilon}_{0,\tau}\mathcal{L}^{*,\epsilon}_{0,\tau}\Phi + \mathcal{L}^{*,\epsilon}_{1,\tau}\mathcal{L}^{\epsilon}_{1,\tau}\Phi.$$

$\Lambda^{\epsilon}_{\tau}$ is a closed positive self-adjoint operator.

For $\Phi = (a, b)$ let

$$\frac{\partial \Phi}{\partial \bar{z}_\ell} := \left(\frac{\partial a}{\partial \bar{z}_\ell}, \frac{\partial b}{\partial \bar{z}_\ell}\right), \quad \frac{\partial \Phi}{\partial \bar{z}} = \left(\frac{\partial \Phi}{\partial \bar{z}_\ell}\right)_{\ell=1,\dots,n}, \quad P\Phi := \frac{(\langle \partial\rho, a\rangle, 0)}{-\rho}.$$

By the same method as in section 2 one obtains the following basic estimates.

Proposition 1. *Let $\Phi \in \mathrm{dom}(\mathcal{L}^{\epsilon}_{1,\tau}) \cap \mathrm{dom}(\mathcal{L}^{*,\epsilon}_{0,\tau})$. Then $P\Phi \in W^{0,\epsilon}_{\tau}$ and $\frac{\partial \Phi}{\partial \bar{z}_\ell} \in W^{1,\epsilon}_{\tau}, \ell = 1, 2, \dots, n$. More precisely we have*

$$\mathcal{Q}^{\epsilon}_{\tau}(\Phi, \Phi) \geq t\|\Phi\|^2_{\epsilon,t,\tau} + (5t^2 + \tau)\|P\Phi\|^2_{\epsilon,t,\tau} + \|\frac{\partial \Phi}{\partial \bar{z}}\|^2_{\epsilon,t,\tau}.$$

Corollary 1. $\Lambda^{\epsilon}_{\tau}$: $\mathrm{dom}(\Lambda^{\epsilon}_{\tau}) \to W^{1,\epsilon}_{\tau}$ *is bijective and*

$$(\Lambda^{\epsilon}_{\tau}\Phi, \Phi)_{\epsilon,t,\tau} \geq t\|\Phi\|^2_{\epsilon,t,\tau}.$$

Therefore we can define a bounded self-adjoint operator

$$\mathcal{N}^{\epsilon}_{\tau} := (\Lambda^{\epsilon}_{\tau})^{-1} : W^{1,\epsilon}_{\tau} \to \mathrm{dom}(\Lambda^{\epsilon}_{\tau}) \subset W^{1,\epsilon}_{\tau}.$$

Analogously to the Neumann operator $\mathcal{N}$ for $\square$, we get for $\alpha \in W^{1,\epsilon}_{\tau}$, with $\mathcal{L}^{\epsilon}_{1,\tau}\alpha = 0$, that a solution of $\mathcal{U} := \mathcal{L}^{*,\epsilon}_{0,\tau}\mathcal{N}^{\epsilon}_{\tau}\alpha$ is

$$\mathcal{L}^{\epsilon}_{0,\tau}\mathcal{U} = \alpha.$$

When choosing $\alpha = (0, 0, \dots, 0, \sqrt{1+t})$, this solves a decomposition problem. The additional term $\|P\Phi\|^2$ in Proposition 1 makes it possible to compare $\mathcal{Q}^{\epsilon}_{\tau}(\Phi, \Phi)$ for $\tau = 0$ with the same term for some positive τ. This is necessary for the following reasons. We intend to construct solutions of the decomposition problem which are holomorphic with respect to the z variable. They are allowed to blow up when z approaches the boundary. If we keep the other parameters fixed, then the rate of growth is controlled by τ. A larger τ results in a larger growth. If we consider a solution for say $\tau = 0$ and differentiate it with respect to ζ, then it will fall out of the defined function space or at least we will lose uniform control with respect to ζ and ϵ. Moreover, if we want to control higher derivatives we are obliged to apply $\mathcal{N}$, defined with respect to the same parameters, again and again to the derivatives. But these derivatives are no longer elements of the original space. This is somewhat illuminated by the following formulae. We have

$$\Lambda\mathcal{N} = id.$$

This implies that if $D_\zeta\Lambda$ denotes a first order differentiation applied to the coefficients of Λ, then we have

$$D_\zeta\mathcal{N} = -\mathcal{N}(D_\zeta\Lambda)\mathcal{N} + \mathcal{N}D_\zeta.$$

$D_\zeta\Lambda$ is essentially given by $\frac{\partial}{\partial \overline{z}_i}$, with coefficients which increase the growth. For $\alpha = (0, 0, \ldots, \sqrt{1+t})$ we are interested in controlling $\mathcal{L}_0^*\mathcal{N}\alpha$. The main problem here is to control the derivatives of $\mathcal{N}\alpha$. But then we obtain

$$D_\zeta(\mathcal{N}\alpha) = -\mathcal{N}(D_\zeta\Lambda)(\mathcal{N}\alpha).$$

So in order to control $\mathcal{N}$ applied to $(D_\zeta\Lambda)(\mathcal{N}\alpha)$, by estimates which are uniform in ζ and ϵ, we need the operator $\mathcal{N} = \mathcal{N}_{\zeta,t,0,\epsilon}$ to be also continuous on the space where $(D_\zeta\Lambda)(\mathcal{N}\alpha)$ is living.

We learned about this kind of continuity on a second space of an operator, which has been defined elsewhere, from a paper of Berndtsson [2]. The above described considerations, however, are hidden behind technical lemmas in our original paper on this subject.

Our next goal is to compare $\mathcal{Q}^\epsilon(= \mathcal{Q}_0^\epsilon)$, Λ^ϵ, $\mathcal{N}^\epsilon$ with $\mathcal{Q}_\tau^\epsilon$, Λ_τ^ϵ and $\mathcal{N}_\tau^\epsilon$ for $\tau > 0$. Occurring constants should be independent of ζ and ϵ. Clearly one has $\mathcal{L}_{q,\tau}^\epsilon = \mathcal{L}_q^\epsilon$. Let $\alpha \in L_{0,1}^2(\Omega_\epsilon, \varphi_1) \cap \operatorname{dom}(\overline{\partial}_\tau^{*,\epsilon})$ and let ϑ denote the formal adjoint of $\overline{\partial}$. Since $\varphi_1(z,\zeta,t,\tau) = \varphi_1(z,\zeta,t,0) + \tau\psi(z)$ one gets

$$\overline{\partial}_\tau^{*,\epsilon}\alpha = e^{\varphi_1(z,\zeta,t,\tau)}\vartheta(e^{-\varphi_1(z,\zeta,t,\tau)}\alpha) = \overline{\partial}^{*,\epsilon}\alpha + \tau\frac{\langle\partial\rho, \alpha\rangle}{-\rho}.$$

Therefore we have, for $\Phi = (a, b) \in \operatorname{dom}(\mathcal{L}_{0,\tau}^{*,\epsilon})$,

$$\mathcal{L}_{0,\tau}^{*,\epsilon}\Phi = \mathcal{L}_0^{*,\epsilon}\Phi + \tau P\Phi.$$

Starting with Proposition 1, one finally arrives at the following lemma by a delicate and long induction procedure in which commutators of the $\mathcal{L}$-operators with differentiations arise.

Lemma 1. *Let $k \geq 0$, $t \geq 4$, $t \geq \tau \geq 6\nu k$. Let $\Phi \in W^{1,\epsilon}$ such that for all J, with $|J| \leq k$, $D_\zeta^J\mathcal{L}_0^\epsilon\mathcal{L}_0^{*,\epsilon}\Phi \in W^{1,\epsilon}$, $D_\zeta^J\mathcal{L}_1^\epsilon\Phi \in W^{2,\epsilon}$. Then for all J, with $|J| \leq k$, $D_\zeta^J\Phi \in \operatorname{dom}(\mathcal{L}_1^\epsilon) \cap \operatorname{dom}(\mathcal{L}_0^{*,\epsilon})$, $\mathcal{L}_0^{*,\epsilon}D_\zeta^J\Phi \in \operatorname{dom}(\mathcal{L}_0^\epsilon)$, $D_\zeta^J\frac{\partial\Phi}{\partial\overline{z}} \in W^{1,\epsilon}$,*

$PD_\zeta^J \Phi \in W^{0,\epsilon}$. *Moreover for I with $|I| = k$*

$$(t-2)\|D_\zeta^I \Phi\|^2_{\epsilon,t,\tau} + (5t^2 + \tau - 4\tau^2)\|PD_\zeta^I \Phi\|^2_{\epsilon,t,\tau} + \|D_\zeta^I \frac{\partial \Phi}{\partial \bar{z}}\|^2_{\epsilon,t,\tau}$$

$$\lesssim \sum_{r=0}^{k} \sum_{\substack{p+|L|\le k-r \\ p\ge 0}} \Big[\|D_\zeta^L \mathcal{L}_0^\epsilon \mathcal{L}_0^{*,\epsilon} \Phi\|^2_{\epsilon,t,\tau-2\nu(p+3r)}$$

$$+ \|D_\zeta^L \mathcal{L}_1^\epsilon \Phi\|^2_{\epsilon,t,\tau-2\nu(p+3r)} \Big].$$

The coefficients of the $\mathcal{L}$-operators are explicitly known. More twists give

Lemma 2. *Let $k \ge 0$, $t \ge 4$, $t \ge \tau \ge 2\nu(1+3k)$. Let $\Phi \in W^{1,\epsilon}$ such that for all J, with $|J| \le k$, $D_\zeta^J \mathcal{L}_0^\epsilon \mathcal{L}_0^{*,\epsilon} \Phi \in W^{1,\epsilon}$, $D_\zeta^J \mathcal{L}_1^\epsilon \Phi \in W^{2,\epsilon}$. Then for all I, with $|I| = k$, $D_\zeta^I \mathcal{L}_0^{*,\epsilon} \Phi \in W^{0,\epsilon}$ and*

$$\|D_\zeta^I \mathcal{L}_0^{*,\epsilon} \Phi\|^2_{\epsilon,t,\tau} \lesssim \sum_{\substack{p+|L|\le k-r \\ p\ge 0}} \Big[\|D_\zeta^L \mathcal{L}_0^\epsilon \mathcal{L}_0^{*,\epsilon} \Phi\|^2_{\epsilon,t,\tau-2\nu(1+3r)}$$

$$+ \|D_\zeta^L \mathcal{L}_1^\epsilon \Phi\|^2_{\epsilon,t,\tau-2\nu(1+3r)} \Big].$$

This result implies easily

Lemma 3. *Let $k \ge 0$, $t \ge 4$, $t \ge 2\nu(1+3k)$. Let $\alpha = (0, \sqrt{1+t})$ and $(w^\epsilon, 0) = \mathcal{L}_0^{*,\epsilon} \mathcal{N}^\epsilon \alpha$. Then there exists a constant $C(t,\tau,k)$, not depending on ϵ and ζ, such that for all I and τ, with $|I| = k' \le k$, $t \ge \tau \ge 2\nu(1+3k')$,*

$$\|D_\zeta^I w^\epsilon\|_{\epsilon,t,\tau} \le C(t,\tau,k')\|\alpha\|_{\epsilon,t,\tau-2\nu(1+3k')},$$

and $D_\zeta^I w^\epsilon(z,\zeta)$ is continuous for $(z,\zeta) \in \Omega_\epsilon \times (U \setminus \Omega)$.

Since $\mathcal{L}_0^\epsilon(w^\epsilon, 0) = (0, \sqrt{1+t})$, $w^\epsilon = (w_1^\epsilon, \cdots, w_n^\epsilon)$ is holomorphic with respect to the z variable and satisfies

$$\sum_{i=1}^{n} w_i^\epsilon(z,\zeta)(\zeta_i - z_i) = 1$$

for $z \in \Omega^\epsilon$ and $\zeta \in U \setminus \Omega$.

If the L^2-estimates are transformed into pointwise estimates by applying Cauchy inequalities one obtains Theorem 4.

3. The $\overline{\partial}_b$-problem on submanifolds of pseudoconvex boundaries

The results of this section appeared in [28]. As a first application of the two barriers, we shall construct a homotopy formula for the tangential Cauchy–Riemann operator $\overline{\partial}_b$ on an open domain of the boundary of a weakly pseudoconvex domain. The homotopy operators will be given by explicit integral formulae. Here integration takes place on some wedge-like subsets of the ambient space. In order to do this we are obliged to extend classes $[f]$ of forms on the submanifold to differential forms $E[f]$ of $\mathbb{C}^n$. We have to assume high regularity of f in order to make sure that $E[\overline{\partial}_b f] - \overline{\partial}E[f]$ vanishes of very high order on the submanifold. In particular, if the class $[f]$ is $\overline{\partial}_b$-closed, then $\overline{\partial}E[f]$ will vanish there. The integral kernels will contain terms coming from our barriers and which blow up in the edges of the wedges. The geometry of the wedges and the vanishing of $E[\overline{\partial}_b f] - \overline{\partial}E[f]$, however, will guarantee the existence of the integral operators. But the loss of regularity is considerable and there is no explicit control on it. So our results work most naturally in the C^∞-category.

After the description of the extension problem we give the definitions of the integral operators and the geometric situation. Finally we mention the case of $(0, n-2)$ forms, which is somewhat special because of a disturbing term in the homotopy formula. That this term actually vanishes can be shown by solving a $\overline{\partial}$-equation on piecewise smooth weakly pseudoconvex domains as described in [28]. Then this solvability gives rise to a holomorphic approximation result by using an Oka map. Finally the holomorphic approximation replaces the polynomial convexity which was needed in the strictly pseudoconvex case by Henkin (see below).

The first results will be mainly formulated in the C^k-class with finite k. Now a standard regularization gives the solvability of the $\overline{\partial}_b$-equation with solutions, which are C^∞-smooth up to the boundary.

Solutions for the $\overline{\partial}_b$-equation in the strictly pseudoconvex case have been given by Henkin [12] and Webster [44]. Ma and Michel [12, 20] have given the optimal C^k-estimates for strictly pseudoconvex submanifolds of the above type. L^p-estimates were shown by Shaw [37].

For $M = b\Omega$, where Ω is weakly pseudoconvex, the existence of C^∞ solutions has been shown by Rosay [33]. Local solvability for $\overline{\partial}_b$ on weakly pseudoconvex boundaries of finite type has been studied by Shaw [38,39]. C^k-estimates for open submanifolds of the boundary of complex pseudo-ellipsoids were given by K. Schaal [34].

We want to apply the integral formula approach to give homotopy formulae for certain open submanifolds of $b\Omega$. In order to do this we need two barrier maps $w_+ = (w_{+,1}, \ldots, w_{+,n})$ and $w_- = (w_{-,1}, \ldots, w_{-,n})$, depending on $z, \zeta \in \mathbb{C}^n$ and having values in $\mathbb{C}^n$, such that $\sum_{i=1}^n w_{\pm,i}(z,\zeta)(\zeta_i - z_i) = 1$. One has $z \in \overline{\Omega}$, $\zeta \notin \overline{\Omega}$ for w_+, $z \notin \Omega$, $\zeta \in \Omega$ for w_-, and $\overline{\partial}_z w_+ = 0, \overline{\partial}_\zeta w_- = 0$. If ζ approaches $b\Omega$, $w_\pm$ blows up of some finite order in terms of the boundary distance $\delta(\zeta)$. Weak information about the derivatives of $w_\pm$ will be enough to show our results. We use an integral formulae method, which was developed for $\overline{\partial}$ and later for $\overline{\partial}_b$ by Lieb and Range [19], Peters [30], Chaumat and Chollet [3,4], Michel [22,23] Michel and Perotti [25] and Ma and Michel [20], and which we generalize here.

The domain of integration will always contain some simplices, which are parametrized by the variable λ. The starting point is the well-known Bochner–Martinelli–Koppelman integral representation formula for differential forms. Here the range of λ is a single point. By passing over to more complicated Stokes chains with higher dimensional ranges for λ using Stokes formula, we obtain an integral representation formula which can be used to solve the $\overline{\partial}_b$-equation on M. In our case there are no barriers with good regularity properties up to the boundary in both variables as in the strictly pseudoconvex case. Therefore in order to make the calculations work we have to multiply the kernels $\mathcal{D}_{n,q}(z,\zeta,\lambda)$, which explode if ζ approaches the boundary of Ω, by the damping factor $\overline{\partial}E[f] - E(\overline{\partial}_b[f])$. This will lead to linear operators $[T_q([f])]$ and $[T_{q+1}(\overline{\partial}_b[f])]$ satisfying the following homotopy formula

$$[f] = \overline{\partial}_b[T_q([f])] + [T_{q+1}(\overline{\partial}_b[f])].$$

Therefore if $[f]$ is $\overline{\partial}_b$-closed we obtain solutions for the $\overline{\partial}_b$-equation.

I. Extension of tangential forms

Let $\Omega \subset \mathbb{C}^n$, $n \geq 2$, be a bounded pseudoconvex domain given by a C^∞ defining function $\rho\colon \mathcal{U} \to \mathbb{R}$, where $\mathcal{U} \supset \overline{\Omega}$ is an open neighborhood of $\overline{\Omega}$, such that $d\rho|_{b\Omega} \neq 0$ and

$$\Omega = \{z \in \mathcal{U} | \rho(z) < 0\}.$$

Let h be a holomorphic function on an open set $\mathcal{U}_1 \subset \mathcal{U}$ such that for $H := \{z \in \mathcal{U}_1 | \mathrm{Re}h(z) = 0\}$ we have the following properties:

i) on $\mathcal{U}_1$ there exists a Hefer decomposition

$$h(\zeta) - h(z) = \sum_{i=1}^n h_i(z,\zeta)(\zeta_i - z_i),$$

with h_i holomorphic on $\mathcal{U}_1 \times \mathcal{U}_1$.

ii) $d\mathrm{Re}h \neq 0$ on H, $d\mathrm{Re}h \wedge d\rho \neq 0$ on $H \cap b\Omega$; H is a relatively closed real hypersurface of $\mathcal{U}$ which decomposes $\mathcal{U}$ into 2 connected parts $\mathcal{U}^+ = \{z \in \mathcal{U}_1 | \mathrm{Re}h(z) \geq 0\}$ and $\mathcal{U}^- = \mathcal{U} \setminus \{z \in \mathcal{U}_1 | \mathrm{Re}h(z) > 0\}$. Moreover $M = \{z \in b\Omega | \mathrm{Re}h(z) \geq 0\}$ is a connected hypersurface with boundary.

By shrinking the neighborhoods a little bit, obviously one can assume that $M_+ := \{z \in \mathcal{U}_1 \setminus \Omega | \rho(z) + \mathrm{Re}h(z) = 0\}$ is a smooth relatively closed hypersurface of $\mathcal{U} \setminus \Omega$, which intersects $b\Omega$ and H transversally, and that $M_- := \{z \in \overline{\Omega} \cap \mathcal{U}_1 | \gamma\rho(z) - \mathrm{Re}h(z) = 0\}$, for a given fixed small $\gamma > 0$, is a smooth closed hypersurface in $\overline{\Omega}$. Then we have

i) $M_- \cap b\Omega = H \cap b\Omega, M_- \cap H = b\Omega \cap H$ and these intersections are transversal;

ii) $\overline{\Omega}$ is decomposed by M_- into two parts R_- and $\overline{\Omega} \setminus R_-$, where R_- is the one containing $H \cap \overline{\Omega}$.

We set finally $R_+ := \{z \in \mathcal{U}_1 \setminus \Omega | \rho(z) + \mathrm{Re}h(z) \geq 0\}$ and $K := \overline{\mathcal{U}_1 \setminus (R_+ \cup R_-)} \cap \mathcal{U}_1$.

If $[f]$ denotes the equivalence class of f, we set $C^k_{p,q}(M) = \{[f] | f \in C^k_{p,q}(\mathbb{C}^n)\}$. Because of the expected loss of regularity we suppose that $k >> 1$.

We set $\overline{\partial}_b[f] := [\overline{\partial} f]$. If $|g|_{k,\mathcal{U}^+}$ denotes the C^k-norm on $\mathcal{U}^+$ we set

$$|[f]|_{k,M} := \inf\{|g|_{k,\mathcal{U}^+} |\ g \sim f\},$$

where $g \sim f$ indicates that f and g are in the same class.

An elementary extension now gives the two following lemmas.

Lemma 4. *Let $[f] \in C^k_{0,q}(M)$ be $\overline{\partial}_b$-closed. Then for every $0 \leq r \leq (k-2)/2$ there exist forms $E_r[f] \in C^{k-2r}_{0,q}(\mathcal{U})$, $g_r \in C^{k-2r-1}_{0,q+1}(\mathcal{U})$, $c_r \in C^{k-2r-1}_{0,q+1}(U)$, $X_r \in C^{k-2r-2}_{0,q}(\mathcal{U})$, such that*

i) *$E_r[f]$ has compact support in $\mathcal{U}_1$, $[E_r[f]] = [f]$, $c_r = 0$ on $\mathcal{U}^+$ and*

$$\overline{\partial} E_r[f] = \rho^r g_r + c_r;$$

ii) *$E_{r+1}[f] - E_r[f] = \rho^{r+1} X_r$;*

iii) *there exist constants K_r not depending on $[f]$, with*

$$|E_r[f]|_{k-2r,\mathcal{U}} + |g_r|_{k-2r-1,\mathcal{U}} + |c_r|_{k-2r-1,\mathcal{U}} \leq K_r |[f]|_{k,M}.$$

Lemma 5. *Let* $[f] \in C^k_{0,q}(M)$ *with* $\overline{\partial}_b[f] \in C^k_{0,q+1}(M)$. *Let for* $0 \leq r \leq (k-2)/2$ $E_r\overline{\partial}_b[f]$ *be the extension of* $\overline{\partial}_b[f]$ *which was constructed in Lemma 4. Then there exist forms* $E_r[f] \in C^{k-2r}_{0,q}(\mathcal{U})$, $G_r \in C^{k-2r-1}_{0,q+1}(\mathcal{U})$, $C_r \in C^{k-2r-1}_{0,q+1}(\mathcal{U})$, $Y_r \in C^{k-2r-2}_{0,q}(\mathcal{U})$, *with*

i) $E_r[f]$ *has compact support in* $\mathcal{U}_1$, $[E_r[f]] = [f]$, $C_r = 0$ *on* $\mathcal{U}^+$ *and*

$$\overline{\partial}E_r[f] - E_r\overline{\partial}_b[f] = \rho^r G_r + C_r;$$

ii) $E_{r+1}[f] - E_r[f] = \rho^{r+1}Y_r$;

iii) *there exist constants* K_r *not depending on* $[f]$, *with*

$$|E_r[f]|_{k-2r,\mathcal{U}} + |G_r|_{k-2r-1,\mathcal{U}} + |C_r|_{k-2r-1,\mathcal{U}} \leq K_r(|[f]|_{k,M} + |\overline{\partial}_b[f]|_{k,M}).$$

II. Homotopy formulae for $\overline{\partial}_b$

Now let $w_+ = (w_{+,1}, \ldots, w_{+,n})\colon \overline{\Omega} \times (\mathcal{U} \setminus \overline{\Omega}) \to \mathbb{C}^n$, be a C^2 barrier mapping, as constructed in section 1, which solves

$$\overline{\partial}_z w_+(z,\zeta) = 0, \quad \sum_{j=1}^n w_{+,j}(z,\zeta)(\zeta_j - z_j) = 1$$

and which fulfills the following C^S-estimates for all integers $S \geq 0$, with constants C_S not depending on ζ:

$$|w_+(\cdot,\zeta)|_{S,\Omega} + |\mathrm{grad}_\zeta w_+(\cdot,\zeta)|_{S,\Omega} \leq \frac{C_S}{\delta(\zeta)^{t_S}}.$$

Denote the second barrier by w_-, with $\zeta \in \Omega$ and $z \in \mathcal{U} \setminus \Omega$. Then much more precise estimates have been obtained. But in the solution operators for $\overline{\partial}_b$, which we want to study in this paper, w_+ and w_- are mixed. So we cannot use the better information on w_-. Therefore we shall not give here all the shown properties of w_- and we prefer instead to treat both barriers in a more streamlined way. w_- is holomorphic with respect to the variable ζ, acting here as a parameter, but it is only of class C^S with respect to z (S is finite but can be chosen arbitrarily). So when it is necessary we shall write $w_{-,S}$ instead of w_-.

So, more precisely, for any $S \geq 0$ there exists a C^S map $w_- = w_{-,S} = (w_{-,1}, \cdots, w_{-,n})\colon (\mathcal{U} \setminus \Omega) \times \Omega \to \mathbb{C}^n$ which solves

$$\overline{\partial}_\zeta w_{-,S}(z,\zeta) = 0, \quad \sum_{j=1}^n w_{-,j}(z,\zeta)(\zeta_j - z_j) = 1$$

and which fulfills the following estimates, with constants C_S not depending on ζ and z:

$$|D_z^I w_{-,S}(z,\zeta)| \leq \frac{C_S}{\delta(\zeta)^{A(S)}},$$

for all multi-indices I with $|I| \leq S$. (Here D_z^I denotes a differentiation with respect to z of order $|I|$ and $A(S)$ is a positive integer).

For $2n$-dimensional sets we carry over the orientation of $\mathbb{C}^n$. We give the induced orientation to their boundaries. We orient M, M_+ and M_- in such a way that

$$bR_+ = -M - M_+, \quad bR_- = M - M_-, \quad bK = M_+ + M_-.$$

Set

$$\Delta_{01+-} = \{(\lambda_0, \lambda_1, \lambda_+, \lambda_-) \in \mathbb{R}^4 | \lambda_\nu \geq 0, \lambda_0 + \lambda_1 + \lambda_+ + \lambda_- = 1\}.$$

For an ordered subset $\emptyset \neq A \subset \{0,1,+,-\}$ we set

$$\Delta_A = \{\lambda \in \Delta_{01+-} | \lambda_\nu = 0 \text{ for } \nu \notin A\}.$$

We orient Δ_A, with $A = \{a_1, \ldots, a_\nu\}, \nu \geq 2$, in such a way that

$$b\Delta_A = \Delta_{a_2,\ldots,a_\nu} - \Delta_{a_1,a_3,\ldots,a_\nu} + \cdots + (-1)^{\nu-1}\Delta_{a_1,\ldots,a_{\nu-1}}.$$

For $a = 0, 1, +, -$ we define the following barrier forms:

$$\eta_a = \sum_{i=1}^n w_{a,i}(z,\zeta) d\zeta_i,$$

with $w_{0,i}(z,\zeta) = \dfrac{\overline{\zeta}_i - \overline{z}_i}{|\zeta - z|^2}, w_{1,i}(z,\zeta) = \dfrac{h_i(z,\zeta)}{h(\zeta) - h(z)}$.

For $\lambda \in \Delta_A$ we set

$$\eta = \sum_{a \in A} \lambda_a \eta_a$$

and for $q = 0, 1, 2, \ldots, n$ and a certain constant $c_{n,q}$

$$\mathcal{D}_{n,q}(z,\zeta,\lambda) = c_{n,q}\eta \wedge ((\overline{\partial}_\zeta + d_\lambda)\eta)^{n-q-1} \wedge (\overline{\partial}_z \eta)^q.$$

Here d_λ denotes the total differential on Δ_A. With $\mathcal{D}_{n,-1} = \mathcal{D}_{n,n+1} = 0$ one has the Koppelman formula for Cauchy–Fantappiè forms

$$\overline{\partial}_z \mathcal{D}_{n,q-1} = (-1)^q (\overline{\partial}_\zeta + d_\lambda) \mathcal{D}_{n,q}.$$

Moreover we set

$$\sum = R_+ \times \Delta_{0+} + R_- \times \Delta_{0-} + K \times \Delta_{01} + M_+ \times \Delta_{01+} + M_- \times \Delta_{01-}.$$

Now let $[f] \in C^k_{0,q}(M)$, with $\overline{\partial}_b[f] \in C^k_{0,q+1}(M)$. We choose extensions $E_r[f]$ and $E_r\overline{\partial}_b[f]$ according to Lemmas 4 and 5. Then for k sufficiently large and $r \leq (k-2)/2$ appropriately chosen with respect to S, we obtain the decomposition

$$\overline{\partial} E_r[f] - E_r\overline{\partial}_b[f] = \rho^{Ms+1}G + (\mathrm{Re}h)^{Ms+S+2}H,$$
$$\overline{\partial} E_r\overline{\partial}_b[f] = \rho^{Ms+1}g + (\mathrm{Re}h)^{Ms+S+2}h,$$

with C^1 forms g, h, G, H, such that h and H vanish on $\mathcal{U}^+$. This will imply the existence of the following operators:

$$T_q([f]) := \int_{\Sigma} (\overline{\partial} E_r[f] - E_r\overline{\partial}_b[f]) \wedge \mathcal{D}_{n,q-1} + \int_{\mathcal{U}\times\Delta_0} E_r[f] \wedge \mathcal{D}_{n,q-1},$$

$$T_{q+1}((\overline{\partial}_b[f]) := \int_{\Sigma} \overline{\partial} E_r\overline{\partial}_b[f] \wedge \mathcal{D}_{n,q} + \int_{\mathcal{U}\times\Delta_0} E_r\overline{\partial}_b[f] \wedge \mathcal{D}_{n,q}.$$

If $\overline{\partial}_b[f] = 0$ we choose $E_r\overline{\partial}_b[f] = 0$. Hence $T_{q+1}(\overline{\partial}_b[f]) = 0$ in this case. Note that on $\mathcal{U} \times \Delta_0$, $\mathcal{D}_{n,q}$ is the well-known Bochner–Martinelli–Koppelman kernel. Now the Koppelman formula and Stokes theorem lead to the following theorem.

Theorem 5. *For every integer $S \geq 0$ there exist integers $A(S) \geq 0$ with the following properties:*
For every $[f] \in C^{A(S)}_{0,q}(M)$ with $\overline{\partial}_b[f] \in C^{A(S)}_{0,q+1}(M)$, $q \geq 1$, the forms

$$T_q([f]) = \int_{\Sigma} (\overline{\partial} E_r[f] - E_r\overline{\partial}_b[f]) \wedge \mathcal{D}_{n,q-1} + \int_{\mathcal{U}\times\Delta_0} E_r[f] \wedge \mathcal{D}_{n,q-1},$$

$$T_{q+1}(\overline{\partial}_b[f]) = \int_{\Sigma} \overline{\partial} E_r\overline{\partial}_b[f] \wedge \mathcal{D}_{n,q} + \int_{\mathcal{U}\times\Delta_0} E_r\overline{\partial}_b[f] \wedge \mathcal{D}_{n,q}$$

are C^S on M.
If $1 \leq q \leq n-3$ one has

$$\overline{\partial}_b[T_q([f])] + [T_{q+1}(\overline{\partial}_b[f])] = [f].$$

If $q = n-2$, $\overline{\partial}_b[f] = 0$, $E_r\overline{\partial}_b[f] = 0$ one has

$$\overline{\partial}_b[T_{n-2}([f])] = [f].$$

In all these cases there exist constants C_S, which are independent of $[f]$, with

$$|[T_q([f])]|_{S,M} + |[T_{q+1}(\overline{\partial}_b[f])]|_{S,M} \le C_S(|[f]|_{A(S),M} + |\overline{\partial}_b[f]|_{A(S),M}).$$

In the proof one starts with the Martinelli–Bochner–Koppelman formula. Then one arrives at the following result.

$$\overline{\partial}T_q([f]) + T_{q+1}(\overline{\partial}_b[f]) = E_r[f] + \int_{M_- \times \Delta_{1-}} (\overline{\partial}E_r[f] - E_r\overline{\partial}_b[f]) \wedge \mathcal{D}_{n,q}.$$

If $q \le n-3$, the integral on the right-hand side vanishes. Let $q = n-2$ and $\overline{\partial}_b[f] = 0$. Then the integral in question also vanishes but this is not obvious. It is a consequence of the following lemma. For a compact set L, let $A^\infty(L)$ denote the space of C^∞ functions on L which are holomorphic in the interior of L.

Lemma 6. *Every function holomorphic in a neighborhood of $H \cap \overline{\Omega}$ can be uniformly approximated on $H \cap \overline{\Omega}$ by functions of $A^\infty(\{z \in \overline{\Omega} | Reh(z) \ge 0\})$.*

Remark. The vanishing result is new even in the strictly pseudoconvex case. In the proof one uses an Oka map argument. It is based on the solvability of the $\overline{\partial}$-equation on the intersection of weakly pseudoconvex domains. We cannot give the details here.

An analogous calculation as for Theorem 5 shows, if the starting point is $T_1'([f])$ (T_1' will be defined in Theorem 6 and $\Sigma' = R_+ \times \Delta_+ - M_+ \times \Delta_{1+}$), the following theorem:

Theorem 6. *Let $n \ge 2$. For every integer $S \ge 0$ there exist integers $A(S) \ge 0$ with the following properties:*

For every function $[f] \in C^{A(S)}(M)$ with $\overline{\partial}_b[f] \in C^{A(S)}(M)$ the forms

$$T_0'([f]) = \int_{\Sigma'} (\overline{\partial}E_r[f] - E_r\overline{\partial}_b[f]) \wedge \mathcal{D}_{n,0},$$

$$T_1'(\overline{\partial}_b[f]) = \int_{\Sigma} \overline{\partial}E_r\overline{\partial}_b[f] \wedge \mathcal{D}_{n,0} + \int_{\mathcal{U} \times \Delta_0} E_r\overline{\partial}_b[f] \wedge \mathcal{D}_{n,0}$$

are C^S on M. $T_0'([f])$ is holomorphic on $\mathcal{U}^+ \cap \Omega$ and C^S on $\mathcal{U}^+ \cap \overline{\Omega}$. If $\overline{\partial}_b[f] = 0$ we can assume that $T_1'(\overline{\partial}_b[f]) = 0$.

If $n \ge 3$ we have

$$[T_0'([f])] + [T_1'(\overline{\partial}_b[f])] = [f].$$

If $n = 2$ and $\overline{\partial}_b[f] = 0$ we have

$$[T_0'([f])] = [f].$$

In all these cases there exists a constant C_S, which is independent of $[f]$, with

$$\|[T_0'([f])]\|_{S,\mathcal{U}^+\cap\overline{\Omega}} + \|[T_1'(\overline{\partial}_b[f])]\|_{S,M} \le C_S\big(\|[f]\|_{A(S),M} + |\overline{\partial}_b[f]|_{A(S),M}\big).$$

If $[f] \in C^\infty(M)$, with $\overline{\partial}_b[f] = 0$, there exists an extension $E_\infty[f] \in C^\infty(\mathcal{U})$ with compact support in $\mathcal{U}_1$, such that $\overline{\partial}E_\infty[f]$ vanishes of infinite order on M. w_+ is C^∞ on $\overline{\Omega}$ with respect to z. Therefore we obtain the following result.

Corollary 2. *Under the same conditions as in Theorem 6 there exists for every CR function $f \in C^\infty(M)$ an extension*

$$\widetilde{T}_0([f]) = \int_{\Sigma'} \overline{\partial}E_\infty[f] \wedge \mathcal{D}_{n,0}$$

of f, which is C^∞ on $\mathcal{U}^+ \cap \overline{\Omega}$ and holomorphic on $\mathcal{U}^+ \cap \Omega$.

When we consider solutions which are more and more regular, we obtain by a standard method

Theorem 7. *Let $[f] \in C^\infty_{0,q}(\overline{M})$ with $\overline{\partial}_b[f] = 0$, $1 \le q \le n-2$. Then there exists $[u] \in C^\infty_{0,q-1}(\overline{M})$ with*

$$\overline{\partial}_b[u] = [f].$$

4. The $\overline{\partial}$-problem on annuli

Let Ω be a bounded domain in $\mathbb{C}^n$ such that Ω has a piecewise smooth boundary. In this section we study the solvability of the Cauchy–Riemann equation

$$\overline{\partial}u = \alpha \quad \text{in} \quad \Omega,$$

where α is a smooth $\overline{\partial}$-closed (p, q) form with coefficients C^∞ up to the boundary of Ω, $0 \le p \le n$ and $1 \le q \le n$. In particular, we describe its solvability with u smooth up to the boundary (for appropriate degree q) if Ω satisfies one of the following conditions:

i) $\Omega = \Omega_1 \setminus \overline{\Omega}_2$, $\Omega_2 \subset\subset \Omega_1$, where Ω_2 is the union of bounded smooth pseudoconvex domains and Ω_1 is a pseudoconvex convex domain with a piecewise smooth boundary.

ii) $\Omega = \Omega_1 \setminus \overline{\Omega}_2\ \Omega_2 \subset\subset \Omega_1$, where Ω_2 is the intersection of bounded smooth pseudoconvex domains and Ω_1 is a pseudoconvex convex domain with a piecewise smooth boundary.

If Ω is the transversal intersection of bounded smooth pseudoconvex domains, then related results are given in [24].

When Ω is an annulus between two bounded pseudoconvex domains with C^∞ boundaries, regularity results have been obtained in Shaw [36].

The $\overline{\partial}$ problem on piecewise smooth domains is not only interesting in itself, it also arises from the local solvability of tangential Cauchy–Riemann equations. Let M be an open subset of the boundary of a bounded smooth pseudoconvex domain in $\mathbb{C}^n$. We consider the equation

$$\overline{\partial}_b u = \alpha \quad \text{in} \quad M,$$

where α is a smooth $\overline{\partial}_b$-closed (p,q) form on $\overline{M}$, $1 \leq q \leq n-3$. We show that when the boundary bM lies in the transversal intersection of $b\Omega$ with k Levi-flat hypersurfaces, then one can find a smooth solution u, for $1 \leq q \leq n-k-2$, provided these k hypersurfaces satisfy some global conditions. Previous results (cf. Henkin [12], Shaw [37,38], Ma and Michel [20], Michel–Shaw [28]) all require that the boundary is smooth and bM lies in some Levi-flat hypersurface.

The plan of this section is as follows: In I we construct a homotopy formula on an annulus such that Ω_2 is the union of finitely many smooth pseudoconvex domains. The proof depends on the barrier functions constructed earlier. We then use induction to construct a solution for the $\overline{\partial}$-equation when Ω_2 is the transversal intersection of finitely many smooth pseudoconvex domains. In II we prove the solvability of the $\overline{\partial}_b$-equation, with regularity up to the boundary, on open submanifolds M of the boundary of a weakly pseudoconvex domain.

I. Boundary regularity for $\overline{\partial}$ on piecewise smooth annuli.

Let Ω be a bounded piecewise smooth pseudoconvex domain in $\mathbb{C}^n$. Let $D_i \subset\subset \Omega$, $i = 1, \ldots, k$, such that each D_i is a bounded pseudoconvex domain with C^2 boundary bD_i defined by $\{\rho_i = 0\}$. We assume that $d\rho_{i_1} \wedge \cdots \wedge d\rho_{i_\ell} \neq 0$ on $\rho_{i_1} = \cdots = \rho_{i_\ell} = 0$ for every $I = (i_1, \ldots, i_\ell)$, $1 \leq i_1 < \cdots < i_\ell \leq k$. Let

$$D = \Omega \setminus \overline{(\cup_{i=1}^k D_i)}.$$

Then D is the annulus between a pseudoconvex domain Ω and the union of finitely many bounded pseudoconvex domains with C^2 boundary. We consider $\overline{\partial}$ on D with solutions smooth up to the boundary and shall construct a homotopy formula for $\overline{\partial}$ on D.

Let $w_-^{(i)}$ be the second barrier for D_i as constructed in section 2. Then $w_-^{(i)} = (w_{-,1}^{(i)}, \ldots, w_{-,n}^{(i)})\colon \mathcal{U}_i \times D_i \to \mathbb{C}^n$ is holomorphic in $\zeta \in D_i$ for each fixed $z \in \mathcal{U}_i$.

We set $w_{0,\mu}(z,\zeta) = (\overline{\zeta}_\mu - \overline{z}_\mu)/ \mid \zeta - z \mid^2$ for $\mu = 1, \ldots, n$ and we define

$$w_{-,\mu}(z,\zeta,\lambda) = \lambda_0 w_{0,\mu}(z,\zeta) + \sum_{j=1}^{k} \lambda_j w_{-,\mu}^{(j)}(z,\zeta),$$

$$\eta_-(z,\zeta,\lambda) = \sum_{\mu=1}^{n} \lambda_\mu w_{-,\mu}(z,\zeta) d\zeta_\mu,$$

wherever it is defined with $\lambda_j \geq 0$ and $\lambda_0 + \lambda_1 + \cdots + \lambda_k = 1$. In particular, if $\lambda_{i_1} + \cdots + \lambda_{i_r} = 1$, $z \in D$, $\zeta \in \cap_{\nu=1}^{r} D_{i_\nu}$, then η_- is well defined and holomorphic in ζ if $\lambda_0 = 0$. For $0 \leq q \leq n-1$, we set, with some constant $c_{n,q}$,

$$\Omega_{n,q}^0 = c_{n,q} \eta_- \wedge ((\overline{\partial}_\zeta + d_\lambda)\eta_-))^{n-1-q} \wedge (\overline{\partial}_z \eta_-)^q,$$

which is of degree q in z and of degree $2n-q-1$ in (ζ,λ). Set $\Omega_{n,n}^0 = \Omega_{n,-1}^0 = 0$ and $\cup_{i=1}^k D_i = R^0$.
For each increasing index $I = (i_1, \ldots, i_\ell)$, $1 \leq \ell \leq k$, we define for small $\epsilon_0 > 0$,

$$\begin{aligned} R_I^0 = \{z \in \cap_{\nu=1}^{\ell} D_{i_\nu} \mid -\epsilon_0 \leq \rho_{i_1}(z) = \cdots = \rho_{i_\ell}(z) \leq 0, \\ \rho_j(z) \geq \rho_{i_1}(z) \text{ for } j \notin I \text{ and } z \in D_j\}. \end{aligned}$$

We require that the orientation on R_I^0 be skew symmetric in the components of I and we define

$$\begin{aligned} S_I &= \{z \in b \cup_{i=1}^k D_i \mid \rho_i(z) = 0, i \in I\}, \\ S_I^{\epsilon_0} &= \{z \in b \cup_{i=1}^k D_i \mid \rho_i(z) = -\epsilon_0,\ i \in I\} \end{aligned}$$

and for each S_I and $S_I^{\epsilon_0}$ the natural induced orientation is given. Then we have

$$bR_I^0 = \sum_{j=1}^{k} R_{Ij}^0 + S_I - S_I^{\epsilon_0},$$

$$\begin{aligned} b\Big(\sum_I (-1)^{|I|}(R_I^0 \times \Delta_{0I})\Big) = \sum_I R_I^0 \times \Delta_I - R^0 \times \Delta_0 + \sum_I (-1)^{|I|} S_I \times \Delta_{0I} \\ - \sum_I (-1)^{|I|} S_I^{\epsilon_0} \times \Delta_{0I}, \end{aligned}$$

where the summation is over all ordered increasing subsets of $\{1, \ldots, k\}$.

Define

$$S_q^{(m)}\alpha = C_n\left\{ \int_{\sum_I (-1)^{|I|} R_I^0 \times \Delta_{0I}} (\bar\partial E\alpha - E\bar\partial\alpha) \wedge \Omega^0_{n,q-1} - \int_{\mathbb{C}^n \times \Delta_0} E\alpha \wedge \Omega^0_{n,q-1} \right\}$$

for $1 \le q \le n-k-1$.

Again, the Stokes theorem and the Koppelman formula give for $z \in D$,

$$\alpha = S_1^{(m)} \bar\partial\alpha, \quad \alpha \in C^\infty_{(0,0)}(\overline{D})$$

and

$$\alpha = \bar\partial S_q^{(m)}\alpha + S_{q+1}^{(m)}\bar\partial\alpha, \quad \alpha \in C^\infty_{(0,q)}(\overline{D}) \quad \text{and } 1 \le q \le n-k-1.$$

Here we supposed that α vanishes on $\mathbb{C}^n \setminus \Omega$. For the general case we need to modify the above construction. One has to also plug in the finitely many barriers coming from $b\Omega$. The construction does not pose any further problems. Let $A^\infty(\overline{D}) = C^\infty(\overline{D}) \cap \mathcal{O}(D)$, where $\mathcal{O}(D)$ is the set of holomorphic functions in D. Then we can show the following theorem.

Theorem 8. *Let Ω be a bounded pseudoconvex domain in $\mathbb{C}^n$ with piecewise C^∞-smooth boundary. Let $D_i \subset\subset \Omega, i = 1, \ldots, k$ be pseudoconvex domains with C^2 boundary and $D = \Omega \setminus \overline{(\cup_{i=1}^k D_i)}$. We assume that the $\{D_i\}_{i=1}^k$ intersect transversally. For $1 \le q \le n-k$ and every nonnegative integer m there exist linear operators $\tilde{S}_q^{(m)}: C^\infty_{(0,q)}(\overline{D}) \to C^m_{(0,q-1)}(\overline{D})$, such that for every $\alpha \in C^\infty_{(0,q)}(\overline{D})$, $z \in D$, we have*

$$\alpha = \bar\partial \tilde{S}_q^{(m)}\alpha + \tilde{S}_{q+1}^{(m)}\bar\partial\alpha, \quad \textit{where } 1 \le q \le n-k-1.$$

When $q = 0$, $k \le n$, there exists an operator $\tilde{S}_0^{(m)}: C^\infty(\overline{D}) \to A^m(\overline{D})$ such that for every $\alpha \in C^\infty(\overline{D})$ we have on D,

$$\alpha = \tilde{S}_0^{(m)}\alpha + \tilde{S}_1^{(m)}\bar\partial\alpha.$$

Corollary 3. *Let D be the same as in Theorem 8. If $\alpha \in C^\infty_{(0,q)}(\overline{D})$ and $\bar\partial\alpha = 0$ where $1 \le q \le n-k-1$, then there exists a $u \in C^\infty_{(0,q-1)}(\overline{D})$ satisfying $\bar\partial u = \alpha$ in D.*

Corollary 3 follows from Theorem 8 by a standard regularizing method.

Next we want to describe the boundary regularity for $\bar\partial$ on an annulus between a pseudoconvex domain and an intersection of smooth pseudoconvex domains. For every increasing multi-index $I = (i_1, \ldots, i_\mu), 1 \le$

$i_1 < \cdots < i_\mu \le k$, we define $D_I = \cap_{i\in I} D_i$. We set for fixed $1 \le \gamma \le k$, $1 \le j_\nu \le k$,

$$\Omega_2 = \cup_{\nu=1}^{\gamma} D_{Ij_\nu} \quad \text{and} \quad A = \Omega \setminus \Omega_2,$$

where $I = (i_1, \ldots, i_\mu)$, $\gamma + \mu \le k$, and $j_\nu \notin I$. We set $D_\phi = \mathbb{C}^n$ and we assume Ω_2 and A are connected. We also assume that each $\Omega \setminus \{\cup_{\nu=1}^{\mu+\gamma-\tilde{\mu}} (D_1 \cap \cdots \cap D_{\tilde{\mu}}) \cap D_{\tilde{\mu}+\nu}\}$ is connected for $0 \le \tilde{\mu} \le \mu$. In a first step one shows the regularity for $\overline{\partial}$ on A.

Theorem 9. *For every $f \in C^\infty_{(0,q)}(\overline{A})$, where $1 \le q \le n-1-\gamma$ such that $\overline{\partial} f = 0$ in A, there exists a $g \in C^\infty_{(0,q-1)}(\overline{A})$ satisfying $\overline{\partial} g = f$ in A.*

By choosing $\gamma = 1$, $\mu = k-1$, one obtains

Corollary 4. *Let Ω be a bounded pseudoconvex domain in $\mathbb{C}^n$ with piecewise smooth C^∞ boundary. Let $D_i \subset\subset \Omega$, $i = 1, \ldots, k$ be pseudoconvex domains with C^2 boundary and $G = \Omega \setminus (\cap_{i=1}^k D_i)$. For every $\overline{\partial}$-closed $f \in C^\infty_{(0,q)}(\overline{G})$, $1 \le q \le n-2$, there exists a $u \in C^\infty_{(0,q-1)}(\overline{G})$ such that $\overline{\partial} u = f$ in G.*

Let $D^0 = \cap_{i=1}^k D_i$. Corollary 4 implies the following:

Corollary 5. *For every $\alpha \in C^\infty_{(0,q)}(\mathbb{C}^n)$ such that $\overline{\partial}\alpha = 0$ in $\mathbb{C}^n$ and $\operatorname{supp}\alpha \subset \overline{D}^0$, where $1 \le q \le n-1$, there exists a $u \in C^\infty_{(0,q-1)}(\mathbb{C}^n)$ satisfying $\overline{\partial} u = \alpha$ in $\mathbb{C}^n$ and $\operatorname{supp} u \subset \overline{D}^0$.*

To prove Theorem 9, one needs the following lemma.

Lemma 7. *Let $I = (i_1, \ldots, i_\mu)$ and $0 < \gamma \le n-1$, $\mu+\gamma \le n-1$, $\mu, \gamma \le k$. If $1 \le q \le n-1-\gamma$, $f \in C^\infty_{(0,q)}(\overline{A})$ such that $\overline{\partial} f = 0$ in A and $f = 0$ in $\Omega \setminus D_I$, there exists a $g \in C^\infty_{(0,q-1)}(\overline{A})$ such that $\overline{\partial} g = f$ in A and $g = 0$ in $\Omega \setminus D_I$.*

Then the proof of Theorem 9 is an induction on μ for all $0 < \gamma \le n-1-\mu$. For $\mu = 0$, this is proved in Theorem 8 (since $1 \le q \le n-1-\gamma$ and $D_\phi = \mathbb{C}^n$).

II. Applications to the local solvability of $\overline{\partial}_b$.

Let Ω be a bounded smooth pseudoconvex domain in $\mathbb{C}^n$. Let M be a connected open subset of $b\Omega$ with piecewise smooth boundary bM. By this we mean that there exist bounded domains D_i, $i = 1, \ldots, k$, with smooth boundary bD_i such that

$$M = b\Omega \cap (\cap_{i=1}^k D_i),$$

where bD_i and M intersect transversally wherever they intersect.

Definition. *M is called a domain with admissible boundary bM if the following conditions hold:*

i) *$\Omega_I \equiv \Omega \cap (\cap_{i=1}^k D_i)$ is a bounded piecewise smooth pseudoconvex domain.*

ii) *For each $1 \le i \le k$, the set $\Omega_i^c \equiv \Omega \cap (\mathbb{C}^n \setminus \overline{D}_i)$ is equal to Ω intersected with a bounded smooth pseudoconvex domain.*

We note that i) and ii) imply that bM consists of smooth pieces which lie in Levi-flat hypersurfaces. Examples of admissible boundaries are those defined by real hyperplanes in $\mathbb{C}^n$.

Theorem 10. *Let $M \subset b\Omega$ be a domain with admissible boundary. For every $\overline{\partial}_b$-closed form $\alpha \in C^\infty_{(0,q)}(\overline{M})$, $1 \le q \le n-k-2$, there exists a $u \in C^\infty_{(0,q-1)}(\overline{M})$ such that*

$$\overline{\partial}_b u = \alpha$$

in M.

The proof uses the solvability of $\overline{\partial}$ on the auxiliary domains of I.

Remark. We note that the condition on the degree q and k cannot be relaxed. Let $S = \{z | \, |z_1|^2 + \cdots + |z_n|^2 = 1\}$ be the unit sphere in $\mathbb{C}^n$, $n \ge 3$. Let

$$M = S \cap \left(\cap_{i=3}^{i=n} \{z | \, |z_i|^2 < \frac{1}{2(n-2)} \} \right).$$

Then M is a domain with admissible boundary with $k = n-2$. We shall show that the $\overline{\partial}_b$-equation is not solvable for $q = 1$ in M. Let

$$\alpha = \frac{\overline{z}_1 d\overline{z}_2 - \overline{z}_2 d\overline{z}_1}{(|z_1|^2 + |z_2|^2)^2}.$$

Then $\alpha \in C^\infty_{0,1}(\overline{M})$ and $\overline{\partial}_b \alpha = 0$ in M. Let $S_0 = S \cap \{z_3 = \frac{1}{\sqrt{2(n-2)}}, \ldots, z_n = \frac{1}{\sqrt{2(n-2)}}\}$. If $\alpha = \bar{\partial}_b u$ for some $u \in C^\infty(\overline{M})$, then we would have

$$\int_{S_0} \alpha dz_1 \wedge dz_2 = \int_{S_0} \bar{\partial} u \wedge dz_1 \wedge dz_2 = 0.$$

On the other hand, we have that

$$\begin{aligned} \int_{S_0} \alpha dz_1 \wedge dz_2 &= \frac{1}{4} \int_{|z_1|^2 + |z_2|^2 = \frac{1}{2}} \{\bar{z}_1 d\bar{z}_2 - \bar{z}_2 d\bar{z}_1\} dz_1 \wedge dz_2 \\ &= \frac{1}{2} \int_{|z_1|^2 + |z_2|^2 < \frac{1}{2}} d\bar{z}_1 \wedge d\bar{z}_2 \wedge dz_1 \wedge dz_2 \\ &\ne 0. \end{aligned}$$

Thus the condition on the number of intersections k cannot be removed.

On the other hand, if we take

$$M = S \cap \left(\cap_{i=4}^{i=n} \{z \mid |z_i|^2 < \frac{1}{2(n-2)}\} \right),$$

where $n > 4$, then using Theorem 10 we can find a $u \in C^\infty(\overline{M})$ satisfying $\bar{\partial}_b u = \alpha$ in M.

References

[1] Aizenberg, L.A., Dautov, Sh.A., *Differential forms orthogonal to holomorphic functions or forms and their properties,* Providence: AMS (1983).

[2] Berndtsson, B., *$\overline{\partial}_b$ and Carleson type inequalities,* Complex Analysis II, Lecture Notes in Math. **1276,** Springer-Verlag, 42–54.

[3] Chaumat, J., Chollet, A.M., *Estimations Hölderiennes pour les équations de Cauchy-Riemann dans les convexes compacts de $\mathbb{C}^n$,* Math Zeit. **207** (1991), 501–534.

[4] Chaumat, J., Chollet, A.M., *Nogaux pour résoudre l'équation $\overline{\partial}$ dans des classes ultradifférentiables sur des compacts irréguliers de $\mathbb{C}^n$,* Proceedings on the special year on Complex Analysis, 87–88 Mittag-Leffler, Princeton University Press (1993), 205–226.

[5] Cumenge, A., *Sharp Hölder estimates on convex domains of finite type,* (preliminary paper).

[6] Derridj, M., *Le problème de Cauchy pour $\bar{\partial}$ et application,* Ann Scient. Ecole Norm. Sup. **17** (1984), 439–449.

[7] Diederich, K., Fornaess, J.E., *Pseudoconvex domains: Bounded strictly plurisubharmonic exhaustion functions,* Invent. Math. **39** (1977), 129–141.

[8] Diederich, K., Fornaess, J. E., *Pseudoconvex domains: an example with non trivial Nebenhuelle,* Math. Ann. **225** (1977), 275–292.

[9] Diederich,K., Ohsawa, T., *On the parameter dependence of solutions to the $\overline{\partial}$-equations,* Math. Ann.**289** (1991), 581–587.

[10] Dufresnoy, A., *Sur L'opérateur d'' et les fonctions différentiables au sens de Whitney,* Ann. Inst. Fourier, Grenoble, **29** (1979), 229–238.

[11] Folland, G.B., Kohn, J.J., *The Neumann problem for the Cauchy–Riemann Complex,* Annals of Math. Studies **75**, Princeton University Press, 1972.

[12] Henkin, G.M., *The H. Lewy equation and analysis on pseudoconvex manifolds,* Russian Math. Surveys **32** (1977), 59–130.

[13] Hörmander, L., *L^2-estimates and existence theorems for the $\overline{\partial}$ operator,* Acta Math. **113** (1965), 89–152.

[14] Hörmander, L., *An Introduction to Complex Analysis in Several Variables, 3rd edition,* North-Holland, 1994.

[15] Kohn, J.J., *Global regularity for $\overline{\partial}$ on weakly pseudoconvex manifolds,* Trans. Am. Math. Soc. **181** (1973), 273–292.

[16] Kohn, J.J., *Methods of partial differential equations in Complex Analysis,* Proc. Symp. Pure Math. **30** (1977), 215–236.

[17] Kohn, J.J., Nirenberg, L., *Non-coercive boundary value problems,* Comm. Pure Appl. Math. **18** (1965), 443–492.

[18] Kohn, J.J., Nirenberg, L., *A pseudoconvex domain not admitting a holomorphic support function,* Math. Ann. **201** (1973), 265–268.

[19] Lieb, I., Range, R.M., *Lösungsoperatoren für den Cauchy-Riemann Komplex mit C^k-Abschätzungen,* Math. Ann. **253** (1980), 145–164.

[20] Ma, L., Michel, J., *Local regularity for the tangential Cauchy-Riemann Complex,* J. Reine Angew. Math **442** (1993), 63–90.

[21] Ma, L., Michel, J., *Regularity of local embeddings of strictly pseudoconvex CR structures,* J. Reine Angew. Math. **447** (1994), 147–164.

[22] Michel, J., *Randregularität des $\overline{\partial}$-Problems für die Halbkugel in $\mathbb{C}^n$,* Man. Math. **55** (1986), 239–268.

[23] Michel, J., *Randregularität des $\overline{\partial}$-Problems für stückweise streng pseudokonvexe Gebiete in $\mathbb{C}^n$,* Math. Ann. **280** (1988), 46–68.

[24] Michel, J., *Integral representations on weakly pseudoconvex domains,* Math. Zeit. **208** (1991), 437–462.

[25] Michel, J., Perotti, A., *C^k-regularity for the $\overline{\partial}$-equation on strictly pseudoconvex domains with piecewise smooth boundaries,* Math. Zeit. **203** (1990), 414–427.

[26] Michel, J., Perotti, A., *C^k-regularity for the $\overline{\partial}$-equation on a piecewise smooth union of strictly pseudoconvex domains in $\mathbb{C}^n$,* Ann. Sc. Norm. Sup. Pisa **21** (1994), 483–495.

[27] Michel, J., Shaw, M.-C., *A decomposition problem on weakly pseudoconvex domains,* Math. Zeit. **230** (1999), 1–19.

[28] Michel, J., Shaw, M.-C., *C^∞-regularity of solutions of the tangential CR-equations on weakly pseudoconvex manifolds,* Math. Ann. **311** (1998), 147–162.

[29] Michel, J., Shaw, M.-C., *The $\overline{\partial}$-problem on domains with piecewise smooth boundaries with applications,* to appear Transactions of the AMS.

[30] Peters, K., *Lösungsoperatoren für die $\overline{\partial}$-Gleichung auf nichttransversalen Durchschnitten von streng pseudokonvexen Gebieten,* Diss. A, Humboldt Universität Berlin, 1990.

[31] Range, R.M. *Holomorphic functions and integral representations in several complex variables,* Graduate Texts in Math. **108,** Springer-Verlag, 1986.

[32] Range, R.M. and Siu, Y.T., *Uniform estimates for the $\overline{\partial}$-equation on domains with piecewise smooth strictly pseudoconvex boundaries,* Math. Ann. **83** (1973), 325–354.

[33] Rosay, J.-P., *Equation de Lewy-Résolubilité globale de l'équation $\overline{\partial}_b u = f$ sur la frontière de domaines faiblement pseudo-convexes de $\mathbb{C}^2$ (où $\mathbb{C}^n$),* Duke Math. J. **49** (1982), 121–128.

[34] Schaal, K., *Lokale Regularität für die tangentialen Cauchy-Riemann Gleichungen auf Rändern komplexer Ellipsoide,* Dissertation, Universität Bonn, 1995.

[35] Seeley, R.T., *Extension of C^∞-functions defined in a half space,* Proc. Amer. Math. Soc. **15** (1964), 625–626.

[36] Shaw, M.-C., *Global solvability and regularity for $\overline{\partial}$ on an annulus between two weakly pseudo-convex domains,* Trans. Amer. Math. Soc. **291** (1985), 255–267.

[37] Shaw, M.-C., *L^p estimates for local solutions of $\bar{\partial}_b$ on strongly pseudo-convex CR manifolds,* Math. Ann. **288** (1990), 35–62.

[38] Shaw, M.-C., *Local existence theorems with estimates for $\overline{\partial}_b$ on weakly pseudoconvex boundaries,* Math. Ann. **294** (1992), 677–700.

[39] Shaw, M.-C., *Semi-global existence theorems of $\overline{\partial}_b$ for $(0, n-2)$ forms on pseudo-convex boundaries in $\mathbb{C}^n$,* Astérisque, Société Mathématiqe de France, Colloque d'Analyse Complexe et Géométrie, Marseille (1993), 227–240.

[40] Skoda, H., *Application de techniques L^2 à la théorie des idéaux d'une algèbre de fonctions holomorphes avec poids,* Ann. Sc. Ec. Norm. Sup. **5** (1972), 545-579.

[41] Stein E.M., *Singular integrals and differentiability properties of functions,* Princeton University Press, Princeton, New Jersey, 1970.

[42] Thilliez, V., *Hilbert-valued forms, and barriers on weakly pseudoconvex domains,* (preprint).

[43] Vassiliadou, S., *Homotopy fomulas for* $\overline{\partial}_b$ *and subelliptic estimates for the* $\overline{\partial}$*-Neumann problem,* Thesis, Notre Dame, 1997.

[44] Webster, S., *On the local solution of the tangential Cauchy–Riemann equations,* Ann. Inst. H. Poincaré (ANL) **6** (1989), 167–182.

[45] Webster, S., *On the proof of Kuranishi's embedding theorem,* Ann. Inst. H. Poincaré (ANL) **6** (1989), 183–207.

UNIVERSITE DU LITTORAL
CENTRE UNIVERSITAIRE DE LA MI-VOIX
F-62228 CALAIS, FRANCE.
E-mail: michel@lma.univ-littoral.fr

DEPARTMENT OF MATHEMATICS
UNIVERSITY OF NOTRE DAME
NOTRE DAME, IN, 46556, USA
E-mail: mei-chi.shaw.1@nd.edu

CHAPTER IX

A Note on the Closed Rangeness of Vector Bundle Valued Tangential Cauchy–Riemann Operator

Kimio Miyajima[†]

Abstract

We prove the vector bundle value version of J. J. Kohn's closed range theorem over three dimensional strongly pseudoconvex CR manifolds. That closed range theorem is a crucial step toward CR construction of the semi-universal family of normal isolated surface singularities (cf. [11]).

Introduction

Let M be a compact real C^∞-manifold of $\dim_{\mathbf{R}} M = 2n-1 \geq 3$. A CR structure on M is a subbundle $T^{(1,0)}M \subset \mathbf{C}TM$ of $\operatorname{rank}_{\mathbf{C}} T^{(1,0)}M = n-1$ such that

$$T^{(1,0)}M \cap T^{(0,1)}M = \{0\} \text{ where } T^{(0,1)}M = \overline{T^{(1,0)}M}, \tag{0.1}$$

$$T^{(1,0)}M \text{ is closed under the bracket operation.} \tag{0.2}$$

Since $\mathbf{C}TM/\left(T^{(1,0)}M \oplus T^{(0,1)}M\right)$ is a $\mathbf{C}$-line bundle compatible with the complex conjugate operation of $\mathbf{C}TM$, there exists a global real vector field ξ such that ξ is nowhere vanishing in $\mathbf{C}TM/\left(T^{(1,0)}M \oplus T^{(0,1)}M\right)$. If F denotes the subbundle of $\mathbf{C}TM$ generated by ξ, then we have a splitting as differentiable vector bundles

$$\mathbf{C}TM = F \oplus T^{(1,0)}M \oplus T^{(0,1)}M. \tag{0.3}$$

An Hermitian form $\mathcal{L}_p$ on $T_p^{(1,0)}M$ $(p \in M)$ defined by

$$\sqrt{-1}\mathcal{L}_p(Z,W)\xi(p) \equiv [\tilde{Z}, \overline{\tilde{W}}](p) \bmod T_p^{(1,0)}M \oplus T_p^{(0,1)}M$$

[†]Partially supported by Grant-in-Aid for Scientific Research (No. 09640123), the Ministry of Education and Culture of Japan

is called the *Levi-form* associated with the CR structure $T^{(1,0)}M$ (and the vector field ξ), where $\tilde{Z}$ and $\tilde{W}$ are local sections of $T^{(1,0)}M$ such that $\tilde{Z}(p) = Z$ and $\tilde{W}(p) = W$. A CR structure is called *strongly pseudoconvex* if the Levi-form has a definite sign at each point. (Note that this property is independent of the choice of ξ.) Throughout this paper, we consider only strongly pseudoconvex CR structures.

For M with CR structure $T^{(1,0)}M$, the tangential Cauchy–Riemann complex is induced:

$$0 \to A_b^0 \xrightarrow{\bar{\partial}_b} A_b^{0,1} \xrightarrow{\bar{\partial}_b} A_b^{0,2} \xrightarrow{\bar{\partial}_b} \cdots$$

where we denote

$$A_b^{0,q} = \Gamma(M, \wedge^q (T^{(0,1)}M)^*),$$
$$\bar{\partial}_b \phi = d\phi_{|T^{(0,1)}M \times \cdots \times T^{(0,1)}M}.$$

If M is a real hypersurface of a complex manifold X, a natural CR structure ${}^\circ T'$ is induced from the complex structure on X by ${}^\circ T' = T^{1,0}X_{|M} \cap \mathbf{C}TM$. In this case, there exists an analytic restriction map

$$\mu \colon A_X^{0,q} \to A_b^{0,q} \quad \text{given by} \quad \mu(\phi) = \phi_{|{}^\circ T'' \times \cdots \times {}^\circ T''}$$

where we denote ${}^\circ T'' = T^{0,1}X_{|M} \cap \mathbf{C}TM$. Then we have a commutative diagram of Cauchy–Riemann complexes;

$$\begin{array}{ccccccccc}
0 & \longrightarrow & A_X^0 & \xrightarrow{\bar{\partial}} & A_X^{0,1} & \xrightarrow{\bar{\partial}} & A_X^{0,2} & \xrightarrow{\bar{\partial}} & \cdots \\
 & & \downarrow{\scriptstyle\mu} & & \downarrow{\scriptstyle\mu} & & \downarrow{\scriptstyle\mu} & & \\
0 & \longrightarrow & A_b^0 & \xrightarrow{\bar{\partial}_b} & A_b^{0,1} & \xrightarrow{\bar{\partial}_b} & A_b^{0,2} & \xrightarrow{\bar{\partial}_b} & \cdots .
\end{array}$$

A CR structure induced from an embedding as a real hypersurface is called an *embeddable* CR structure.

J. J. Kohn's closed range theorem is the following:

Theorem. ([8], Theorem 5.2.) *If a strongly pseudoconvex CR structure is embeddable, then the Cauchy–Riemann operator $\bar{\partial}_b$ has closed range in $L^2_{(0,q)}(M)$ $(1 \le q \le n-1)$ where we denote by $L^2_{(0,q)}(M)$ the space of L^2-sections of $\wedge^q (T^{(0,1)}M)^*$ and by the same symbol $\bar{\partial}_b$ the L^2-closure of $\bar{\partial}_b$.*

(In [8], Theorem 5.2, the closed range theorem is proved under some weaker condition than strongly pseudoconvexity. Since our interest is in the links of isolated singularities, we only consider the case of smooth boundaries of strongly pseudoconvex domains.)

By [2], it is shown that the L^2-closed rangeness implies the embeddability of strongly pseudoconvex CR structures. A major difference between three dimensional CR structures and higher dimensional ones is that there exist nonembeddable strongly pseudoconvex three dimensional CR structures while all strongly pseudoconvex CR structures of $\dim_{\mathbf{R}} \geq 5$ are embeddable ([3]).

The subject of this paper is to generalize the above J. J. Kohn's closed range theorem to the vector bundle-valued Cauchy–Riemann complex. The motivation of that generalization is in the deformation theory of CR structures. Construction of the semi-universal family of normal isolated singularities by means of deformations of strongly pseudoconvex CR structures on its link is a natural idea originated with M. Kuranishi ([10]). Since the deformation theory of CR structures is in analogy to the one of complex structures due to K. Kodaira, D. C. Spencer and M. Kuranishi (e.g. [7]), we need the harmonic analysis for vector bundle-valued tangential Cauchy–Riemann complexes. If $\dim_{\mathbf{R}} M \geq 5$, it is established for general holomorphic vector bundles by directly showing the basic estimate ([5]). However, it has not been discussed yet in the three dimensional case. Hence, we will concentrate on the vector bundle-valued tangential Cauchy–Riemann complexes on three dimensional strongly pseudoconvex CR manifolds. (See §12 of [11], for the application of the result of this paper.)

Let $\mathbf{E}$ be a holomorphic vector bundle over a strongly pseudoconvex three dimensional CR manifold $(M, T^{(1,0)}M)$; that is, there exists a differential operator

$$\overline{\partial}_{\mathbf{E}} \colon A_b^{0,0}(\mathbf{E}) \to A_b^{0,1}(\mathbf{E})$$

such that

$$\overline{\partial}_{\mathbf{E}}(fu) = f\overline{\partial}_{\mathbf{E}}u + (\overline{\partial}_b f)u,$$
$$\overline{\partial}_{\mathbf{E}}(u)([\overline{Z}, \overline{W}]) = \overline{\partial}_{\mathbf{E}}(\overline{\partial}_{\mathbf{E}}(u)(\overline{W}))(\overline{Z}) - \overline{\partial}_{\mathbf{E}}(\overline{\partial}_{\mathbf{E}}(u)(\overline{Z}))(\overline{W})$$

hold for $u \in A_b^0(\mathbf{E})$, $f \in C^\infty(M)$ and $\overline{Z}$, $\overline{W} \in \Gamma(M, T^{(0,1)}M)$ (cf. [12]), where we denote

$$A_b^{0,q}(\mathbf{E}) := \Gamma(M, \mathbf{E} \otimes \wedge^q (T^{(0,1)}M)^*).$$

If $T^{(1,0)}M$ is induced from an embedding of M into a complex surface and $\mathbf{E} = E_{|M}$ for a holomorphic vector bundle E over the ambient complex surface, a natural $\overline{\partial}_{\mathbf{E}}$ is induced from $\overline{\partial}_b$ as above. For this $\overline{\partial}_{\mathbf{E}}$, we have a commutative diagram induced from the analytic restriction μ as above;

$$\begin{array}{ccccccc} 0 & \longrightarrow & A_X^0(E) & \xrightarrow{\overline{\partial}_E} & A_X^{0,1}(E) & \xrightarrow{\overline{\partial}_E} & A_X^{0,2}(E) & \longrightarrow & 0 \\ & & \downarrow \mu & & \downarrow \mu & & \downarrow \mu & & \\ 0 & \longrightarrow & A_b^0(\mathbf{E}) & \xrightarrow{\overline{\partial}_{\mathbf{E}}} & A_b^{0,1}(\mathbf{E}) & \longrightarrow & 0 & & \end{array}$$

The main result of this paper is as follows. Throughout this paper, we assume that M is a three dimensional real manifold with a strongly pseudoconvex CR structure $T^{(1,0)}M$ and denote by $\mathbf{E}$ a holomorphic vector bundle on M.

Theorem 3.1. *Suppose that M is a smooth boundary of a strongly pseudoconvex domain of a complex surface X and that $\mathbf{E} = E_{|M}$ for a holomorphic vector bundle E on X. Then* $\operatorname{Im}\left\{\overline{\partial}_{\mathbf{E}} \colon L^2_{(0,0)}(\mathbf{E}) \to L^2_{(0,1)}(\mathbf{E})\right\}$ *is closed where $L^2_{(0,q)}(\mathbf{E})$ denotes the space of L^2-sections of $\mathbf{E} \otimes \wedge^q (T^{(0,1)}M)^*$ and we denote the L^2-closure of $\overline{\partial}_{\mathbf{E}}$ by the same symbol $\overline{\partial}_{\mathbf{E}}$.*

The closed rangeness of $\overline{\partial}_{\mathbf{E}}$ implies the closed rangeness of $\overline{\partial}^*_{\mathbf{E}}$, $\square^{(0,0)}_{\mathbf{E}} = \overline{\partial}^*_{\mathbf{E}}\overline{\partial}_{\mathbf{E}}$ and $\square^{(0,1)}_{\mathbf{E}} = \overline{\partial}_{\mathbf{E}}\overline{\partial}^*_{\mathbf{E}}$, and then we have the orthogonal projections

$$\rho^0_{\mathbf{E}} \colon L^2_{(0,0)}(\mathbf{E}) \to \operatorname{Ker} \square^{(0,0)}_{\mathbf{E}}, \qquad \rho^1_{\mathbf{E}} \colon L^2_{(0,1)}(\mathbf{E}) \to \operatorname{Ker} \square^{(0,1)}_{\mathbf{E}}$$

and the partial inverses of $\square^{(0,0)}_{\mathbf{E}}$ and $\square^{(0,1)}_{\mathbf{E}}$ respectively

$$N^0_{\mathbf{E}} \colon L^2_{(0,0)}(\mathbf{E}) \to \operatorname{Dom} \square^{(0,0)}_{\mathbf{E}}, \qquad N^1_{\mathbf{E}} \colon L^2_{(0,1)}(\mathbf{E}) \to \operatorname{Dom} \square^{(0,1)}_{\mathbf{E}}.$$

The next task is to obtain the estimates for these operators. We denote by $\| \ \|_{(k)}$ and $| \ |_{(\alpha)}$ the Folland-Stein norm of order k and the nonisotropic Lipschitz norm of order α respectively. (See [6] for these norms.)

Proposition 4.1. *Let $k \geq 0$ and $\alpha > 0$.*

$$\|\rho^0_{\mathbf{E}}\mathbf{u}\|_{(k)} \leq C\|\mathbf{u}\|_{(k)}, \qquad |\rho^0_{\mathbf{E}}\mathbf{u}|_{(\alpha)} \leq C|\mathbf{u}|_{(\alpha)} \tag{1}$$
$$\|\rho^1_{\mathbf{E}}\mathbf{u}\|_{(k)} \leq C\|\mathbf{u}\|_{(k)}, \qquad |\rho^1_{\mathbf{E}}\mathbf{u}|_{(\alpha)} \leq C|\mathbf{u}|_{(\alpha)} \tag{2}$$
$$\|N^0_{\mathbf{E}}\mathbf{u}\|_{(k+2)} \leq C\|\mathbf{u}\|_{(k)}, \qquad |N^0_{\mathbf{E}}\mathbf{u}|_{(\alpha+2)} \leq C|\mathbf{u}|_{(\alpha)}. \tag{3}$$

The arrangement of this paper is as follows. Though the vector bundle value version seems to be a trivial corollary of the J. J. Kohn's closed range theorem, it fails in general even if the base CR manifold is embeddable.

In §1, we present a simple counter example which is given by L. Lempert. In §2, we obtain a vector bundle version of the J. J. Kohn's microlocal estimate which played a crucial roles in his proof. We will prove Theorem 3.1 in §3. The proof is a vector bundle-valued analogue of [8]. Since our purpose of this work is to establish a harmonic analysis from the viewpoint of deformation theory, we need the estimates in Proposition 4.1. In §4, we obtain them as a consequence of the Heisenberg calculus in [1] and [6].

The author would like to thank Professors J. Bland, C. Epstein, L. Lempert and the participants of the Taniguchi Symposium for helpful discussions in this work, especially to Professor L. Lempert for showing him the counterexample presented in §1.

1. Counterexample

The closed rangeness for $\overline{\partial}_{\mathbf{E}}$ fails in general even if the CR structure $T^{(1,0)}M$ is embeddable. We present an example given by L. Lempert.

We consider the situation that M is a strongly pseudoconvex embeddable CR manifold and M' an unramified double cover of M such that the CR structure on M' induced via π is unembeddable. The following is a simple example of this situation: Let V be a subvariety of $\mathbf{C}^3$ defined by $x_1x_2 - x_3^2 = 0$. Then V has only normal singularity at the origin 0. We denote by U its regular part. Then the map $\mathbf{C}^2 \to \mathbf{C}^3$ given by $(z, w) \mapsto (x_1, x_2, x_3) = (z^2, w^2, zw)$ induces a unramified double cover $\pi\colon \mathbf{C}^2 \setminus \{(0,0)\} \to U$. We take a smoothing V' of the singularity $(V, 0)$. Then the complex structure of V' induces a new complex structure on U and then on $\mathbf{C}^2 \setminus \{(0,0)\}$. If we take a strongly pseudoconvex real hypersurface M in U and set $M' = \pi^{-1}(M)$, then by Proposition 1.3 of [4], M' is not embeddable.

Next, let $\mathbf{L}$ be a line bundle over M given by

$$\mathbf{L} = M' \times \mathbf{C}/\mathbf{Z}_2$$

where the operation $\mathbf{Z}_2$ is generated by

$$g\colon (p, \zeta) \mapsto (g(p), -\zeta)$$

for the generator g of the covering transformation group.

We remark that

$$g^*\overline{\partial}_b u = \overline{\partial}_b g^* u$$

holds for $u \in A_b^{0,q}(M')$. The following lemma is obvious.

Lemma 1.1

$$A_b^{(0,q)}(M, \mathbf{L}) \simeq \left\{ u \in A_b^{(0,q)}(M') \,|\, g^*u = -u \right\} \ (q = 0, 1)$$

and $\overline{\partial}_{\mathbf{L}}$ is given by $\overline{\partial}_b$ with respect to the above isomorphism.

We define an inner product in $A_b^{(0,q)}(M, \mathbf{L})$ using the inner product in $A_b^{(0,q)}(M')$ via the isomorphism in Lemma 1.1. We remark that it defines a topology in $A_b^{(0,q)}(M, \mathbf{L})$ which is equivalent to the one defined in the standard manner (using the fibre metric of $\mathbf{L}$).

Proposition 1.2 $\operatorname{Im}\left\{\overline{\partial}_{\mathbf{L}} \colon L^2_{(0,0)}(M, \mathbf{L}) \to L^2_{(0,1)}(M, \mathbf{L})\right\}$ *is not closed.*

Proof. Since $\operatorname{Im}\left\{\overline{\partial}_b \colon L^2_{(0,0)}(M') \to L^2_{(0,1)}(M')\right\}$ is not closed, there exist $\alpha \in L^2_{(0,1)}(M')$ and $u_i \in L^2_{(0,0)}(M')$ such that $\overline{\partial}_b u_i \to \alpha$ and $\alpha \notin \operatorname{Im}\overline{\partial}_b$. Let

$$u_i^+ = \frac{1}{2}(u_i + g^* u_i), \qquad \alpha^+ = \frac{1}{2}(\alpha + g^*\alpha),$$
$$u_i^- = \frac{1}{2}(u_i - g^* u_i), \qquad \alpha^- = \frac{1}{2}(\alpha - g^*\alpha).$$

Then, we have

$$u_i^+ \in L^2_{(0,0)}(M), \qquad \alpha^+ \in L^2_{(0,1)}(M), \qquad \overline{\partial}_b u_i^+ \to \alpha^+,$$
$$u_i^- \in L^2_{(0,0)}(M, \mathbf{L}), \qquad \alpha^- \in L^2_{(0,1)}(M, \mathbf{L}), \qquad \overline{\partial}_{\mathbf{L}} u_i^- \to \alpha^-.$$

Since M is embeddable, $\alpha^+ = \overline{\partial}_b u^+$ holds for $u^+ \in L^2_{(0,0)}(M)$. If, further, $\alpha^- = \overline{\partial}_{\mathbf{L}} u^-$ holds for $u^- \in L^2_{(0,0)}(M, \mathbf{L})$, we have

$$\alpha = \overline{\partial}_b(u^+ + u^-), \quad u^+ + u^- \in L^2_{(0,0)}(M').$$

This contradicts $\alpha \notin \operatorname{Im}\overline{\partial}_b$. □

2. E-valued version of J. J. Kohn's microlocal estimate

The key estimate of the proof of the J. J. Kohn's closed range theorem is the microlocal estimate [8], Theorem 3.3. In this section, we obtain the vector bundle-valued version of that estimate.

Let E be a holomorphic vector bundle on a complex manifold X. Then we have the E-valued Cauchy–Riemann complex;

$$0 \to A_X^0(E) \xrightarrow{\overline{\partial}_E} A_X^{0,1}(E) \xrightarrow{\overline{\partial}_E} A_X^{0,2}(E) \xrightarrow{\overline{\partial}_E} \cdots.$$

If we fix an inner product h along the fibres of E, an inner product

$$(\phi, \psi) = \int_X \langle \phi, \psi \rangle dV$$

is defined for $\phi,\ \psi \in A_X^{0,q}(E)$ with compact supports where $\langle \phi, \psi \rangle dV = \sum_{\alpha,\beta} h_{\alpha\overline{\beta}}\phi^\alpha \wedge *\overline{\psi^\beta}$ for $\phi = \sum_{\alpha=1}^r \phi^\alpha \mathbf{e}_\alpha$ and $\psi = \sum_{\beta=1}^r \psi^\beta \mathbf{e}_\beta$ with denoting by $*$ the Hodge $*$-operator, $\{\mathbf{e}_1, \ldots, \mathbf{e}_r\}$ a local frame of E and $h_{\alpha\overline{\beta}} = h(\mathbf{e}_\alpha, \mathbf{e}_\beta)$. Then the formal adjoint operator

$$\vartheta_E \colon A_X^{0,q}(E) \to A_X^{0,q-1}(E)$$

is induced by

$$(\phi, \overline{\partial}_E \psi) = (\vartheta_E \phi, \psi) \text{ for all } \psi \in A_X^{0,q-1}(E) \text{ with compact supports.}$$

Now, let Ω is a strongly pseudoconvex bounded domain of X defined by $r < 0$ with $r \in C^\infty(X)$ such that $dr \neq 0$ on the boundary $M = \partial\Omega$. We consider the Cauchy–Riemann complex

$$0 \to A_{\overline{\Omega}}^{0}(E) \overset{\overline{\partial}_E}{\to} A_{\overline{\Omega}}^{0,1}(E) \overset{\overline{\partial}_E}{\to} A_{\overline{\Omega}}^{0,2}(E) \overset{\overline{\partial}_E}{\to} \cdots .$$

We denote by the same symbol $\overline{\partial}_E$ the L^2-closure of $\overline{\partial}_E$ and its Hilbert space adjoint by $\overline{\partial}_E^*$. Then

$$\mathrm{Dom}(\overline{\partial}_E^*) \cap A_{\overline{\Omega}}^{0,q}(E) = \left\{ \phi \in A_{\overline{\Omega}}^{0,q}(E) \,|\, \sigma(\vartheta_E, dr)\phi = 0 \right\}.$$

We denote by $\mathcal{D}_E^{0,q}(\overline{\Omega})$ the right-hand side. $\overline{\partial}_E$-harmonic theory on a strongly pseudoconvex domain $\overline{\Omega}$ relies on the basic estimate (cf. [5]);

$$\sum_{\alpha,J,k} \|\frac{\partial \phi_J^\alpha}{\partial \overline{z}_k}\|^2 + \int_M |\phi|^2 + \|\phi\|^2 \leq C Q_E(\phi, \phi) \text{ for } \phi \in \mathcal{D}_E^{0,q}(\overline{\Omega}) \ (q \geq 1), \tag{2.1}$$

where we denote

$$Q_E(\phi, \phi) = (\overline{\partial}_E \phi, \overline{\partial}_E \phi) + (\vartheta_E \phi, \vartheta_E \phi) + (\phi, \phi)$$

and $\|\phi\|$ (resp. $\sqrt{\int_M |\phi|^2}$) the L^2-norm of ϕ (resp. $\phi_{|M}$).

The proof of J. J. Kohn's closed range theorem relies on a microlocal estimate for zero-th order tangential pseudodifferential operators ([8], Theorem 3.3) as well as the basic estimate for the $\overline{\partial}$-Neumann problem with weight on a (not necessarily strong) pseudoconvex domain. (In the case of strongly pseudoconvex domains, it is enough to use the above basic estimate (2.1) neglecting the weight.) In this section, we shall obtain the following $\overline{\partial}_E$-version of Theorem 3.3 of [8] for a complex surface X.

Let $p_0 \in M$ and U be a neighbourhood of p_0 in X such that ${}^\circ T'_{|U\cap M}$ is trivialized. We denote by L and T the local frames of ${}^\circ T'_{|U\cap M}$ and $F_{|U\cap M}$

respectively such that $\overline{T} = -T$. We choose a local coordinate (x_1, x_2, x_3) of $U \cap M$ such that

$$L(p_0) = \frac{1}{2}\left(\frac{\partial}{\partial x_1} - \sqrt{-1}\frac{\partial}{\partial x_2}\right),$$

$$\frac{\partial}{\partial x_3} = \sqrt{-1}gT,\ g \in C^\infty(U \cap M),\ g > 0.$$

Denote by (ξ_1, ξ_2, ξ_3) the dual coordinate of (x_1, x_2, x_3). Then the symbols of L, T are given by

$$\sigma(L) = -\frac{1}{2}(\sqrt{-1}\xi_1 + \xi_2) + \sum_{k=1}^{3} a^k(x, r)\xi_k,$$

$$\sigma(T) = \frac{1}{g(x)}\xi_3 + \sum_{k=1}^{3} b^k(x, r)\xi_k$$

where $a^k(x, r)$, $b^k(x, r) \in C^\infty(U \cap \overline{\Omega})$ such that $a^k(0, 0) = b^k(x, 0) = 0\,(1 \leq k \leq 3)$.

A tangential pseudodifferential operator of order zero is an operator $P\colon C_0^\infty(U \cap \overline{\Omega}) \to C^\infty(U \cap \overline{\Omega})$ given by

$$Pu(x, r) = \int_{\mathbf{R}^3} e^{\sqrt{-1}x\cdot\xi} p(x, r, \xi)\tilde{u}(\xi, r)d\xi$$

for some $p(x, r, \xi) \in C^\infty(\mathbf{R}^3 \times \mathbf{R} \times \mathbf{R}^3)$ such that there exists a compact subset $K \subset U \cap \overline{\Omega}$ such that $p(x, r, \xi) = 0$ for $(x, r) \notin K$ and there exists a constant $C = C(\alpha, j, \beta) > 0$ such that

$$|D_x^\alpha D_r^j D_\xi^\beta p(x, r, \xi)| \leq C(1 + |\xi|)^{-|\beta|}$$

holds, where we denote by $\tilde{u}(\xi, r)$ the tangential Fourier transform of $u(x, r)$.

Let $\mathcal{P}$ be the set of all tangential pseudodifferential operators of order zero and let

$$\mathcal{P}^0 = \left\{P \in \mathcal{P} \,|\, p(x, 0, \xi) = 0 \text{ if } \xi_3^2 \geq 2(\xi_1^2 + \xi_2^2) \text{ and } |\xi| > 1\right\},$$

$$\mathcal{P}^+ = \left\{P \in \mathcal{P} \,|\, p(x, 0, \xi) = 0 \text{ if } \xi_3 \leq \sqrt{\xi_1^2 + \xi_2^2} \text{ and } |\xi| > 1\right\},$$

$$\mathcal{P}^- = \left\{P \in \mathcal{P} \,|\, p(x, 0, \xi) = 0 \text{ if } \xi_3 \geq -\sqrt{\xi_1^2 + \xi_2^2} \text{ and } |\xi| > 1\right\}.$$

We note that, after fixing a trivialization of E on U, $P \in \mathcal{P}$ naturally induces an operator $P\colon C_0^\infty(U \cap \overline{\Omega}, E \otimes \wedge^q (T^{0,1}X)^*) \to C^\infty(U \cap \overline{\Omega}, E \otimes \wedge^q (T^{0,1}X)^*)$.

Theorem 2.1 *Let $\Omega \subset X$ and U be as above and $s \in \mathbf{R}$ and $m \in \mathbf{Z}_{\geq 0}$ be given. Let Λ^s be the tangential pseudodifferential operator with symbol $(1+|\xi|^2)^{\frac{s}{2}}$. Suppose $P \in \mathcal{P}^0 \cup \mathcal{P}^-$ and $\zeta, \zeta' \in C_0^\infty(\widetilde{U \cap \overline{\Omega})}$ are given such that $\zeta' = 1$ in a neighbourhood of* Supp ζ. *Then, if $\phi \in \widetilde{\mathcal{D}^{0,1}_{\overline{\Omega}}}(E)$ and $\alpha \in H^s_{(0,1)}(\Omega, E)$ satisfy*

$$Q_E(\phi, \psi) = (\alpha, \psi) \text{ for all } \psi \in \mathcal{D}^{0,1}_{\overline{\Omega}}(E),$$

the following estimate holds:

$$\|\left(\frac{\partial}{\partial r}\right)^m \Lambda^{s+2-m} P(\zeta\phi)\| \leq$$
$$\begin{cases} C(\|\Lambda^s\zeta'\alpha\| + \|\alpha\|_{s-1}) & \text{if } m \leq 2 \\ C\left(\sum_{j=0}^{m-2} \|(\frac{\partial}{\partial r})^j \Lambda^{s-j}\zeta'\alpha\| + \|\alpha\|_{s-1}\right) & \text{if } m \geq 2, \end{cases}$$

where $\| \ \|$ (resp. $\| \ \|_s$) denotes the L^2-norm (resp. the Sobolev norm of order s), $H^s_{(0,1)}(\Omega, E)$ the closure of $A^{0,1}_{\overline{\Omega}}(E)$ with respect to $\| \ \|_s$ and $\widetilde{\mathcal{D}^{0,1}_{\overline{\Omega}}}(E)$ the closure of $\mathcal{D}^{0,1}_{\overline{\Omega}}(E)$ with respect to $Q_E(\phi,\phi)$. Furthermore, we have

$$\|\left(\frac{\partial}{\partial r}\right)^m \Lambda^{s+1-m} P(\overline{\partial}\zeta\phi)\| + \|\left(\frac{\partial}{\partial r}\right)^m \Lambda^{s+1-m} P(\overline{\partial}^*\zeta\phi)\| \leq$$
$$\begin{cases} C(\|\Lambda^s\zeta'\alpha\| + \|\alpha\|_{s-1}) & \text{if } m \leq 1 \\ C\left(\sum_{j=0}^{m-1} \|(\frac{\partial}{\partial r})^j \Lambda^{s-j}\zeta'\alpha\| + \|\alpha\|_{s-1}\right) & \text{if } m \geq 1. \end{cases}$$

Theorem 2.1 implies the following.

Corollary 2.2 *Under the same assumption of Theorem 2.1, for $s \geq 0$, we have*

(1) $\|P(\zeta\phi)\|_s \leq C\left(\|\zeta'\alpha\|_{s-2} + \|\alpha\|_{s-3}\right)$,

(2) $\|P(\overline{\partial}\zeta\phi)\|_s + \|P(\overline{\partial}^*\zeta\phi)\|_s \leq C\left(\|\zeta'\alpha\|_{s-1} + \|\alpha\|_{s-2}\right)$.

For the case of $E = 1_X$, Theorem 2.1 is proved in Theorem 3.3 of [8] for a more general pseudoconvex domain Ω. The proof relies on the basic estimate for the scalar-valued $\overline{\partial}$-Neumann problem with weight on a pseudoconvex domain and the following strong inequality for $P \in \mathcal{P}^0 \cup \mathcal{P}^-$;

$$\|Pu\|_1 \leq C\left(\sum_k \|\frac{\partial Pu}{\partial \overline{z}_k}\| + \|Pu\|\right) \quad \text{for } u \in C_0^\infty(U \cap \overline{\Omega}) \qquad (2.2)$$

where $\| \;\|$ (resp. $\| \;\|_1$) denotes the L^2-norm (resp. the Sobolev 1-norm).

Theorem 2.1 is proved by an argument parallel to [8], using the basic estimate (2.1) and the same inequality (2.2). (Note that (2.2) still works in the E-valued case since $\Gamma(U \cap \overline{\Omega}, E)$ is naturally identified with $\oplus^r C^\infty(U \cap \overline{\Omega})$).

3. E-valued version of the closed range theorem

Let Ω be a strongly pseudoconvex bounded domain of a complex surface X defined by $r < 0$ with a C^∞ function r such that $dr \neq 0$ on the boundary $M = \partial\Omega$ and E is a holomorphic vector bundle on X.

Theorem 3.1 *Let M and E be as above and $\mathbf{E} = E_{|M}$. Then*

$$\operatorname{Im}\left\{\overline{\partial}_{\mathbf{E}} \colon L^2_{(0,0)}(M, \mathbf{E}) \to L^2_{(0,1)}(M, \mathbf{E})\right\}$$

is closed.

The proof of Theorem 3.1 is parallel to [8]. It is enough to prove the following assertion:

(3.1) There exists a constant $C > 0$ such that given $\mathbf{u} \in C^\infty(M, \mathbf{E})$ there exists $\mathbf{v} \in C^1(M, \mathbf{E})$ such that $\overline{\partial}_{\mathbf{E}}\mathbf{v} = \overline{\partial}_{\mathbf{E}}\mathbf{u}$ and $\|\mathbf{v}\|_{L^2(M)} \leq C\|\overline{\partial}_{\mathbf{E}}\mathbf{u}\|_{L^2(M)}$, where we denote by $C^1(M, \mathbf{E})$ the space of all C^1-sections of $\mathbf{E}$ and by $\| \;\|_{L^2(M)}$ the L^2-norm on M.

$\mathbf{v}$ in (3.1) is obtained by the following three steps. Let $\mathbf{a} = \overline{\partial}_{\mathbf{E}}\mathbf{u}$.

(i) We construct an extension operator

$$A\colon \ (\operatorname{Ker} \overline{\partial}_{\mathbf{E}}) \cap A_b^{0,1}(\mathbf{E}) \to (\operatorname{Ker} \overline{\partial}_E) \cap A_{\overline{\Omega}}^{0,1}(E).$$

Since we need a modification only in this step, we repeat the argument parallel to [8]: Let $\{U_i\}$ be a coordinate neighbourhood covering of $\overline{\Omega}$ such that $E_{|U_i}$ is trivialized as holomorphic vector bundles. Choose a partition of unity $\{\zeta_i\}$ subordinate to the covering $\{U_i\}$ and $\zeta_i' \in C_0^\infty(U_i)$ such that $\zeta_i' = 1$ in a neighbourhood of Supp ζ_i. First we can extend $\zeta_i'\mathbf{a}$ to $\alpha_i' \in \oplus^r A^{0,1}_{U_i \cap \overline{\Omega}}(E)$ such that

$$\|\alpha_i'\|_{H^s(\Omega)} \leq C\|\zeta_i'\mathbf{a}\|_{H^{s-\frac{1}{2}}(M)}$$

using the local Fourier transform (cf. [8]) where $\| \;\|_{H^t(M)}$ (resp. $\| \;\|_{H^t(\Omega)}$) denotes the Sobolev norm of order t on M (resp. on Ω). Then we have an extension of $\mathbf{a}$, $\alpha' = \sum_i \zeta_i \alpha_i' \in A_{\overline{\Omega}}^{0,1}(E)$ such that

$$\|\alpha'\|_{H^s(\Omega)} \leq C'\|\mathbf{a}\|_{H^{s-\frac{1}{2}}(M)}.$$

We proceed with the argument in p.540 of [8] using the following two properties (cf. [5], Ch. V and [9] for these properties):

(3.2) Let $\phi \in A^{p,q}_{\overline{\Omega}}(E)$. Then $\mu(\phi) = 0$ if and only if $\sigma(\overline{\partial}, dr)\phi = \overline{\partial} r \wedge \phi = 0$ on M.

(3.3) Let $\#\colon A^{p,q}_X(E) \to A^{2-p,2-q}_X(E^*)$ be an anti-linear isomorphism given by

$$\theta \wedge \phi = h(\theta, \#\phi)dV,\ \phi \in A^{p,q}_X(E),\ \theta \in A^{2-p,2-q}_X(E^*)$$

where h is an Hermitian inner product induced from the fixed fibre metric of E^*. Then

$$\vartheta_{E^*}\# = (-1)^{pq+1}\#\overline{\partial}_E,$$
$$\#\mathcal{C}^{p,q}_{\overline{\Omega}}(E) = \mathcal{D}^{2-p,2-q}_{\overline{\Omega}}(E^*)$$

where $\mathcal{C}^{p,q}_{\overline{\Omega}}(E) = \left\{\phi \in A^{p,q}_{\overline{\Omega}}(E) \mid \mu(\alpha) = 0\right\}$.

Then the argument is parallel to [8], p.540. Since $\overline{\partial}_{\mathbf{E}}\mathbf{u} = \mathbf{a}$, by extending $\mathbf{u}$ to $u' \in A^{0,0}_{\overline{\Omega}}(E)$ we have $\overline{\partial} r \wedge \left(\alpha' - \overline{\partial}_E u'\right) = 0$ on M. Let $\theta \in A^{2,0}_{\overline{\Omega}}(E^*)$. Then, since $\iota^*(\partial r + \overline{\partial} r) = 0$, we have $\iota^*\left(\theta \wedge (\alpha' - \overline{\partial}_E u')\right) = 0$ where $\iota\colon M \hookrightarrow X$ denotes the natural inclusion map. Hence

$$\int_\Omega \theta \wedge \overline{\partial}_E \alpha' = \int_M \iota^*\left(\theta \wedge \alpha'\right) - \int_\Omega \overline{\partial}_{E^*}\theta \wedge \alpha'$$
$$= -\int_M \iota^*\left(\overline{\partial}_{E^*}\theta \wedge u'\right) - \int_\Omega \overline{\partial}_{E^*}\theta \wedge \alpha'.$$

This implies

$$(\theta, \#\overline{\partial}_E\alpha') = 0 \ \text{ for all } \theta \in A^{2,0}_{\overline{\Omega}}(E^*) \ \text{ with } \overline{\partial}_{E^*}\theta = 0. \tag{3.4}$$

Hence, if

$$\gamma = N_{E^*}\overline{\partial}_{E^*}\#\overline{\partial}_E\alpha' \in A^{2,1}_{\overline{\Omega}}(E^*),$$

then we have

$$\#^{-1}\gamma \in \mathcal{C}^{0,1}_{\overline{\Omega}}(E),$$
$$\overline{\partial}_E\#^{-1}\gamma = -\#^{-1}\vartheta_{E^*}\gamma = -\#^{-1}\overline{\partial}^*_{E^*}N_{E^*}\overline{\partial}_{E^*}\#\overline{\partial}_E\alpha' = -\overline{\partial}_E\alpha'$$

by (3.3) and (3.4). We set

$$A\mathbf{a} = \alpha' + \#^{-1}\gamma. \tag{3.5}$$

(ii) By the Hodge decomposition theorem for $L^2_{(0,1)}(\overline{\Omega}, E)$ (cf. [5]), we have a decomposition

$$A\mathbf{a} = \rho_E A\mathbf{a} + \overline{\partial}_E \overline{\partial}_E^* N_E A\mathbf{a}, \tag{3.6}$$

where ρ_E (resp. N_E) denotes the orthogonal projection onto the harmonic space (resp. the Neumann operator). Let $v' = \overline{\partial}_E^* N_E A\mathbf{a}$.

(iii) We restrict the decomposition (3.6) onto M. Since $\mathbf{a}$ is $\overline{\partial}_{\mathbf{E}}$-exact, there exists $\mathbf{v}_0 \in A_b^{0,0}(\mathbf{E})$ such that $\mu(\rho_E A\mathbf{a}) = \overline{\partial}_{\mathbf{E}} \mathbf{v}_0$. Then, if we set

$$\mathbf{v} = \mathbf{v}_0 + \mu(v') \in A_b^{0,0}(\mathbf{E}),$$

we have $\overline{\partial}_{\mathbf{E}} \mathbf{v} = \mathbf{a}$.

The main analytical work is to estimate $\|\mathbf{v}\|_{L^2(M)}$ by $\|\mathbf{a}\|_{L^2(M)}$. Since the harmonic space is finite dimensional, it is reduced to estimate $\|v'\|_{H^{\frac{1}{2}}(\Omega)}$ by $\|\alpha'\|_{H^{\frac{1}{2}}(\Omega)}$. It is done by decomposing

$$\zeta v' = P^0(\zeta v') + P^+(\zeta v') + P^-(\zeta v') + R(\zeta v')$$

where $\zeta = \zeta_i$, $P^0 \in \mathcal{P}^0$, $P^+ \in \mathcal{P}^+$, $P^- \in \mathcal{P}^-$ and R is of order -1.

$\|P^0(\zeta v') + P^-(\zeta v')\|_{H^{\frac{1}{2}}(\Omega)}$ is estimated by $\|A\mathbf{a} - \rho_E A\mathbf{a}\|_{H^{-\frac{1}{2}}(\Omega)}$ and then by $\|A\mathbf{a}\|_{H^{-\frac{1}{2}}(\Omega)}$, using Theorem 2.1 and the fact that the harmonic space is finite dimensional. By (3.5), $\|A\mathbf{a}\|_{H^{-\frac{1}{2}}(\Omega)}$ is estimated by $\|\alpha'\|_{H^{\frac{1}{2}}(\Omega)}$.

Instead of estimating $\|P^+(\zeta v')\|_{H^{\frac{1}{2}}(\Omega)}$, we estimate $\|P^+(\zeta \#^{-1}\gamma)\|_{H^{\frac{1}{2}}(\Omega)}$. This is also done by Theorem 2.1 and Proposition 2.17 of [8].

Lemma 3.2 *If $P \in \mathcal{P}^+$ then there exists $P' \in \mathcal{P}^-$ such that*

$$\#P\alpha = P'\#\alpha \quad \textit{holds for } \alpha \in A^{p,q}_{\overline{\Omega}}(E) \textit{ with } \operatorname{Supp} \alpha \subset U.$$

Proof. We may assume that $E_{|U}$ and $E^*_{|U}$ are trivialized as holomorphic vector bundles. Then $\#$ is represented by a composition of the complex conjugation and a $C^\infty(U)$-linear map. Then Lemma 3.2 follows from the following assertion which is obvious from the definitions of $\mathcal{P}^\pm$: For $P \in \mathcal{P}^+$ then there exists $P' \in \mathcal{P}^-$ such that $\overline{Pu} = P'\overline{u}$ holds for $u \in C_0^\infty(U)$. □

By Lemma 3.2 using Theorem 2.1 again, $\|P^+(\zeta \#^{-1}\gamma)\|_{H^{\frac{1}{2}}(\Omega)}$ is estimated by $\|\#\overline{\partial}_E \alpha'\|_{H^{-\frac{1}{2}}(\Omega)}$ and then by $\|\alpha'\|_{H^{\frac{1}{2}}(\Omega)}$. An estimate of $\|P^+(\zeta v')\|_{H^{\frac{1}{2}}(\Omega)}$ follows from this estimate of $\|P^+(\zeta \#^{-1}\gamma)\|_{H^{\frac{1}{2}}(\Omega)}$.

Thus we proved (3.1) and hence Theorem 3.1.

Theorem 3.1 implies the following.

Corollary 3.3 *(Cf. [8], Corollary 4.14) Under the same assumption as Theorem 3.1,*

(1) $\mathrm{Im}\left\{\overline{\partial}^*_{\mathbf{E}}\colon L^2_{(0,1)}(M,\mathbf{E}) \to L^2_{(0,0)}(M,\mathbf{E})\right\}$ *is closed,*

(2) $\mathrm{Im}\left\{\Box^{(0,0)}_{\mathbf{E}} = \overline{\partial}^*_{\mathbf{E}}\overline{\partial}_{\mathbf{E}}\colon L^2_{(0,0)}(M,\mathbf{E}) \to L^2_{(0,0)}(M,\mathbf{E})\right\}$ *is closed,*

(3) $\mathrm{Im}\left\{\Box^{(0,1)}_{\mathbf{E}} = \overline{\partial}_{\mathbf{E}}\overline{\partial}^*_{\mathbf{E}}\colon L^2_{(0,1)}(M,\mathbf{E}) \to L^2_{(0,1)}(M,\mathbf{E})\right\}$ *is closed.*

The proof is same as Corollary 4.14 of [8].
By Corollary 3.3, we have the strong decomposition;

$$L^2_{(0,q)}(\mathbf{E}) = \mathrm{Ker}\ \Box^{(0,q)}_{\mathbf{E}} \oplus \mathrm{Im}\ \Box^{(0,q)}_{\mathbf{E}} \quad (q = 0, 1).$$

We denote

$$H^{(0,q)}(\mathbf{E}) = \mathrm{Ker}\ \Box^{(0,q)}_{\mathbf{E}}$$

and by $\rho^q_{\mathbf{E}}$ and $N^q_{\mathbf{E}}$ the orthogonal projection onto $H^{(0,q)}(\mathbf{E})$ and the partial inverse of $\Box^{(0,q)}_{\mathbf{E}}$ respectively.

Proposition 3.4 (1) $\overline{\partial}_{\mathbf{E}}\rho^0_{\mathbf{E}} = \rho^1_{\mathbf{E}}\overline{\partial}_{\mathbf{E}} = 0$, $\overline{\partial}^*_{\mathbf{E}}\rho^1_{\mathbf{E}} = \rho^0_{\mathbf{E}}\overline{\partial}^*_{\mathbf{E}} = 0$,

(2) $\overline{\partial}^*_{\mathbf{E}}\overline{\partial}_{\mathbf{E}}N^0_{\mathbf{E}} = 1 - \rho^0_{\mathbf{E}}$, $N^0_{\mathbf{E}}\overline{\partial}^*_{\mathbf{E}}\overline{\partial}_{\mathbf{E}} = 1 - \rho^0_{\mathbf{E}}$ *on* Dom $\overline{\partial}_{\mathbf{E}}$,

(3) $\overline{\partial}_{\mathbf{E}}N^0_{\mathbf{E}}\overline{\partial}^*_{\mathbf{E}} = 1 - \rho^1_{\mathbf{E}}$ *on* Dom $\overline{\partial}^*_{\mathbf{E}}$,

(4) $\overline{\partial}_{\mathbf{E}}\overline{\partial}^*_{\mathbf{E}}N^1_{\mathbf{E}} = 1 - \rho^1_{\mathbf{E}}$, $N^1_{\mathbf{E}}\overline{\partial}_{\mathbf{E}}\overline{\partial}^*_{\mathbf{E}} = 1 - \rho^1_{\mathbf{E}}$ *on* Dom $\overline{\partial}^*_{\mathbf{E}}$.

Proof. (1), (2) and (4) are clear from the definitions of $\rho^q_{\mathbf{E}}$ and $N^q_{\mathbf{E}}$. Proof of (3): For $v \in$ Dom $\overline{\partial}^*_{\mathbf{E}}$, $\overline{\partial}^*_{\mathbf{E}}\left(1 - \overline{\partial}_{\mathbf{E}}N^0_{\mathbf{E}}\overline{\partial}^*_{\mathbf{E}}\right)v = \overline{\partial}^*_{\mathbf{E}}v - \Box_{\mathbf{E}}N^0_{\mathbf{E}}\overline{\partial}^*_{\mathbf{E}}v = 0$. If $\overline{\partial}^*_{\mathbf{E}}v = 0$, then $\left(1 - \overline{\partial}_{\mathbf{E}}N^0_{\mathbf{E}}\overline{\partial}^*_{\mathbf{E}}\right)v = v$. Therefore we have $1 - \overline{\partial}_{\mathbf{E}}N^0_{\mathbf{E}}\overline{\partial}^*_{\mathbf{E}} = \rho^1_{\mathbf{E}}$ on Dom $\overline{\partial}^*_{\mathbf{E}}$. □

4. Estimates of operators

Let $\rho^q_{\mathbf{E}}\colon L^2_{(0,q)}(\mathbf{E}) \to H^{(0,q)}(\mathbf{E})$ and $N^q_{\mathbf{E}}\colon L^2_{(0,q)}(\mathbf{E}) \to \mathrm{Dom}\left(\Box^{(0,q)}_{\mathbf{E}}\right)$ be operators defined in §3. Let $\|\ \|_{(k)}$ and $|\ |_{(\alpha)}$ be the Folland-Stein norm of order k and the nonisotropic Lipschitz norm of order α respectively (cf. [6] for these norms).

Proposition 4.1 *Let $k \geq 0$ and $\alpha > 0$.*

$$\|\rho^0_{\mathbf{E}}\mathbf{u}\|_{(k)} \leq C\|\mathbf{u}\|_{(k)}, \qquad |\rho^0_{\mathbf{E}}\mathbf{u}|_{(\alpha)} \leq C|\mathbf{u}|_{(\alpha)}. \tag{1}$$
$$\|\rho^1_{\mathbf{E}}\mathbf{u}\|_{(k)} \leq C\|\mathbf{u}\|_{(k)}, \qquad |\rho^1_{\mathbf{E}}\mathbf{u}|_{(\alpha)} \leq C|\mathbf{u}|_{(\alpha)}. \tag{2}$$
$$\|N^0_{\mathbf{E}}\mathbf{u}\|_{(k+2)} \leq C\|\mathbf{u}\|_{(k)}, \qquad |N^0_{\mathbf{E}}\mathbf{u}|_{(\alpha+2)} \leq C|\mathbf{u}|_{(\alpha)}. \tag{3}$$

We fix the Levi metric on M, then by Theorems 15.19 and 15.20 of [6], (1) and (3) of Proposition 4.1 follow from the following assertion (4.1) ((2) follows from (3)):

(4.1) $N^0_{\mathbf{E}}$ and $\rho^0_{\mathbf{E}}$ are operators of order -2 and 0 respectively in the Heisenberg calculus on M relative to the contact plane field underlying the CR structure.

Assertion (4.1) is a consequence of the Heisenberg calculus in §25 of [1] where (4.1) is proved for $\mathbf{E} = 1_M$. We follow the argument of [1].

(i) Let $y \in M$. For the local Heisenberg model operator $\Box^y_{\mathbf{E}}$ of $\Box^{(0,0)}_{\mathbf{E}}$, there exist a partial inverse $Q^y_{\mathbf{E}}$ of $\Box^y_{\mathbf{E}}$ and the projection $S^y_{\mathbf{E}}$ onto the complement of the image of $\Box^y_{\mathbf{E}}$ such that

$$\Box^y_{\mathbf{E}} Q^y_{\mathbf{E}} + S^y_{\mathbf{E}} = I. \tag{4.2}$$

Theorem 23.9 of [1] proves the assertion (i) for $\mathbf{E} = 1_M$. It also provides that $Q^y_{\mathbf{E}}$ and $S^y_{\mathbf{E}}$ since $\Box^y_{\mathbf{E}} = \oplus^r \Box^y_b$. (4.2) implies

$$\Box^{(0,0)}_{\mathbf{E}} Q^y_{\mathbf{E}} + S^y_{\mathbf{E}} - I \ \text{ is of order } -1. \tag{4.3}$$

(ii) There exists a local Heisenberg operator (a $\mathcal{V}$-operator, in the terminology of [1]) $\tilde{S}^*_{\mathbf{E}}$ of order 0 such that

$$\begin{aligned} \tilde{S}^*_{\mathbf{E}} \vartheta_{\mathbf{E}} &\sim 0, \\ \tilde{S}^*_{\mathbf{E}} - (S^y_{\mathbf{E}})^* &\ \text{ is of order } -1. \end{aligned} \tag{4.4}$$

Assertion (ii) is proved by Theorem 25.59 of [1] for $\mathbf{E} = 1_M$ by constructing the kernel of $\tilde{S}^*$ asymptotically using an asymptotic expansion of ϑ_b. The same calculation is still valid for $\vartheta_{\mathbf{E}}$ since the asymptotic expansion of $\vartheta_{\mathbf{E}}$ has the same leading term as $\oplus^r \vartheta_b$. If $\tilde{S}_{\mathbf{E}}$ is the adjoint of $\tilde{S}^*_{\mathbf{E}}$, then

$$\begin{aligned} \Box^{(0,0)}_{\mathbf{E}} \tilde{S}_{\mathbf{E}} &\sim 0, \\ \tilde{S}_{\mathbf{E}} - S^y_{\mathbf{E}} &\ \text{ is of order } -1. \end{aligned} \tag{4.5}$$

Then using Proposition 25.4 of [1], we can adjust $Q^y_{\mathbf{E}}$ and $\tilde{S}_{\mathbf{E}}$ to local Heisenberg operators $Q_{\mathbf{E}}$ and $S_{\mathbf{E}}$ having the same order and the properties

$$\begin{gathered}\Box^{(0,0)}_{\mathbf{E}} Q_{\mathbf{E}} + S_{\mathbf{E}} \sim I,\\ S_{\mathbf{E}} \sim S^*_{\mathbf{E}} \sim S^2_{\mathbf{E}},\ \Box^{(0,0)}_{\mathbf{E}} S_{\mathbf{E}} \sim S_{\mathbf{E}} \Box^{(0,0)}_{\mathbf{E}} \sim 0,\\ S_{\mathbf{E}} Q_{\mathbf{E}} \sim Q_{\mathbf{E}} S_{\mathbf{E}} \sim 0,\\ Q_{\mathbf{E}} \sim Q^*_{\mathbf{E}},\ \Box^{(0,0)}_{\mathbf{E}} Q_{\mathbf{E}} \sim Q_{\mathbf{E}} \Box^{(0,0)}_{\mathbf{E}}.\end{gathered} \tag{4.6}$$

By patching the local operators as above using the partition of unity, we can globalize them to global Heisenberg operators having the same order and satisfying (4.6). Finally, (4.1) follows from Theorem 25.20 of [1].

References

[1] Beals, R. and Greiner P., "Calculus on Heisenberg manifolds", Annals of Mathematics Studies 119, Princeton Univ. Press, 1988.

[2] Burns, D., *Global behavior of some tangential Cauchy–Riemann equations*, in "Partial Differential Equations and Geometry", Dekker, 1979, 51–56.

[3] Boutet de Monvel, *Intégration des equations de Cauchy–Riemann induites formelles*, Séminaire Goulaouic-Lions-Schwartz, Exposé (1974-1975).

[4] Falbel, E., *Non-embeddable CR-manifolds and surface singularities*, Invent. Math. **108** (1992), 49–65.

[5] Folland, G. B. and Kohn, J. J., "The Neumann problem for the Cauchy–Riemann complex", Annals of Mathematics Studies 75, Princeton Univ. Press, 1972.

[6] Folland, G. B. and Stein, E. M., *Estimates for the $\overline{\partial}_b$ complex and analysis on the Heisenberg group*, Comm. Pure Appl. Math. **27** (1974), 429–522.

[7] Kodaira, K., "Complex Manifolds and Deformation of Complex Structures", Die Grundlehren der mathematischen Wissenschaften 283, Springer-Verlag, 1986.

[8] Kohn, J. J., *The range of the tangential Cauchy–Riemann operator*, Duke Math. J. **53** (1986), 525–545.

[9] Kohn, J. J. and Rossi, H., *On the extension of holomorphic functions from the boundary of a complex manifold*, Ann. of Math. **81** (1965), 451–472.

[10] Kuranishi, M., *Application of $\overline{\partial}_b$ to deformation of isolated singularities*, Proc. Symp. Pure Math. 30, A.M.S., 1977, 97–106.

[11] Miyajima, K., *CR construction of the flat deformations of normal isolated singularities*, to appear in J. Alg. Geom.

[12] Tanaka, N., "A differential-geometric study on strongly pseudoconvex manifolds", Lectures in Mathematics, Kyoto University, No. 9, Kinokuniya Shoten, 1975.

Department of Mathematics and Computer Science
Faculty of Science
Kagoshima University
Kagoshima, 890-0065, JAPAN

E-mail: miyajima@sci.kagoshima-u.ac.jp

CHAPTER X

Discrete Groups of Complex Hyperbolic Isometries and Pseudo-Hermitian Structures

Shin Nayatani†

Abstract

For a discrete isometry group of complex hyperbolic space, we construct a distinguished contact form on the quotient of the domain of discontinuity (contained in the sphere at infinity) by the group, compatible with its strongly pseudoconvex CR structure. We compute the pseudo-Hermitian curvature and torsion of the contact form, and indicate an application.

Introduction

Complex hyperbolic space $H_{\mathbb{C}}^{n+1}$ is a complete, simply-connected Kähler manifold of complex dimension $n+1$ whose holomorphic sectional curvature is identically -4. Each automorphism (= holomorphic isometry) of $H_{\mathbb{C}}^{n+1}$ extends to its boundary at infinity $\partial_\infty H_{\mathbb{C}}^{n+1} = S^{2n+1}$, and preserves the standard strongly pseudoconvex CR structure on it. The automorphism group of $H_{\mathbb{C}}^{n+1}$ is thus identified with the CR automorphism group of S^{2n+1}, and we may use the common notation $G_{\mathbb{C}}(n+1)$ to denote both these groups.

Let Γ be a torsion-free, discrete subgroup of $G_{\mathbb{C}}(n+1)$. Then Γ acts on $H_{\mathbb{C}}^{n+1}$ properly discontinuously and freely, and the quotient manifold $H_{\mathbb{C}}^{n+1}/\Gamma$ inherits the local Kähler geometry of $H_{\mathbb{C}}^{n+1}$. Turning our sight to the boundary S^{2n+1}, the domain of discontinuity $\Omega(\Gamma)$ is the largest open subset of S^{2n+1} on which Γ acts properly discontinuously and freely, and the quotient manifold $X = \Omega(\Gamma)/\Gamma$ comes equipped with a natural CR structure, locally equivalent to the standard one of S^{2n+1}. Since this CR structure on X is again strongly pseudoconvex, the underlying corank one

†Partly supported by the Grant-in-Aid for Scientific Research, The Ministry of Education, Science, Sports and Culture, Japan.

subbundle gives a contact structure, and a choice of compatible contact form gives rise to a pseudo-Hermitian geometry on X.

In this paper, we propose a distinguished contact form θ on X compatible with the CR structure. The standard contact form θ_0 on S^{2n+1} is not invariant but rather transforms "conformally" under the action of $G_{\mathbb{C}}(n+1)$, and our idea is to multiply θ_0 by a positive function λ on $\Omega(\Gamma)$ so that the resulting contact form $\theta = \lambda\theta_0$ be Γ-invariant. In fact, following the construction [12] in the conformal category, we display such a λ explicitly in terms of the Green function for the CR Yamabe operator of (S^{2n+1}, θ_0) and the Patterson-Sullivan measure on the limit set $\Lambda(\Gamma) = S^{2n+1} \setminus \Omega(\Gamma)$.

Upon computing the pseudo-Hermitian curvature and torsion of θ, it turns out that they are closely related to the critical exponent $\delta(\Gamma)$ of Γ, which coincides with the Hausdorff dimension of $\Lambda(\Gamma)$ with respect to the Carnot distance of S^{2n+1} for convex cocompact Γ [5]. For example, it is roughly true that if $\delta(\Gamma) <$ (resp. $=, >$) n, the pseudo-Hermitian scalar curvature of θ is positive (resp. zero, negative) everywhere (see Theorem 2.4 for the precise statement).

We then discuss a possible application; a differential-geometric proof of M. Bourdon and C.-B. Yue's recent result on the Hausdorff dimension of the limit set of a "complex quasi-Fuchsian group" [2, 18]. In fact, we reduce this result to a conjecture on vanishing of the cohomology of X, which we believe could be proved by applying (possibly an appropriate improvement of) M. Rumin's general pseudo-Hermitian vanishing theorem [14] to our contact form θ. While this conjecture is left unproved, we prove a related vanishing result for $H^2(X;\mathbb{R})$ (Theorem 3.5).

This paper is organized as follows. In §1 we review basic concepts, formulas and examples in CR and pseudo-Hermitian geometry. In §2, for a discrete complex-hyperbolic group, we construct a contact form on the quotient of the domain of discontinuity by the group, and compute its pseudo-Hermitian curvature and torsion. In §3 we indicate an application of our contact form, and prove a vanishing result.

We have learned that C. B. Yue [19] constructed the same contact form as ours, and computed its pseudo-Hermitian scalar curvature.

1. Preliminaries

Let M be an orientable manifold of real dimension $2n+1$. A *CR structure* on M is given by a corank one subbundle Q of TM, the tangent bundle of M, together with a complex structure $J : Q \to Q$. We shall assume throughout that the CR structure is *integrable*; that is, it satisfies the formal Frobenius condition $[Q^{1,0}, Q^{1,0}] \subset Q^{1,0}$, where

$$Q^{1,0} = \{Z \in Q \otimes \mathbb{C} \mid JZ = \sqrt{-1}\, Z\}.$$

Note that $Q^{1,0}$ is a complex rank n subbundle of $TM \otimes \mathbb{C}$ satisfying $Q^{1,0} \cap \overline{Q^{1,0}} = \{0\}$, and recovers Q and J by $Q = \mathrm{Re}(Q^{1,0} \oplus \overline{Q^{1,0}})$ and $J(Z + \overline{Z}) = \sqrt{-1}(Z - \overline{Z})$ for $Z \in Q^{1,0}$, respectively.

Let θ be a one-form on M whose kernel is the bundle of hyperplanes Q. Such a θ exists globally, since we assume M is orientable, and Q is oriented by its complex structure. Associated with θ is the real Hermitian (i.e., J-invariant, symmetric) form L_θ on Q:

$$L_\theta(X, Y) = d\theta(X, JY), \quad X, Y \in Q,$$

called the *Levi form* of θ. If θ is replaced by $\theta' = \lambda\theta$, $\lambda > 0$, then L_θ changes conformally by $L_{\theta'} = \lambda L_\theta$. Two such forms θ and θ' are called *pseudoconformal* to each other. We shall assume that M is *strongly pseudoconvex*, that is, that L_θ is positive definite for a suitable choice of θ. In this case, Q gives a contact structure on M, and we call θ a *contact form*.

The most important example of an integrable CR structure is that induced by an embedding of M in a complex manifold Ω of complex dimension $n + 1$, in which case $Q^{1,0} = T^{1,0}\Omega \cap (TM \otimes \mathbb{C})$. If ρ is a defining function for M, then one choice for the contact form is $\theta = \dfrac{\sqrt{-1}}{2}(\bar{\partial} - \partial)\rho$.

A *pseudo-Hermitian structure* on M is a CR structure together with a choice of contact form θ. Corresponding to such a choice, there is a unique vector field (Reeb field) $T = T_\theta$ on M transverse to Q, defined by

$$d\theta(T, \cdot) = 0, \quad \theta(T) = 1.$$

This defines T uniquely because $d\theta$ is nondegenerate on Q and thus has precisely one null direction transverse to Q. Also, M carries a natural volume form $\theta \wedge (d\theta)^n$. The Levi form L_θ will be denoted by g when it is regarded as a Hermitian metric on Q. For p, $q \in M$, the *Carnot distance* $d(p, q)$ between p and q is defined as the infimum of g-length of curves from p to q whose tangent vectors lie in the contact subbundle Q. By Mitchell's calculation [11], the Hausdorff dimension of M with respect to the distance d is $2n + 2$.

On a pseudo-Hermitian manifold (M, θ) there is a natural affine connection $D = D_\theta$, known as the *Tanaka-Webster connection* [16, 17]. The connection D is characterized by the following conditions:

(i) the contact subbundle Q is D-parallel;

(ii) the tensor fields $g = L_\theta$, T and J are all D-parallel;

(iii) the torsion tensor Tor of D satisfies

$$\mathrm{Tor}(X,Y) = d\theta(X,Y)T,$$

$$\mathrm{Tor}(T,JX) = -J(\mathrm{Tor}(T,X))$$

for all $X, Y \in Q$.

It follows from the conditions (i), (ii) that θ and $d\theta$ are also D-parallel.

Explicitly, for sections X, Y of $Q^{1,0}$, the covariant derivatives $D_{\overline{X}}Y$, D_XY and D_TY are the sections of $Q^{1,0}$ given by

$$D_{\overline{X}}Y = \text{the } Q^{1,0}\text{-component of } [\overline{X},Y], \tag{1.1}$$

$$g(D_XY,\overline{Z}) = X \cdot g(Y,\overline{Z}) - g(Y,\overline{D_{\overline{X}}Z}) \text{ for all sections } Z \text{ of } Q^{1,0}, \tag{1.2}$$

$$D_TY = \text{the } Q^{1,0}\text{-component of } [T,Y], \tag{1.3}$$

respectively. Moreover, $D_X\overline{Y} = \overline{D_{\overline{X}}Y}$, $D_{\overline{X}}\overline{Y} = \overline{D_XY}$, $D_T\overline{Y} = \overline{D_TY}$, and $DT = 0$. It follows that

$$\mathrm{Tor}(T,Y) = -(\text{the } \overline{Q^{1,0}}\text{-component of } [T,Y]). \tag{1.4}$$

The symmetric bilinear form τ on Q defined by

$$\tau(X,Y) = g(\mathrm{Tor}(T,X),JY), \quad X,Y \in Q,$$

is equivalent to Webster's torsion one-forms of (M,θ) [17]. By the condition (iii) above, τ is J-anti-invariant:

$$\tau(JX,JY) = -\tau(X,Y).$$

Henceforth, we shall refer to τ as the *pseudo-Hermitian torsion* of (M,θ).

The pseudo-Hermitian curvature tensor R is part of the curvature tensor of D which is characterized by

$$\begin{aligned} g(R(JX,JY)Z,W) &= g(R(X,Y)Z,W), \\ g(R(X,Y)Z,W) &= g(R(Z,W)X,Y) \end{aligned}$$

for all $X, Y, Z, W \in Q$.

The *pseudo-Hermitian Ricci tensor* is the real Hermitian form Ric on Q defined by

$$\mathrm{Ric}(X,Y)=\sum_{i=1}^{2n} g(R(X,e_i)e_i,Y), \quad X,Y\in Q,$$

where $\{e_1,\dots,e_{2n}\}$ is a g-orthonormal basis for Q. The *pseudo-Hermitian scalar curvature* is $S=\frac{1}{2}\mathrm{tr}_g\mathrm{Ric}$, where we set $\mathrm{tr}_g A=\sum_{i=1}^{2n} A(e_i,e_i)$ for a bilinear form A on Q. For a function u, $d_b u$ denotes the differential of u restricted to Q, and the Hessian of u with respect to D is given by

$$\begin{aligned} Dd_b u(X,Y) &= (D_Y d_b u)(X)\\ &= YXu-(D_Y X)u, \quad X,Y\in Q. \end{aligned}$$

Note that $Dd_b u$ is *not* symmetric since the connection D has torsion. We denote by $Dd_b u^{\mathrm{Sym}}$ the symmetrization of $Dd_b u$:

$$Dd_b u^{\mathrm{Sym}}(X,Y)=\frac{1}{2}(Dd_b u(X,Y)+Dd_b u(Y,X)).$$

The *sublaplacian* Δ_b, which is subelliptic, is defined on u by $\Delta_b u = -\mathrm{tr}_g Dd_b u$.

We now recall the transformation law for the pseudo-Hermitian curvature and torsion under a change of contact form [9, 10]. Let $\theta'=e^{2f}\theta$ be a new choice of contact form for the CR manifold M. The pseudo-Hermitian Ricci tensor, scalar curvature and torsion associated with θ' are

$$\mathrm{Ric}'=\mathrm{Ric}-2(n+2)\left(Dd_b f^{\mathrm{Sym}}\right)^{(1)}+\left(\Delta_b f-2(n+1)|d_b f|^2\right)g, \tag{1.5}$$

$$S'=e^{-2f}\left(S+2(n+1)\Delta_b f-2n(n+1)|d_b f|^2\right), \tag{1.6}$$

$$\tau'=\tau-2(Dd_b f-2d_b f\otimes d_b f)^{(2)}, \tag{1.7}$$

respectively, where $A^{(1)}$ (resp. $A^{(2)}$) is the J-invariant (resp. J-anti-invariant) part of a bilinear form A on Q; that is,

$$A^{(1)}(X,Y)=\frac{1}{2}(A(X,Y)+A(JX,JY)),$$

$$A^{(2)}(X,Y)=\frac{1}{2}(A(X,Y)-A(JX,JY))$$

for $X, Y \in Q$. Letting $f = \frac{1}{n} \log u$ so that $e^{2f} = u^{2/n}$, we may rewrite (1.6) as

$$S' = u^{-\frac{n+2}{n}} \left(\frac{2(n+1)}{n} \Delta_b u + Su \right).$$

The operator $\mathcal{L} = \frac{2(n+1)}{n} \Delta_b + S$ is called the *CR Yamabe operator* [8].

Example. Let $S^{2n+1} \subset \mathbb{C}^{n+1}$ be the unit sphere:

$$S^{2n+1} = \{ z = (z_1, \ldots, z_{n+1}) \in \mathbb{C}^{n+1} \mid |z|^2 = z \cdot \bar{z} = 1 \},$$

where $v \cdot w = \sum_{i=1}^{n+1} v_i w_i$ for $v = (v_1, \ldots, v_{n+1})$, $w = (w_1, \ldots, w_{n+1}) \in \mathbb{C}^{n+1}$. As a real hypersurface in $\mathbb{C}^{n+1}$, S^{2n+1} is equipped with an integrable CR structure. Explicitly, the subbundle Q and the complex structure J are given by

$$Q_z = \{ X \in \mathbb{C}^{n+1} \mid X \cdot \bar{z} = 0 \}, \quad z \in S^{2n+1},$$

and

$$JX = \sqrt{-1}\, X, \quad X \in Q,$$

respectively. With $\rho(z) = |z|^2 - 1$, a defining function for S^{2n+1}, let

$$\begin{aligned} \theta_0 &= \frac{\sqrt{-1}}{2} (\bar{\partial} - \partial) \rho \\ &= \frac{\sqrt{-1}}{2} \sum_{i=1}^{n+1} (z_i d\bar{z}_i - \bar{z}_i dz_i). \end{aligned}$$

It is easy to verify that the Levi form L_{θ_0} is twice the standard Riemannian metric of S^{2n+1} restricted to Q. Hence the CR structure of S^{2n+1} is strongly pseudoconvex. We shall refer to θ_0 as the *standard contact form* of S^{2n+1}.

Example. The *Heisenberg group* $\mathcal{H}^{2n+1}$ is the Lie group whose underlying manifold is $\mathbb{C}^n \times \mathbb{R}$ with coordinates $(\zeta, t) = (\zeta_1, \ldots, \zeta_n, t)$ and whose (nonabelian) group law is given by

$$(\zeta, t) + (\zeta', t') = (\zeta + \zeta', t + t' + 2\mathrm{Im}\,(\zeta \cdot \bar{\zeta}')).$$

The Heisenberg norm of $x = (\zeta, t) \in \mathcal{H}^{2n+1}$ is

$$\|x\| = \left(|\zeta|^4 + t^2 \right)^{1/4},$$

and the Heisenberg dilations are the mappings

$$x \mapsto ax = \left(a\zeta, a^2 t\right), \quad a > 0.$$

We have the identities

$$a(x+y) = ax + ay, \quad ||ax|| = a||x||$$

for $a > 0$ and $x, y \in \mathcal{H}^{2n+1}$.

The complex vector fields

$$Z_\alpha = \frac{\partial}{\partial \zeta_\alpha} + \sqrt{-1}\,\bar{\zeta}_\alpha \frac{\partial}{\partial t}, \quad \alpha = 1, \ldots, n,$$

are left-invariant, and then $Q^{1,0} = \mathrm{span}_{\mathbb{C}}\{Z_1, \ldots, Z_n\}$ gives a left-invariant CR structure on $\mathcal{H}^{2n+1}$. It is also strongly pseudoconvex, and the left-invariant one-form

$$\theta_1 = \frac{1}{2}\left(dt + \sqrt{-1}\sum_{\alpha=1}^{n}\left(\zeta_\alpha d\bar{\zeta}_\alpha - \bar{\zeta}_\alpha d\zeta_\alpha\right)\right)$$

is the standard choice of contact form. Indeed, the Levi form of θ_1 is given by $L_{\theta_1}\left(Z_\alpha, \overline{Z_\beta}\right) = \delta_{\alpha\beta}$, that is, $\{Z_1, \ldots, Z_n\}$ is an orthonormal basis for $Q^{1,0}$ with respect to L_{θ_1}. $T = 2\frac{\partial}{\partial t}$ is the Reeb field associated with θ_1. Since $[\overline{Z_\alpha}, Z_\beta] = \sqrt{-1}\,\delta_{\alpha\beta}T$ and $[T, Z_\alpha] = 0$, it follows from (1.1)-(1.4) that $\{Z_\alpha, \overline{Z_\alpha}, T\}$ is a parallel frame for the Tanaka-Webster connection, the pseudo-Hermitian torsion τ vanishes identically, and so does the curvature. Also, the Hessian and sub-Laplacian associated with θ_1 is given by

$$Dd_b u\left(Z_\alpha, \overline{Z_\beta}\right) = \overline{Z_\beta} Z_\alpha u, \quad Dd_b u\left(Z_\alpha, Z_\beta\right) = Z_\beta Z_\alpha u,$$

$$\Delta_b u = -\sum_{\alpha=1}^{n}\left(\overline{Z_\alpha} Z_\alpha u + Z_\alpha \overline{Z_\alpha} u\right),$$

respectively.

The sphere S^{2n+1}, with one point removed, is CR equivalent to $\mathcal{H}^{2n+1}$. Indeed, the mapping

$$\begin{aligned} F : (z_1, \ldots, z_{n+1}) &\in S^{2n+1} \backslash \{(0, \ldots, 0, -1)\} \\ &\mapsto \left(\frac{z_1}{1+z_{n+1}}, \ldots, \frac{z_n}{1+z_{n+1}}, \frac{\sqrt{-1}\left(\overline{z_{n+1}} - z_{n+1}\right)}{\left(1+z_{n+1}\right)\left(1+\overline{z_{n+1}}\right)}\right) \in \mathcal{H}^{2n+1} \end{aligned}$$

gives a CR equivalence. Hence $F^*\theta_1$ is a contact form for the CR structure of S^{2n+1}, and thus has the form $\lambda\theta_0$, $\lambda > 0$. Explicitly, we have

$$F^*\theta_1 = \frac{1}{|1+z_{n+1}|^2}\theta_0, \quad \text{or} \quad \theta_0 = F^*\left(\frac{4}{(1+|\zeta|^2)^2+t^2}\theta_1\right).$$

From now on, we identify $S^{2n+1}\setminus\{(0,\dots,0,-1)\}$ with $\mathcal{H}^{2n+1}$ through F, and omit F from the formulas.

Since θ_0 is locally pseudoconformally equivalent to θ_1, which is flat, Chern's pseudoconformal curvature associated with θ_0 vanishes, and thus the pseudo-Hermitian curvature of θ_0 is completely determined by the pseudo-Hermitian Ricci tensor [17]. Computation based on the formulas (1.5)-(1.7) gives

$$\mathrm{Ric}_{\theta_0} = (n+1)g_0, \quad S_{\theta_0} = n(n+1), \quad \tau_{\theta_0} = 0.$$

We now determine the Green function for the CR Yamabe operator $\mathcal{L}_{\theta_0}$. Rather than working with the definition of Green function, we exploit the fact that the contact form $G(\cdot, w_0)^{2/n}\theta_0$, where $w_0 = (0,\dots,0,-1)$, coincides with θ_1 (up to a constant multiple). Since $\theta_1 = |1+z_{n+1}|^{-2}\theta_0$, it follows that $G(z,w_0) = |1+z_{n+1}|^{-n}$. By the invariance of G under the action of the unitary group $U(n+1)$, we obtain

$$G(z,w) = |1 - z\cdot\bar{w}|^{-n}$$

(up to a constant multiple).

We denote the group of CR automorphisms of S^{2n+1} by $\mathrm{Aut}_{CR}(S^{2n+1})$. For $\gamma \in \mathrm{Aut}_{CR}(S^{2n+1})$, we define a positive function j_γ on S^{2n+1} by $\gamma^*\theta_0 = {j_\gamma}^2\theta_0$. We have

Lemma 1.1

$$G(\gamma z, \gamma w) = j_\gamma(z)^{-n} j_\gamma(w)^{-n} G(z,w), \quad \gamma \in \mathrm{Aut}_{CR}(S^{2n+1}),\ z, w \in S^{2n+1}.$$

Proof. It is known and easy to verify that the CR Yamabe operator satisfies the transformation law

$$\mathcal{L}_{\theta'}\phi = u^{-\frac{n+2}{n}}\mathcal{L}_\theta(u\phi),$$

where $\theta' = u^{2/n}\theta$.

Clearly, $(z,w) \mapsto G(\gamma z, \gamma w)$ is a Green function for $\mathcal{L}_{\gamma^*\theta_0}$. We shall show that $G'(z,w) = j_\gamma(z)^{-n} j_\gamma(w)^{-n} G(z,w)$ is also a Green function for

$\mathcal{L}_{\gamma^*\theta_0}$. The lemma will then follow from the uniqueness of Green function. For any function ϕ, we compute

$$\begin{aligned}
&\int_{S^{2n+1}} G'(z,w)\left(\mathcal{L}_{\gamma^*\theta_0}\phi\right)(z)\,\gamma^*\theta_0\wedge d\left(\gamma^*\theta_0\right)^n(z)\\
&= \int_{S^{2n+1}} \left(j_\gamma(z)^{-n} j_\gamma(w)^{-n} G(z,w)\right)\left(j_\gamma(z)^{-(n+2)}\mathcal{L}_{\theta_0}\left({j_\gamma}^n\phi\right)(z)\right)\\
&\quad\times j_\gamma(z)^{2n+2}\theta_0\wedge d{\theta_0}^n(z)\\
&= j_\gamma(w)^{-n}\int_{S^{2n+1}} G(z,w)\mathcal{L}_{\theta_0}\left({j_\gamma}^n\phi\right)(z)\,\theta_0\wedge d{\theta_0}^n(z)\\
&= j_\gamma(w)^{-n}\left({j_\gamma}^n\phi\right)(w)=\phi(w).
\end{aligned}$$

We have used the above transformation law in the first equality, with $\theta=\theta_0$ and $u={j_\gamma}^n$. This shows that G' is a Green function for $\mathcal{L}_{\gamma^*\theta_0}$, completing the proof of Lemma 1.1. □

2. Canonical contact form

As a model of complex hyperbolic space $H_{\mathbb{C}}^{n+1}$, we take the ball

$$B_{\mathbb{C}}^{n+1}=\left\{z\in\mathbb{C}^{n+1}\mid |z|<1\right\}$$

endowed with the Bergman metric

$$\frac{1}{1-|z|^2}\left\{\sum_{i=1}^{n+1} dz_i\cdot d\bar{z}_i+\frac{1}{1-|z|^2}\left(\sum_{i=1}^{n+1} z_i d\bar{z}_i\right)\cdot\left(\sum_{j=1}^{n+1}\bar{z}_j dz_j\right)\right\},$$

normalized to have holomorphic sectional curvature -4. Each holomorphic isometry of $B_{\mathbb{C}}^{n+1}$ extends to $S^{2n+1}=\partial B_{\mathbb{C}}^{n+1}$, and gives a CR automorphism of S^{2n+1}. The automorphism group of $B_{\mathbb{C}}^{n+1}$ is thus identified with $\mathrm{Aut}_{CR}(S^{2n+1})$, and we use the common notation $G_{\mathbb{C}}(n+1)$ to denote both these groups. We refer the reader to Goldman's monograph [6] for extensive information on the complex hyperbolic and spherical CR geometries.

Let Γ be a discrete subgroup of $G_{\mathbb{C}}(n+1)$. Its *limit set* $\Lambda(\Gamma)$ is defined as the set of accumulation points in $\overline{B_{\mathbb{C}}^{n+1}}$ of the Γ-orbit of any point in $B_{\mathbb{C}}^{n+1}$. Since Γ acts properly discontinuously on $B_{\mathbb{C}}^{n+1}$, $\Lambda(\Gamma)$ is a (closed) subset of S^{2n+1}. The complement $\Omega(\Gamma)=S^{2n+1}\setminus\Lambda(\Gamma)$ is called the *domain of discontinuity* of Γ. It is the largest open subset of S^{2n+1} on which Γ acts properly discontinuously. If Γ acts on $\Omega(\Gamma)$ freely, which is the case if Γ is torsion-free, then the quotient manifold $X=\Omega(\Gamma)/\Gamma$ comes equipped with a natural CR structure, locally equivalent to the standard one of S^{2n+1}. The *critical exponent* $\delta(\Gamma)$ is defined by

$$\delta(\Gamma) = \inf\left\{ s > 0 \mid \sum_{\gamma\in\Gamma} e^{-sd(z,\gamma z)} < \infty \right\},$$

where $z,\, w \in B^{n+1}_{\mathbb{C}}$ and d is the complex-hyperbolic distance function on $B^{n+1}_{\mathbb{C}}$. Note that $\delta(\Gamma)$ is independent of the particular choice of the points $z,\, w$. It is known that $0 \leq \delta(\Gamma) \leq 2n+2$.

Following S. J. Patterson [13] and D. Sullivan [15], K. Corlette [5] constructed a distinguished family of measures on $\Lambda(\Gamma)$, which we now recall.

Proposition 2.1 *There exists a family of Borel measures μ_z, $z \in B^{n+1}_{\mathbb{C}}$, which has the following properties:*

(i) $\mu_0(\Lambda(\Gamma)) = 1$;

(ii) $\mu_z = e^{-\delta b(z,\cdot)}\mu_0$, *where $\delta = \delta(\Gamma)$ and b is the Busemann function of complex hyperbolic space*;

(iii) $\gamma^*\mu_z = \mu_{\gamma^{-1}z}, \quad \gamma \in \Gamma.$

It follows from (ii) and (iii) of the proposition that μ_0 satisfies

$$\gamma^*\mu_0 = e^{-\delta b(\gamma^{-1}0,0)}\mu_0, \quad \gamma \in \Gamma. \tag{2.1}$$

We call measures μ_z, $z \in B^{n+1}_{\mathbb{C}}$, as in the proposition *Patterson-Sullivan measures.*

We shall now construct a distinguished contact form on $\Omega(\Gamma)$ which is compatible with the CR structure and hence has the form $u^{2/n}\theta_0$, where $u > 0$ and θ_0 is the standard contact form of S^{2n+1}. Let $\mu = \mu_0$ be a Patterson-Sullivan measure with base point at the origin 0. We consider the positive function u on $\Omega(\Gamma)$ of the form

$$u(z) = \left(\int_{\Lambda(\Gamma)} G(z,w)^p d\mu(w) \right)^q, \quad z \in \Omega(\Gamma), \tag{2.2}$$

where G is the Green function for $\mathcal{L}_{\theta_0}$, and determine the exponents p and q so that the contact form $\theta = u^{2/n}\theta_0$ be Γ-invariant. Recall that the one-form θ_0, the function G and the measure μ respectively satisfy the transformation laws

$$\gamma^*\theta_0 = {j_\gamma}^2\theta_0,$$

$$G(\gamma z, \gamma w) = j_\gamma(z)^{-n} j_\gamma(w)^{-n} G(z,w),$$
$$\gamma^* \mu = {j_\gamma}^\delta \mu$$

for $\gamma \in \Gamma$ and $z,\ w \in S^{2n+1}$. The last one follows from (2.1) and the identity $j_\gamma = e^{-b(\gamma^{-1}0,\,\cdot\,)}$. We compute:

$$\begin{aligned} u(\gamma z) &= \left(\int_{\Lambda(\Gamma)} G(\gamma z, w)^p d\mu(w)\right)^q \\ &= \left(\int_{\Lambda(\Gamma)} G(\gamma z, \gamma w)^p d\left(\gamma^* \mu\right)(w)\right)^q \\ &= \left(\int_{\Lambda(\Gamma)} j_\gamma(z)^{-pn} j_\gamma(w)^{-pn} G(z,w)^p j_\gamma(w)^\delta d\mu(w)\right)^q \\ &= j_\gamma(z)^{-q\delta} u(z) \end{aligned}$$

if we choose $p = \delta/n$, and hence

$$\begin{aligned} \gamma^* \theta &= (u \circ \gamma)^{2/n} \gamma^* \theta_0 \\ &= \left({j_\gamma}^{-q\delta} u\right)^{2/n} {j_\gamma}^2 \theta_0 \\ &= \theta \end{aligned}$$

if $\delta > 0$ and $q = n/\delta$ $(= 1/p)$. We thus obtain, assuming $\delta = \delta(\Gamma) > 0$,

$$\begin{aligned} \theta &= \left(\int_{\Lambda(\Gamma)} G(z,w)^{\delta/n} d\mu(w)\right)^{2/\delta} \theta_0 \\ &= \left(\int_{\Lambda(\Gamma)} \varphi(z,w)^{-\delta} d\mu(w)\right)^{2/\delta} \theta_0, \end{aligned} \tag{2.3}$$

where $\varphi(z,w) = |1 - z \cdot \bar{w}|$. Since θ is Γ-invariant, if Γ acts on $\Omega(\Gamma)$ freely, θ projects to a contact form on the quotient manifold $X = \Omega(\Gamma)/\Gamma$ which is compatible with the CR structure. We denote this contact form on X by the same symbol θ, and call it the *canonical contact form* of X.

Remark. If we take $p = q = 1$ in (2.2), then the contact form $u^{2/n}\theta_0$ has vanishing scalar curvature. Indeed, the pseudo-Hermitian scalar curvature is computed by

$$\begin{aligned} S &= u^{-\frac{n+2}{n}} \mathcal{L}_{\theta_0} u \\ &= u^{-\frac{n+2}{n}} \int_{\Lambda(\Gamma)} \mathcal{L}_{\theta_0} G(\cdot, w)\, d\mu(w) \\ &= 0 \qquad \text{on } \Omega(\Gamma). \end{aligned}$$

The above construction of the canonical contact form is a modification of this one, achieving the Γ-invariance of contact form rather than the vanishing of pseudo-Hermitian scalar curvature.

For the computation of the pseudo-Hermitian curvature and torsion of θ, it is more convenient to write θ in the form $v^{2/n}\theta_1$, where $v > 0$ and θ_1 is the standard contact form of $\mathcal{H}^{2n+1} = S^{2n+1} \setminus \{(0, \ldots, 0, -1)\}$. The identification mapping $F^{-1} : \mathcal{H}^{2n+1} \to S^{2n+1}$ is given by

$$F^{-1}(x) = \left(\frac{2\zeta_1}{1 + |\zeta|^2 - \sqrt{-1}\,t}, \ldots, \frac{2\zeta_n}{1 + |\zeta|^2 - \sqrt{-1}\,t}, \frac{1 - |\zeta|^2 + \sqrt{-1}\,t}{1 + |\zeta|^2 - \sqrt{-1}\,t} \right),$$

where $x = (\zeta, t) \in \mathcal{H}^{2n+1}$. By direct computation, we have

$$\varphi\left(F^{-1}(x), F^{-1}(y)\right) = \lambda(x)\lambda(y)\varphi_1(x, y),$$

where

$$\varphi_1(x, y) = \frac{1}{2}\left[|\zeta - \zeta'|^4 + \left(t - t' + 2\mathrm{Im}\zeta \cdot \bar{\zeta}'\right)^2\right]^{1/2}, \quad y = (\zeta', t'),$$

$$\lambda(x) = \left[\frac{4}{\left(1 + |\zeta|^2\right)^2 + t^2}\right]^{1/2}.$$

Let $\mu_1 = \lambda^{-\delta}\left(F^{-1}\right)^* \mu$, and compute

$$\begin{aligned} &\int_{\Lambda(\Gamma)} \varphi\left(F^{-1}(x), w\right)^{-\delta} d\mu(w) \\ &\quad = \int_{F(\Lambda(\Gamma))} \varphi\left(F^{-1}(x), F^{-1}(y)\right)^{-\delta} d\left(F^{-1}\right)^* \mu(y) \\ &\quad = \lambda(x)^{-\delta} \int_{F(\Lambda(\Gamma))} \varphi_1(x, y)^{-\delta} d\mu_1(y). \end{aligned}$$

Since $\left(F^{-1}\right)^* \theta_0 = \lambda^2\theta_1$, we finally obtain

$$\left(F^{-1}\right)^* \theta = \left(\int_{F(\Lambda(\Gamma))} \varphi_1(\cdot, y)^{-\delta} d\mu_1(y)\right)^{2/\delta} \theta_1,$$

or, omitting F,

$$\theta = \left(\int_{\Lambda(\Gamma)} \varphi_1(\cdot, y)^{-\delta} d\mu_1(y)\right)^{2/\delta} \theta_1.$$

We now give a few illustrative examples, where the canonical contact forms turn out to coincide with the standard ones.

Example. Suppose $\Lambda(\Gamma)$ is a single point, say $w_0 = (0, \ldots, 0, -1)$. For any such Γ, $\delta(\Gamma) > 0$ and μ has to be the Dirac measure at w_0. Therefore,

$$\begin{aligned} \theta &= \varphi(\cdot, w_0)^{-2}\theta_0 \\ &= \frac{1}{|1+z_{n+1}|^2}\theta_0 \\ &= \theta_1, \end{aligned}$$

the standard contact form of $\mathcal{H}^{2n+1} = S^{2n+1} \setminus \{w_0\}$.

Example. Suppose $\Lambda(\Gamma)$ consists of two points, say $w_\pm = (0, \ldots, 0, \pm 1)$. For example, a Heisenberg dilation, transplanted on S^{2n+1}, generates such a Γ. Then $\delta(\Gamma) = 0$, and the formula (2.3) does not make sense. A natural choice of contact form is

$$\begin{aligned} \theta &= \varphi(\cdot, w_+)^{-1}\varphi(\cdot, w_-)^{-1}\theta_0 \\ &= \frac{1}{|1-{z_{n+1}}^2|}\theta_0 \\ &= \frac{1}{||x||^2}\theta_1. \end{aligned}$$

Indeed, this θ is invariant by all the CR automorphisms of S^{2n+1} which preserve $\{w_\pm\}$. By direct computation, using the formulas (1.5)-(1.7), we obtain

$$\mathrm{Ric}_\theta = \frac{(n+1)|\zeta|^2}{(|\zeta|^4+t^2)^{1/2}}\,g, \quad S_\theta = \frac{n(n+1)|\zeta|^2}{(|\zeta|^4+t^2)^{1/2}},$$

$$\tau_\theta(Z_\alpha, Z_\beta) = \frac{-\bar{\zeta}_\alpha\bar{\zeta}_\beta}{\left(|\zeta|^2-\sqrt{-1}\,t\right)^2}.$$

Example. For $0 \le k \le n-1$, let $B_{\mathbb{C}}^{k+1}$ be a complex geodesic subspace of complex dimension $k+1$ in $B_{\mathbb{C}}^{n+1}$, and let $S_{\mathbb{C}}^{2k+1} = \partial\, B_{\mathbb{C}}^{k+1}$. We call such a $S_{\mathbb{C}}^{2k+1}$ a $\mathbb{C}$-*sphere*, while it is called a *chain* when $k=0$. For example, we may choose

$$B_{\mathbb{C}}^{k+1} = \left\{(z_1, \ldots, z_{n+1}) \in B_{\mathbb{C}}^{n+1} \mid z_1 = \cdots = z_{n-k} = 0\right\}.$$

Then, under the identification $S^{2n+1} = \mathcal{H}^{2n+1} \cup \{\infty\}$,

$$S_{\mathbb{C}}^{2k+1} = \left\{(\zeta_1, \cdots, \zeta_n, t) \in \mathcal{H}^{2n+1} \mid \zeta_1 = \cdots = \zeta_{n-k} = 0\right\} \cup \{\infty\}.$$

Let Γ be a discrete subgroup of $G_{\mathbb{C}}(n+1)$ such that $\Lambda(\Gamma) = S_{\mathbb{C}}^{2k+1}$ and $\delta(\Gamma) = 2k+2$. (It is very likely that the latter condition would follow from the former.) Then, as a Patterson-Sullivan measure μ, we may take the measure associated with the volume form $\theta_0 \wedge (d\theta_0)^k$ of $S^{2k+1}_{\mathbf{C}}$, normalized so as to have unit total mass. Let H denote the group of all the CR automorphisms of S^{2n+1} which preserve $S^{2k+1}_{\mathbf{C}}$. It is known that H acts on $S^{2n+1} \setminus S^{2k+1}_{\mathbf{C}}$ transitively [4]. Since $\gamma^*\mu = j_\gamma{}^{2k+2}\mu$ for every $\gamma \in H$, the contact form θ is H-invariant. On the other hand, with the choice of $S^{2k+1}_{\mathbf{C}}$ as above, it is easy to verify that the contact form

$$\theta' = \frac{1}{\sum_{i=1}^{n-k} z_i \bar{z}_i}\,\theta_0 = \frac{1}{\sum_{\alpha=1}^{n-k} \zeta_\alpha \bar{\zeta}_\alpha}\,\theta_1$$

is also H-invariant, and that θ and θ' agree at $(1,0,\ldots,0) \in S^{2n+1} \setminus S^{2k+1}_{\mathbf{C}}$ ($\leftrightarrow (1,0,\ldots,0) \in \mathcal{H}^{2n+1}$). It follows that $\theta = \theta'$ on $S^{2n+1} \setminus S^{2k+1}_{\mathbf{C}}$. Again, direct computation gives

$$\mathrm{Ric}_\theta = -(k+2)\, g|_\Pi + (n-k)\, g|_{\Pi^\perp},$$

where $\Pi = \mathrm{span}_{\mathbf{C}} \left\{ \sum_{\rho=1}^{n-k} \zeta_\rho Z_\rho, Z_{n-k+1}, \ldots, Z_n \right\}$ and $\Pi^\perp$ is the orthogonal complement of Π in Q,

$$S_\theta = (n+1)(n-2k-2),$$

$$\tau_\theta = 0.$$

Let $\mathfrak{H}^{2n+1} \subset \mathbf{C}^{n+1}$ be the hyperboloid:

$$\mathfrak{H}^{2n+1} = \left\{ (z_1, \ldots, z_{n+1}) \in \mathbf{C}^{n+1} \mid -z_1\bar{z}_1 + z' \cdot \bar{z}' = -1 \right\},$$

where $z' = (z_2, \ldots, z_{n+1})$. With the CR structure as a real hypersurface in $\mathbf{C}^{n+1}$, $\mathfrak{H}^{2n+1}$ is strongly pseudoconvex, as it is CR equivalent to $S^{2n+1} \setminus S^{2n-1}_{\mathbf{C}}$ by the mapping

$$G : (z_1, z') \in \mathfrak{H}^{2n+1} \mapsto (1/z_1, z'/z_1) \in S^{2n+1} \setminus S^{2n-1}_{\mathbf{C}}.$$

By direct computation, we obtain

$$\begin{aligned} G^*\theta &= G^*\left(\frac{1}{z_1\bar{z}_1}\,\theta_0\right) \\ &= \frac{\sqrt{-1}}{2}\left[-(z_1 d\bar{z}_1 - \bar{z}_1 dz_1) + \sum_{i=2}^{n+1} (z_i d\bar{z}_i - \bar{z}_i dz_i) \right]. \end{aligned}$$

The right-hand side is nothing but the standard contact form of $\mathfrak{H}^{2n+1}$, which is the pseudo-Hermitian counterpart of the hyperbolic metric in Riemannian geometry.

Example. For $0 \le k \le n$, let $B_{\mathbb{R}}^{k+1}$ be a totally-real geodesic subspace of real dimension $k+1$ in $B_{\mathbb{C}}^{n+1}$, and let $S_{\mathbb{R}}^k = \partial B_{\mathbb{R}}^{k+1}$. Such a $S_{\mathbb{R}}^k$ is called an $\mathbb{R}$-*sphere*, and an $\mathbb{R}$-*circle* when $k=1$. One possible choice of $B_{\mathbb{R}}^{k+1}$ is

$$B_{\mathbb{R}}^{k+1} = \left\{ (z_1, \dots, z_{n+1}) \in B_{\mathbb{C}}^{n+1} \;\middle|\; \begin{array}{l} z_1 = \cdots = z_{n-k} = 0, \\ \mathrm{Im} z_{n-k+1} = \cdots = \mathrm{Im} z_{n+1} = 0 \end{array} \right\},$$

and then

$$S_{\mathbb{R}}^k = \left\{ (\zeta_1, \cdots, \zeta_n, t) \in \mathcal{H}^{2n+1} \;\middle|\; \begin{array}{l} \zeta_1 = \cdots = \zeta_{n-k} = t = 0, \\ \mathrm{Im}\zeta_{n-k+1} = \cdots = \mathrm{Im}\zeta_n = 0 \end{array} \right\} \cup \{\infty\}.$$

The $k=0$ case is nothing but the second example above. In what follows, we assume $k \ge 1$.

Let Γ be a discrete subgroup of $G_{\mathbb{C}}(n+1)$ such that $\Lambda(\Gamma) = S_{\mathbb{R}}^k$ and $\delta(\Gamma) = k$. (This latter condition would also be superficial.) Again, the normalized (Riemannian) volume form of $S_{\mathbb{R}}^k$ gives a Patterson-Sullivan measure, and the contact form θ is invariant by the stabilizer H of $S_{\mathbb{R}}^k$. Its action on $S^{2n+1} \setminus S_{\mathbb{R}}^k$ is transitive if and only if $k=n$, when we can obtain an explicit form of θ as in the last example. We use the fact that $S^{2n+1} \setminus S_{\mathbb{R}}^n$ is H-equivariantly diffeomorphic to the unit tangent sphere bundle $S(H_{\mathbb{R}}^{n+1})$ of the real hyperbolic $n+1$-space, where H acts on $S(H_{\mathbb{R}}^{n+1})$ by the tangent maps of isometries of $H_{\mathbb{R}}^{n+1}$ [4]. We pull back the Liouville form of $S(H_{\mathbb{R}}^{n+1})$ to obtain an H-invariant contact form

$$\theta' = \frac{1}{|1 - z \cdot z|} \theta_0$$

on $S^{2n+1} \setminus S_{\mathbb{R}}^n$, and we must have $\theta = \theta'$ (up to a constant multiple). In view of this and the second example, it is natural to expect

$$\begin{aligned} \theta &= \frac{1}{|1 - \sum_{i=n-k+1}^{n+1} z_i{}^2|} \theta_0 \\ &= \frac{1}{\left|\zeta \cdot \bar{\zeta} - \sqrt{-1}\, t - \sum_{\alpha=n-k+1}^{n} \zeta_\alpha{}^2\right|} \theta_1 \end{aligned}$$

for general k. Assuming this, we compute

$$\mathrm{Ric}_\theta = \frac{(n+1) \sum_{\sigma=1}^{n-k} \zeta_\sigma \bar{\zeta}_\sigma}{\left|\zeta \cdot \bar{\zeta} - \sqrt{-1}\, t - \sum_{\rho=n-k+1}^{n} \zeta_\rho{}^2\right|^2} g,$$

$$S_\theta = \frac{n(n+1) \sum_{\sigma=1}^{n-k} \zeta_\sigma \bar{\zeta}_\sigma}{\left|\zeta \cdot \bar{\zeta} - \sqrt{-1}\, t - \sum_{\rho=n-k+1}^{n} \zeta_\rho{}^2\right|^2},$$

$$\tau(Z_\alpha, Z_\beta) = \frac{-1}{\zeta\cdot\bar\zeta - \sqrt{-1}\,t - \sum_{\rho=n-k+1}^{n}\zeta_\rho^{\,2}} \times \left(e_\alpha\delta_{\alpha\beta} + \frac{(\bar\zeta_\alpha - e_\alpha\zeta_\alpha)(\bar\zeta_\beta - e_\beta\zeta_\beta)}{\zeta\cdot\bar\zeta - \sqrt{-1}\,t - \sum_{\rho=n-k+1}^{n}\zeta_\rho^{\,2}} \right),$$

where

$$e_\alpha = \begin{cases} 0, & 1 \le \alpha \le n-k, \\ 1, & n-k+1 \le \alpha \le n. \end{cases}$$

Observe that Ric_θ is positive semidefinite, and vanishes precisely on $S^{2k+1}_{\mathbb{C}} \setminus S^k_{\mathbb{R}}$, where $S^{2k+1}_{\mathbb{C}}$ is the minimal $\mathbb{C}$-sphere containing $S^k_{\mathbb{R}}$. In particular, if $k = n$, Ric_θ vanishes identically, and so does the whole pseudo-Hermitian curvature of θ.

We shall now compute the pseudo-Hermitian curvature and torsion of the contact form θ. Since θ is pseudoconformally equivalent to the flat contact form θ_1, its pseudo-Hermitian curvature is completely determined by the Ricci tensor. Let

$$f = \log\left(\int_{\Lambda(\Gamma)} \varphi_1(\cdot, y)^{-\delta} d\mu_1(y)\right)^{1/\delta}$$

so that $\theta = e^{2f}\theta_1$. By (1.5), (1.7), we have

$$\mathrm{Ric}_\theta = -2(n+2)\left(Dd_bf^{\mathrm{Sym}}\right)^{(1)} + \left(\Delta_b f - 2(n+1)|d_bf|^2\right) g_1, \qquad (2.4)$$

$$\tau_\theta = -2(Dd_bf - 2d_bf\otimes d_bf)^{(2)}, \qquad (2.5)$$

where $g_1 = L_{\theta_1}$, and Dd_bf, $\Delta_b f$ and $|d_bf|^2$ are computed with respect to θ_1.

Recall that

$$\varphi_1(x,y) = \frac{1}{2}\left[|\zeta - \zeta'|^4 + \left(t - t' + 2\mathrm{Im}\zeta\cdot\bar\zeta'\right)^2\right]^{1/2},$$

where $x = (\zeta, t)$, $y = (\zeta', t')$. Letting $\varphi_y(x) = \varphi_1(x,y)$, we first compute the derivatives of φ_y along Q:

$$Z_\alpha\varphi_y = \frac{\Phi}{2|\Phi|}\left(\bar\zeta_\alpha - \bar\zeta'_\alpha\right),$$

$$\overline{Z_\beta}Z_\alpha\varphi_y = \frac{\left(\bar\zeta_\alpha - \bar\zeta'_\alpha\right)\left(\zeta_\beta - \zeta'_\beta\right) + \Phi\delta_{\alpha\beta}}{2|\Phi|},$$

$$Z_\beta Z_\alpha \varphi_y = -\frac{\Phi^2}{2|\Phi|^3}\left(\bar{\zeta_\alpha} - \bar{\zeta'_\alpha}\right)\left(\bar{\zeta_\beta} - \bar{\zeta'_\beta}\right),$$

where

$$\Phi = |\zeta - \zeta'|^2 + \sqrt{-1}\left(t - t' + 2\mathrm{Im}\zeta \cdot \bar{\zeta}'\right).$$

It follows that

$$\begin{aligned} |d_b\varphi_y|^2 &= 2\sum_{\alpha=1}^{n} |Z_\alpha \varphi_y|^2 \\ &= \frac{1}{2}|\zeta - \zeta'|^2, \end{aligned} \tag{2.6}$$

$$\begin{aligned} Dd_b\varphi_y{}^{\mathrm{Sym}}(Z_\alpha, \overline{Z_\beta}) &= \frac{1}{2}\left(\overline{Z_\beta} Z_\alpha \varphi_y + Z_\alpha \overline{Z_\beta}\varphi_y\right) \\ &= \varphi_y{}^{-1}\left[\left(Z_\alpha\varphi_y\right)\left(\overline{Z_\beta}\varphi_y\right) + \frac{1}{2}|d_b\varphi_y|^2\delta_{\alpha\beta}\right], \end{aligned} \tag{2.7}$$

$$\begin{aligned} \Delta_b\varphi_y &= -2\sum_{\alpha=1}^{n} Dd_b\varphi_y{}^{\mathrm{Sym}}(Z_\alpha, \overline{Z_\alpha}) \\ &= -(n+1)\varphi_y{}^{-1}|d_b\varphi_y|^2, \end{aligned} \tag{2.8}$$

$$\begin{aligned} Dd_b\varphi_y(Z_\alpha, Z_\beta) &= Z_\beta Z_\alpha \varphi_y \\ &= -\varphi_y{}^{-1}(Z_\alpha\varphi_y)(Z_\beta\varphi_y). \end{aligned} \tag{2.9}$$

Note that (2.7), (2.9) mean

$$\left(Dd_b\varphi_y{}^{\mathrm{Sym}}\right)^{(1)} = \varphi_y{}^{-1}\left[\left(d_b\varphi_y \otimes d_b\varphi_y\right)^{(1)} + \frac{1}{2}|d_b\varphi_y|^2 g_1\right], \tag{2.10}$$

$$\left(Dd_b\varphi_y\right)^{(2)} = -\varphi_y{}^{-1}\left(d_b\varphi_y \otimes d_b\varphi_y\right)^{(2)}, \tag{2.11}$$

respectively.

We now fix an arbitrary point $x \in \Omega(\Gamma)$ and compute the right-hand sides of (2.4), (2.5) at x. We introduce a measure ν_1 of unit total mass on $\Lambda(\Gamma)$ by

$$\nu_1 = \frac{\varphi_1(x,\cdot)^{-\delta}\mu_1}{||\varphi_1(x,\cdot)^{-\delta}\mu_1||},$$

where $||\cdot||$ denotes the total mass. By direct computation, we have

$$d_b f = -\int_{\Lambda(\Gamma)} \varphi_y{}^{-1} d_b\varphi_y \, d\nu_1(y),$$

$$\begin{aligned} Dd_bf &= -\delta \int_{\Lambda(\Gamma)} {\varphi_y}^{-1} d_b\varphi_y \, d\nu_1(y) \otimes \int_{\Lambda(\Gamma)} {\varphi_y}^{-1} d_b\varphi_y \, d\nu_1(y) \\ &\quad + (\delta+1) \int_{\Lambda(\Gamma)} {\varphi_y}^{-2} d_b\varphi_y \otimes d_b\varphi_y \, d\nu_1(y) \\ &\quad - \int_{\Lambda(\Gamma)} {\varphi_y}^{-1} Dd_b\varphi_y \, d\nu_1(y). \end{aligned}$$

Using these together with (2.8), (2.10) we obtain

$$\begin{aligned} \mathrm{Ric}_\theta &= -2(n+2)\delta \Bigg[\int_{\Lambda(\Gamma)} {\varphi_y}^{-2} d_b\varphi_y \otimes d_b\varphi_y \, d\nu_1(y) \\ &\quad - \int_{\Lambda(\Gamma)} {\varphi_y}^{-1} d_b\varphi_y \, d\nu_1(y) \otimes \int_{\Lambda(\Gamma)} {\varphi_y}^{-1} d_b\varphi_y \, d\nu_1(y) \Bigg]^{(1)} \\ &\quad + (2(n+1)-\delta) \Bigg[\int_{\Lambda(\Gamma)} {\varphi_y}^{-2} |d_b\varphi_y|^2 d\nu_1(y) \\ &\quad - \left| \int_{\Lambda(\Gamma)} {\varphi_y}^{-1} d_b\varphi_y \, d\nu_1(y) \right|^2 \Bigg] g_1. \end{aligned} \tag{2.12}$$

It follows that

$$\begin{aligned} S_\theta &= \frac{1}{2} \mathrm{tr}_g \mathrm{Ric}_\theta \\ &= 2(n+1)(n-\delta)e^{-2f} \\ &\quad \times \left[\int_{\Lambda(\Gamma)} {\varphi_y}^{-2} |d_b\varphi_y|^2 d\nu_1(y) - \left| \int_{\Lambda(\Gamma)} {\varphi_y}^{-1} d_b\varphi_y \, d\nu_1(y) \right|^2 \right], \end{aligned} \tag{2.13}$$

where $g = L_\theta$. We now use (2.11) to obtain

$$\begin{aligned} \tau_\theta &= -2(\delta+2) \Bigg[\int_{\Lambda(\Gamma)} {\varphi_y}^{-2} d_b\varphi_y \otimes d_b\varphi_y \, d\nu_1(y) \\ &\quad - \int_{\Lambda(\Gamma)} {\varphi_y}^{-1} d_b\varphi_y \, d\nu_1(y) \otimes \int_{\Lambda(\Gamma)} {\varphi_y}^{-1} d_b\varphi_y \, d\nu_1(y) \Bigg]^{(2)}. \end{aligned} \tag{2.14}$$

We now introduce a symmetric bilinear form A on Q_x, defined by

$$\begin{aligned} A &= \int_{\Lambda(\Gamma)} {\varphi_y}^{-2} d_b\varphi_y \otimes d_b\varphi_y \, d\nu_1(y) \\ &\quad - \int_{\Lambda(\Gamma)} {\varphi_y}^{-1} d_b\varphi_y \, d\nu_1(y) \otimes \int_{\Lambda(\Gamma)} {\varphi_y}^{-1} d_b\varphi_y \, d\nu_1(y), \end{aligned}$$

which is positive semidefinite, since the right-hand side may be rewritten as

$$\int_{\Lambda(\Gamma)} \left({\varphi_y}^{-1} d_b \varphi_y - \int_{\Lambda(\Gamma)} {\varphi_y}^{-1} d_b \varphi_y \, d\nu_1(y) \right)^{\otimes 2} d\nu_1(y).$$

Then

$$\mathrm{tr}_{g_1} A = \int_{\Lambda(\Gamma)} {\varphi_y}^{-2} |d_b \varphi_y|^2 d\nu_1(y) - \left| \int_{\Lambda(\Gamma)} {\varphi_y}^{-1} d_b \varphi_y \, d\nu_1(y) \right|^2 ,$$

and (2.12)-(2.14) are simplified as in the following

Proposition 2.2 *The pseudo-Hermitian Ricci tensor, scalar curvature and torsion of the canonical contact form θ are given by*

$$\mathrm{Ric}_\theta = -2(n+2)\delta A^{(1)} + (2n+2-\delta)\,(\mathrm{tr}_g A)\, g, \tag{2.15}$$

$$S_\theta = 2(n+1)(n-\delta)\mathrm{tr}_g A, \tag{2.16}$$

$$\tau_\theta = -2(\delta+2)A^{(2)}, \tag{2.17}$$

respectively, where $A^{(1)}$ (resp. $A^{(2)}$) is the J-invariant (resp. J-anti-invariant) part of A.

For later use, we compute here the pull-back of A by the CR equivalence

$$F : S^{2n+1} \backslash \{(0, \dots, 0, -1)\} \to \mathcal{H}^{2n+1}.$$

Noting that

$$\varphi_1(F(z), F(w)) = \frac{\varphi(z,w)}{|1+z_{n+1}||1+w_{n+1}|},$$

$$F^* \nu_1 = \frac{\varphi(z,\cdot)^{-\delta}\mu}{||\varphi(z,\cdot)^{-\delta}\mu||} \quad (=: \nu),$$

we obtain

$$\begin{aligned} F^* A \;=\; & \int_{\Lambda(\Gamma)} {\varphi_w}^{-2} d_b \varphi_w \otimes d_b \varphi_w \, d\nu(w) \\ & - \int_{\Lambda(\Gamma)} {\varphi_w}^{-1} d_b \varphi_w \, d\nu(w) \otimes \int_{\Lambda(\Gamma)} {\varphi_w}^{-1} d_b \varphi_w \, d\nu(w), \end{aligned}$$

where $\varphi_w = \varphi(\cdot, w)$.

It follows from Proposition 2.2 that if $\delta <$ (resp. $=, >$) n then $S_\theta \geq$ (resp. $=, \leq$) 0. To make this assertion more strict, we need the following lemma. Note that if $A^{(1)}$ vanishes on a subspace Π of Q_z, $z \in \Omega(\Gamma)$, then it (or equivalently A) vanishes on the minimal J-invariant subspace of Q_z containing Π.

Lemma 2.3 *Let $z \in \Omega(\Gamma)$, and suppose that A vanishes on a J-invariant $2k$-plane Π in Q_z. Then $\Lambda(\Gamma)$ lies properly in a $\mathbb{C}$-sphere C of codimension $2k$ such that $z \in C$ and $T_zC \cap Q_z$ is orthogonal to Π. The converse is also true. In particular, A vanishes on Q_z if and only if $\Lambda(\Gamma)$ lies properly in a chain through z.*

Proof. We have only to prove the first assertion and its converse when $k = 1$. The general case follows from repeated use of the $k = 1$ case. With $X \in \Pi \setminus \{0\}$ fixed, the assumption is equivalent to the identity $A(X, X) = A(JX, JX) = 0$. This in turn implies that ${\varphi_w}^{-1} d_b \varphi_w(X) = c_1$ and ${\varphi_w}^{-1} d_b \varphi_w(JX) = c_2$ for μ-a.e. w, where c_1 and c_2 are real constants. Since

$$ {\varphi_w}^{-1} d_b \varphi_w(Y) = -\mathrm{Re} \frac{w \cdot \overline{Y}}{1 - \bar{z} \cdot w}, \quad Y \in Q_z, $$

these equations are equivalent to the single equation

$$ -\frac{w \cdot \overline{X}}{1 - \bar{z} \cdot w} = c, \quad \text{or} \quad (\overline{X} - c\bar{z}) \cdot w = -c, $$

where $c = c_1 + \sqrt{-1}\, c_2$. Then $\Lambda(\Gamma)$ lies in $C = \{w \in S^{2n+1} \mid (\overline{X} - c\bar{z}) \cdot w = -c\}$, which is a $\mathbb{C}$-sphere of codimension two, bounding a complex geodesic hyperplane $\{w \in B^{n+1}_{\mathbb{C}} \mid (\overline{X} - c\bar{z}) \cdot w = -c\}$. Since $X \in Q_z = \{Y \in \mathbb{C}^{n+1} \mid Y \cdot \bar{z} = 0\}$, $(\overline{X} - c\bar{z}) \cdot z = \overline{X} \cdot z - c = -c$, that is, $z \in C$. On the other hand, $T_zC = \{Y \in T_z S^{2n+1} \mid (\overline{X} - c\bar{z}) \cdot Y = 0\}$, and hence $T_zC \cap Q_z = \{Y \in Q_z \mid \overline{X} \cdot Y = 0\}$. This is nothing but the orthogonal complement of Π in Q_z. Thus the first assertion is proved. The converse may be proved by tracing the above argument backwards, which is left to the reader. This completes the proof of Lemma 2.3. □

If $\Lambda(\Gamma)$ is a single point, then A, and hence the pseudo-Hermitian curvature and torsion of θ, vanish identically. It is known that the vanishing of pseudo-Hermitian curvature and torsion implies that θ is locally equivalent to the standard contact form θ_1 of $\mathcal{H}^{2n+1}$.

Theorem 2.4 *Let Γ be a discrete subgroup of $G_{\mathbb{C}}(n+1)$ such that $\Lambda(\Gamma)$ is not a single point, and θ the canonical contact form associated with Γ.*
(i) *If $\delta(\Gamma) = n$, then the pseudo-Hermitian scalar curvature of θ vanishes identically.*
(ii) *If $\delta(\Gamma) < $ (resp. $>$) n, then the pseudo-Hermitian scalar curvature of θ is positive (resp. negative) everywhere unless $\Lambda(\Gamma)$ lies properly in a chain of S^{2n+1}; if $\Lambda(\Gamma)$ lies properly in a chain C, then the pseudo-Hermitian scalar curvature of θ is nonnegative (resp. nonpositive) and vanishes precisely on $C \setminus \Lambda(\Gamma)$.*

Proof. The assertion (i) is obvious from the formula (2.16). Suppose that $\delta(\Gamma) \neq n$, and that S_θ vanishes somewhere. Then by the formula (2.16) and Lemma 2.3, $\Lambda(\Gamma)$ lies properly in a chain C, and S_θ vanishes on $C \setminus \Lambda(\Gamma)$. If S_θ vanishes at some $z \in \Omega(\Gamma) \setminus C$, then $\Lambda(\Gamma)$ lies in another chain through z. But two different chains have at most one point in common, and therefore $\Lambda(\Gamma)$ has to be a single point, contradicting the assumption. This proves (ii) and completes the proof of Theorem 2.4. □

3. Cohomology vanishing and rigidity

Let $H^{n+1}_{\mathbb{R}}$ be real hyperbolic space of dimension $n+1$, and let $G_{\mathbb{R}}(n+1)$ denote the group of orientation-preserving isometries of $H^{n+1}_{\mathbb{R}}$. We take, as a model of $H^{n+1}_{\mathbb{R}}$, the ball $B^{n+1}_{\mathbb{R}} = \{x \in \mathbb{R}^{n+1} \mid |x| < 1\}$ endowed with the Klein metric

$$\frac{1}{1-|x|^2}\left\{\sum_{i=1}^{n+1}(dx_i)^2 + \frac{1}{1-|x|^2}\left(\sum_{i=1}^{n+1} x_i dx_i\right)^2\right\}.$$

Then for $k \leq n$, $H^{k+1}_{\mathbb{R}}$ is naturally embedded in $H^{n+1}_{\mathbb{R}}$ as a geodesic subspace, and in $H^{n+1}_{\mathbb{C}}$ as a totally-real geodesic subspace. Correspondingly, there is an embedding of $G_{\mathbb{R}}(k+1)$ into $G_K(n+1)$ for $K = \mathbb{R}, \mathbb{C}$.

Let Γ be a discrete subgroup of $G_K(n+1)$. Throughout this section, we assume that Γ is torsion-free. Let $C(\Gamma)$ denote the hyperbolic (resp. complex-hyperbolic) convex hull of $\Lambda(\Gamma)$ in B^{n+1}_K when $K = \mathbb{R}$ (resp. $\mathbb{C}$). The set $C(\Gamma)$ is invariant under Γ, and we say that Γ is *convex-cocompact* if the quotient $C(\Gamma)/\Gamma$ is compact. It is known that this condition of convex-cocompactness is equivalent to the condition that $Y = (B^{n+1}_K \cup \Omega(\Gamma))/\Gamma$ is compact, and also that if Γ is convex-cocompact, then its critical exponent $\delta(\Gamma)$ coincides with the Hausdorff dimension of $\Lambda(\Gamma)$ (with respect to the Carnot distance of S^{2n+1} when $K = \mathbb{C}$) [5, 15].

Let Γ_0 be a discrete subgroup of $G_{\mathbb{R}}(k+1)$ such that $B^{k+1}_{\mathbb{R}}/\Gamma_0$ is compact. Through the above embedding, Γ_0 may be regarded as a discrete subgroup of $G_K(n+1)$, giving a convex-cocompact group such that $\Lambda(\Gamma_0) = S^k_{\mathbb{R}}$ $(= \partial B^{k+1}_{\mathbb{R}})$ and $\delta(\Gamma_0) = k$. H. Izeki (when $K = \mathbb{R}$) [7], C.-B. Yue [18] and M. Bourdon [2] have proved the following generalization of a theorem of R. Bowen [3]:

Theorem 3.1 *Let Γ_0 be as above, and let $\rho : \Gamma_0 \to G_K(n+1)$ ($K = \mathbb{R}$, $\mathbb{C}$) be an injective homomorphism such that its image $\Gamma = \rho(\Gamma_0)$ is discrete and convex-cocompact. Then the critical exponent of Γ satisfies $\delta(\Gamma) \geq k$ $(= \delta(\Gamma_0))$, and the equality holds if and only if Γ stabilizes a geodesic* *subspace (resp. a totally-real geodesic subspace) B of dimension $k+1$ in $B^{n+1}_{\mathbb{R}}$ (resp. $B^{n+1}_{\mathbb{C}}$) and B/Γ is compact.*

It should be mentioned that the results of Bourdon and Yue are more general than the statement here.

Izeki's proof of the $K = \mathbb{R}$ case of the theorem is based on two results on the cohomology of the manifold $X = \Omega(\Gamma)/\Gamma$; the vanishing theorem which has been proved in [12], and a non-vanishing lemma. This latter result can easily be generalized to the $K = \mathbb{C}$ case, and we obtain

Lemma 3.2 *Suppose that $k \leq n-1$, and let Γ_0 and Γ be as in Theorem 3.1 with $K = \mathbb{C}$. If the critical exponent of Γ satisfies $\delta(\Gamma) \leq k$, then the $(k+1)$-th real cohomology group of $X = \Omega(\Gamma)/\Gamma$ is nontrivial.*

Proof. Since Γ is convex-cocompact, $\delta(\Gamma)$ coincides with the Hausdorff dimension of $\Lambda(\Gamma)$ with respect to the Carnot distance of S^{2n+1}, which is not less than the ordinary Hausdorff dimension of $\Lambda(\Gamma)$. Therefore, the ordinary Hausdorff dimension of $\Lambda(\Gamma)$ is not greater than k. This implies $\pi_i(\Omega(\Gamma)) = 0$ for $i \leq (2n+1) - k - 2$. To conclude the proof, it remains to repeat Izeki's argument in [7, the proof of Theorem 5.1], which we shall include here for the sake of completeness. It follows from the homotopy exact sequence, the relative Hurewicz theorem and the universal coefficient theorem that $H^i(Y, X; \mathbb{R}) = 0$ for $i \leq (2n+1) - k - 1$, where $Y = (B^{n+1}_{\mathbb{C}} \cup \Omega(\Gamma))/\Gamma$. In particular, we have $H^{k+1}(Y, X; \mathbb{R}) = 0$ since $k+1 < (2n+1) - k - 1$. Then by the cohomology exact sequence, $H^{k+1}(Y; \mathbb{R}) \to H^{k+1}(X; \mathbb{R})$ is injective. Now Y is homotopy equivalent to $B^{n+1}_{\mathbb{C}}/\Gamma_0$ since both are $K(\Gamma_0, 1)$-manifolds. The latter manifold has a structure of disk bundle, and hence is homotopy equivalent to its base manifold $B^{k+1}_{\mathbb{R}}/\Gamma_0$. The $(k+1)$-th cohomology group of $B^{k+1}_{\mathbb{R}}/\Gamma_0$ is nontrivial as it is compact, orientable and has no boundary. This completes the proof of Lemma 3.2. □

Conjecture. Let Γ be a discrete subgroup of $G_{\mathbb{C}}(n+1)$ such that $X = \Omega(\Gamma)/\Gamma$ is compact, and let $\delta = \delta(\Gamma)$.
(i) If k is an integer such that $\delta < k < 2n - 1 - \delta$, then $H^{k+1}(X; \mathbb{R}) = 0$.
(ii) If δ is an integer, $\delta \leq n-1$ and $H^{\delta+1}(X; \mathbb{R}) \neq 0$, then $\Lambda(\Gamma)$ is an $\mathbb{R}$-sphere of dimension δ.

We observe that the $K = \mathbb{C}$ case of Theorem 3.1 follows from this conjecture combined with Lemma 3.2. Let Γ be as in Theorem 3.1. Note that there is a natural embedding of $G_{\mathbb{C}}(n+1)$ into $G_{\mathbb{C}}(n+2)$. By regarding Γ as a discrete subgroup of $G_{\mathbb{C}}(n+2)$ if necessary, we may assume

$k \leq n-1$. Suppose $\delta(\Gamma) \leq k$. Then by Lemma 3.2 and the conjecture, we must have $\delta(\Gamma) = k$ and $\Lambda(\Gamma)$ is an $\mathbb{R}$-sphere of dimension k. Then $C(\Gamma)$ is a totally-real geodesic subspace B of dimension $k+1$ in $B_{\mathbb{C}}^{n+1}$, and B/Γ is compact since Γ is assumed to be convex-cocompact.

Though we have not been able to settle the above conjecture, we can prove a related vanishing result for $H^2(X;\mathbb{R})$, characterizing a chain instead of an $\mathbb{R}$-circle. This follows from the general pseudo-Hermitian vanishing theorem due to M. Rumin [14], which we now review. We believe that the above conjecture would also follow by applying (possibly an appropriate improvement of) Rumin's result to our contact form.

Let (M,θ) be a pseudo-Hermitian manifold, and let $\Omega^k Q$ denote the set of k-forms on the contact subbundle Q. By abuse of notation, we denote the corresponding vector bundle by the same symbol. The metric $g = L_\theta$ on Q canonically extends to a fiber metric $\langle\ ,\ \rangle$ on $\Omega^k Q$, while the complex structure J on Q induces a decomposition of forms on Q according to their bidegree, which we write

$$\Omega^k Q \otimes \mathbb{C} = \sum_{p+q=k} \Omega^{p,q} Q.$$

Let $\mathcal{R} : \Omega^k Q \to \Omega^k Q$ be the endomorphism defined on $\omega \in \Omega^{p,q} Q$ by

$$\begin{aligned}\mathcal{R}\omega &= \sum_{\rho,\sigma=1}^{n} \left[e(\theta_\rho) i(e_\sigma) R(\overline{e_\sigma}, e_\rho)\omega + e(\overline{\theta_\rho}) i(\overline{e_\sigma}) R(e_\sigma, \overline{e_\rho})\omega\right] \\ &\quad + \frac{p-q}{n} \sum_{\rho=1}^{n} R(e_\rho, \overline{e_\rho})\omega,\end{aligned}$$

where $\{e_1, \ldots, e_n\}$ is a g-orthonormal basis for $Q^{1,0}$, $\{\theta_1, \ldots, \theta_n\}$ is the dual basis, and $i(\cdot)$ (resp. $e(\cdot)$) denotes the interior (resp. exterior) product. $\mathcal{R}$ is self-adjoint, and preserves the bidegree: $\mathcal{R}(\Omega^{p,q}Q) \subset \Omega^{p,q}Q$. The pseudo-Hermitian torsion τ operates on forms on Q in a natural way, giving rise to $\mathcal{T} : \Omega^k Q \to \Omega^k Q$. This is also self-adjoint, but violates the bidegree in general, satisfying $\mathcal{T}(\Omega^{p,q}Q) \subset \Omega^{p+1,q-1}Q + \Omega^{p-1,q+1}Q$.

Let

$$J^k = \left\{\omega \in \Omega^k Q \mid \Lambda\omega = 0\right\},$$

where $\Lambda : \Omega^k Q \to \Omega^{k-2} Q$ is the adjoint of $L = (d\theta|_Q)\wedge : \Omega^{k-2} Q \to \Omega^k Q$. Again, J^k represents a vector bundle as well as the set of its sections. $\mathcal{R}$ and $\mathcal{T}$ preserve J^k, and may be regarded as quadratic forms on J^k by using the fiber metric. We are now ready to state [14].

Theorem 3.3 *Suppose M is compact and $k \leq n-1$.*
(i) *If $\mathcal{R}+(n-k+2)\mathcal{T}$ is positive semidefinite on J^k, and positive definite at some point, then $H^k(M;\mathbb{R})=0$.*
(ii) *If $\mathcal{R}+(n-k+2)\mathcal{T}$ is positive semidefinite on J^k and $H^k(M;\mathbb{R})\neq 0$, then there exists a global nonzero k-form $\omega \in J^k$ which is D-parallel along Q and satisfies $(\mathcal{R}+(n-k+2)\mathcal{T})\omega = 0$.*

We now assume that θ is locally pseudoconformally equivalent to a flat contact form, so that its pseudo-Hermitian curvature is completely determined by the Ricci tensor. Precisely, we have [17]

$$R(e_\rho,\overline{e_\sigma})e_\alpha = h_{\rho\bar\sigma}e_\alpha + h_{\alpha\bar\sigma}e_\rho + \delta_{\rho\sigma}\sum_{\beta=1}^{n} h_{\alpha\bar\beta}e_\beta + \delta_{\alpha\sigma}\sum_{\beta=1}^{n} h_{\rho\bar\beta}e_\beta, \tag{3.1}$$

where

$$h = \frac{1}{n+2}\left(\mathrm{Ric} - \frac{S}{2(n+1)}g\right)$$

is the pseudo-Hermitian analogue of the Schouten-Ricci tensor in Riemannian geometry, $\{e_1,\ldots,e_n\}$ is a g-orthonormal basis for $Q^{1,0}$ and $h_{\alpha\bar\beta} = h(e_\alpha,\overline{e_\beta})$. If we choose $\{e_\alpha\}$ so that

$$h_{\alpha\bar\beta} = \mu_\alpha\delta_{\alpha\beta}, \quad \mu_1 \leq \cdots \leq \mu_n,$$

then (3.1) is rewritten as

$$R(e_\rho,\overline{e_\sigma})e_\alpha = (\mu_\alpha+\mu_\rho)(\delta_{\rho\sigma}e_\alpha + \delta_{\alpha\sigma}e_\rho).$$

If $\{\theta_\alpha\}$ is the dual basis, then

$$R(e_\rho,\overline{e_\sigma})\theta_\alpha = -(\mu_\alpha+\mu_\sigma)(\delta_{\rho\sigma}\theta_\alpha + \delta_{\alpha\rho}\theta_\sigma). \tag{3.2}$$

We now assume $n \geq 3$ and compute $\mathcal{R}$ on

$$J^2 \otimes \mathbb{C} = \Omega^{2,0}Q \oplus \Omega^{0,2}Q \oplus (\Omega^{1,1}Q \cap \ker\Lambda)\,.$$

$\{\theta_\alpha\wedge\theta_\beta\}_{1\leq\alpha<\beta\leq n}$ (resp. $\{\overline{\theta_\alpha}\wedge\overline{\theta_\beta}\}_{1\leq\alpha<\beta\leq n}$, $\{\theta_\alpha\wedge\overline{\theta_\beta}\}_{1\leq\alpha,\beta\leq n}$) is a basis for $\Omega^{2,0}Q$ (resp. $\Omega^{0,2}Q$, $\Omega^{1,1}Q$). $\theta_\alpha\wedge\overline{\theta_\beta} \in \Omega^{1,1}Q\cap\ker\Lambda$ for $\alpha\neq\beta$, and the orthogonal projection of $\theta_\alpha\wedge\overline{\theta_\alpha}$ to $\Omega^{1,1}Q\cap\ker\Lambda$ is given by

$$\omega_\alpha = \theta_\alpha\wedge\overline{\theta_\alpha} - \frac{1}{n}\sum_{\rho=1}^{n}\theta_\rho\wedge\overline{\theta_\rho}.$$

The linear combination $\sum_{\alpha=1}^{n} z_\alpha\omega_\alpha$, $z_\alpha\in\mathbb{C}$, is zero if and only if $z_1=\cdots=z_n$.

Direct computation using (3.2) gives

$$\mathcal{R}(\theta_\alpha \wedge \theta_\beta) = \left(1 - \frac{2}{n}\right)\left[(n+2)(\mu_\alpha + \mu_\beta) + 2\sum_{\rho=1}^{n} \mu_\rho\right] \theta_\alpha \wedge \theta_\beta, \tag{3.3}$$

$$\begin{aligned} \mathcal{R}(\overline{\theta_\alpha} \wedge \overline{\theta_\beta}) &= \overline{\mathcal{R}(\theta_\alpha \wedge \theta_\beta)} \\ &= \left(1 - \frac{2}{n}\right)\left[(n+2)(\mu_\alpha + \mu_\beta) + 2\sum_{\rho=1}^{n} \mu_\rho\right] \overline{\theta_\alpha} \wedge \overline{\theta_\beta}, \end{aligned} \tag{3.4}$$

$$\mathcal{R}(\theta_\alpha \wedge \overline{\theta_\beta}) = \left[n(\mu_\alpha + \mu_\beta) + 2\sum_{\rho=1}^{n} \mu_\rho\right] \theta_\alpha \wedge \overline{\theta_\beta}, \quad \alpha \neq \beta, \tag{3.5}$$

$$\mathcal{R}\omega_\alpha = 2\left[n\mu_\alpha + \sum_{\rho=1}^{n} \mu_\rho\right] \omega_\alpha - 2\sum_{\beta=1}^{n} \mu_\beta \omega_\beta. \tag{3.6}$$

Let $\omega = \sum_{\alpha=1}^{n} z_\alpha \omega_\alpha$. Since $\langle \omega_\alpha, \omega_\beta \rangle = \delta_{\alpha\beta} - 1/n$, we obtain

$$\langle \mathcal{R}\omega, \omega \rangle = 2\sum_{\alpha,\beta=1}^{n} \left[\left(n\mu_\alpha + \sum_{\rho=1}^{n} \mu_\rho\right)\delta_{\alpha\beta} - (\mu_\alpha + \mu_\beta)\right] z_\alpha \bar{z}_\beta.$$

Here and in the following, we denote the Hermitian extension of $\langle\ ,\ \rangle$ by the same symbol.

Lemma 3.4 *Let* $\omega = \sum_{\alpha=1}^{n} z_\alpha \omega_\alpha$, $z_\alpha \in \mathbb{C}$.

(i) *If* $\mu_1 + \mu_2 \geq 0$, *then* $\langle \mathcal{R}\omega, \omega \rangle \geq 0$.

(ii) *If* $\mu_1 + \mu_2 > 0$ *and* $\langle \mathcal{R}\omega, \omega \rangle = 0$, *then* $\omega = 0$.

Proof. Let $c_{\alpha\beta} = \left(n\mu_\alpha + \sum_{\rho=1}^{n} \mu_\rho\right)\delta_{\alpha\beta} - (\mu_\alpha + \mu_\beta)$, and suppose $\mu_1 + \mu_2 \geq 0$. Then for $\alpha < \beta$, $c_{\alpha\beta} = -(\mu_\alpha + \mu_\beta) \leq 0$. Hence

$$\begin{aligned} \frac{1}{2}\langle \mathcal{R}\omega, \omega \rangle &= \sum_\alpha c_{\alpha\alpha}|z_\alpha|^2 + \sum_{\alpha<\beta} c_{\alpha\beta}(z_\alpha \bar{z}_\beta + \bar{z}_\alpha z_\beta) \\ &\geq \sum_\alpha c_{\alpha\alpha}|z_\alpha|^2 + \sum_{\alpha<\beta} c_{\alpha\beta}\left(|z_\alpha|^2 + |z_\beta|^2\right) \\ &= \sum_\alpha \left(c_{\alpha\alpha} + \sum_{\beta\neq\alpha} c_{\alpha\beta}\right)|z_\alpha|^2 \\ &= 0. \end{aligned}$$

Thus $\langle \mathcal{R}\omega, \omega\rangle \geq 0$, and the equality holds if and only if $c_{\alpha\beta}|z_\alpha - z_\beta|^2 = 0$ for $\alpha < \beta$. If $\mu_1 + \mu_2 > 0$ so that $c_{\alpha\beta} < 0$ for $\alpha < \beta$, then the last equation is equivalent to $z_1 = \cdots = z_n$, or $\omega = 0$. This completes the proof of Lemma 3.4. □

We shall now prove

Theorem 3.5 *Suppose $n \geq 3$, and let Γ be a discrete subgroup of $G_{\mathbb{C}}(n+1)$ such that $X = \Omega(\Gamma)/\Gamma$ is compact. Assume that the pseudo-Hermitian torsion τ_θ of the canonical contact form θ vanishes identically.*
(i) *If $\delta(\Gamma) < 2$, then $H^2(X;\mathbb{R}) = 0$.*
(ii) *If $\delta(\Gamma) = 2$ and $H^2(X;\mathbb{R}) \neq 0$, then $\Lambda(\Gamma)$ is a chain.*

Proof. Let $\lambda_1, \dots, \lambda_n$ denote the eigenvalues of $A^{(1)}$ (viewed as a Hermitian form on $Q^{1,0}$), numbered so that $\lambda_1 \geq \cdots \geq \lambda_n (\geq 0)$. Then by (2.15), (2.16), we have

$$\mu_\alpha = 2\left(-\delta\lambda_\alpha + \sum_{\rho=1}^{n} \lambda_\rho\right),$$

and therefore

$$\mu_1 + \mu_2 = 2\left[(2-\delta)(\lambda_1 + \lambda_2) + 2\sum_{\rho=3}^{n} \lambda_\rho\right], \tag{3.7}$$

$$\sum_{\rho=1}^{n} \mu_\rho = 2(n-\delta)\sum_{\rho=1}^{n} \lambda_\rho. \tag{3.8}$$

If $\delta \leq 2$, these are both nonnegative, and by the formulas (3.3)-(3.5) and Lemma 3.4 (i), $\mathcal{R}$ is positive semidefinite on J^2.

We remark that if $\delta \leq 2$, $\Lambda(\Gamma)$ consists of more than one point. Indeed, if $\Lambda(\Gamma)$ is a single point, then Γ contains a finite-index subgroup which is conjugate to a cocompact Heisenberg lattice Γ' [1]. But then $\delta = \delta(\Gamma') = n+1 > 2$, a contradiction. Now, by the argument in the proof of Theorem 2.4, $A^{(1)}$ vanishes at most on a chain. Hence if $\delta < 2$, we have $\mu_1 + \mu_2 > 0$ almost everywhere. By the formulas (3.3)-(3.5) and Lemma 3.4 (ii), $\mathcal{R}$ is positive definite on J^2 almost everywhere. By Theorem 3.3 (i), $H^2(X;\mathbb{R}) = 0$, proving (i).

We now prove (ii). By the assumption and Theorem 3.3 (ii), there exists a global non-zero 2-form $\omega \in J^2$ which is D-parallel along Q and satisfies $\mathcal{R}\omega = 0$. Since ω is D-parallel along Q, ω has constant length, and in particular, it is nowhere vanishing. We still have $\sum_{\rho=1}^{n} \mu_\rho > 0$ almost everywhere. By the formulas (3.3)-(3.5), $\mathcal{R}$ is positive definite on

$$\Omega^{2,0}Q \oplus \Omega^{0,2}Q \oplus \mathrm{span}_{\mathbb{C}}\{\theta_\alpha \wedge \overline{\theta_\beta}\}_{\alpha\neq\beta}$$

almost everywhere. Hence ω is a linear combination of ω_α at each point. Since ω is real, $\omega = \sum_{\alpha=1}^{n} \sqrt{-1}\, x_\alpha \omega_\alpha$, where x_α are real. Now, by the proof of Lemma 3.4, we must have $c_{\alpha\beta}(x_\alpha - x_\beta)^2 = 0$ for $\alpha < \beta$. Since $\omega \neq 0$, x_α are not all equal, and hence at least one of $c_{\alpha\beta}$ must be zero. This implies $\mu_1 + \mu_2 = 0$, and therefore $\lambda_3 = \cdots = \lambda_n = 0$ by (3.7). Suppose $\lambda_2 > 0$. Then $c_{\alpha\beta} \neq 0$ unless $(\alpha, \beta) = (1, 2)$, which implies that x_α are all equal, a contradiction. Hence $\lambda_2 = \lambda_3 = \cdots = \lambda_n = 0$. This holds everywhere, since ω is nowhere vanishing. Since $\lambda_1 > 0$ almost everywhere, we must have $x_2 = \cdots = x_n (=: y)$. By replacing x_α by $x_\alpha - y$ and multiplying by a constant, we may assume that $x_2 = \cdots = x_n = 0$ and $x_1 = 1$, that is,

$$\omega = \sqrt{-1}\left(\theta_1 \wedge \overline{\theta_1} - \frac{1}{n}\sum_{\rho=1}^{n} \theta_\rho \wedge \overline{\theta_\rho}\right),$$

or

$$\omega + \frac{1}{n} d\theta|_Q = \sqrt{-1}\, \theta_1 \wedge \overline{\theta_1}.$$

It follows that the two-plane field Π_0 spanned by the real and imaginary parts of e_1 is globally well-defined.

On the other hand, by Lemma 2.3, $\Lambda(\Gamma)$ lies in a $\mathbb{C}$-sphere of dimension 3 through any given point of $\Omega(\Gamma)$, and hence lies in a chain C, since two different $\mathbb{C}$-spheres of dimension 3 intersect at most along a chain. It remains to show that $\Lambda(\Gamma)$ fills C. Assume the contrary, and take $z \in C \setminus \Lambda(\Gamma)$ and a J-invariant two-plane Π_z in Q_z such that $\Pi_z \neq (\Pi_0)_z$. Then there is a unique $\mathbb{C}$-sphere S of dimension 3 such that $C \subset S$ and $T_z S \cap Q_z = \Pi_z$. Let $\Pi_w = T_w S \cap Q_w$ for $w \in S \setminus C$. By Lemma 2.3, $A_w^{(1)}$ vanishes precisely on $\Pi_w^\perp$, the orthogonal complement of Π_w in Q_w. This means that $\Pi_w = (\Pi_0)_w$. By continuity, we obtain $\Pi_z = (\Pi_0)_z$, a contradiction. Thus $\Lambda(\Gamma) = C$. This completes the proof of Theorem 3.5. □

References

[1] B. Apanasov and X. Xie, *Discrete actions on nilpotent groups and negatively curved spaces*, preprint.

[2] M. Bourdon, *Sur le birapport au bord des CAT* (-1)*-espaces*, I.H.E.S. Publ. Math. **83** (1996), 95–104.

[3] R. Bowen, *Hausdorff dimension of quasi-circles*, I.H.E.S. Publ. Math. **50** (1979), 11–25.

[4] D. Burns, Jr. and S. Shnider, *Spherical hypersurfaces in complex manifolds*, Invent. Math. **33** (1976), 223–246.

[5] K. Corlette, *Hausdorff dimensions of limit sets. I*, Invent. Math. **102** (1990), 521–541.

[6] W. Goldman, *Complex Hyperbolic Geometry*, Oxford University Press, New York, 1999.

[7] H. Izeki, *Limit sets of Kleinian groups and conformally flat Riemannian manifolds*, Invent. Math. **122** (1995), 603–625.

[8] D. Jerison and J. M. Lee, *The Yamabe problem on CR manifolds*, J. Differential Geometry **25** (1987), 167–197.

[9] J. M. Lee, *The Fefferman metric and pseudo-Hermitian invariants*, Trans. Amer. Math. Soc. **296** (1986), 411–429.

[10] J. M. Lee, *Pseudo-Einstein structures on CR manifolds*, Amer. J. Math. **110** (1988), 157–178.

[11] J. Mitchell, *On Carnot-Carathéodory metrics*, J. Differential Geometry **21** (1985), 35–45.

[12] S. Nayatani, *Patterson-Sullivan measure and conformally flat metrics*, Math. Z. **225** (1997), 115–131.

[13] S. J. Patterson, *The limit set of a Fuchsian group*, Acta Math. **136** (1976), 241–273.

[14] M. Rumin, *Formes différentielles sur les variétés de contact*, J. Differential Geometry **39** (1994), 281–330.

[15] D. Sullivan, *The density at infinity of a discrete group of hyperbolic motions*, I.H.E.S. Publ. Math. **50** (1979), 171–202.

[16] N. Tanaka, A differential geometric study on strongly pseudo-convex manifolds, Lectures in Mathematics, Kyoto University, No. 9, Kinokuniya, Tokyo, 1975.

[17] S. Webster, *Pseudo-Hermitian structures on a real hypersurface*, J. Differential Geometry **13** (1978), 25–41.

[18] C.-B. Yue, *Dimension and rigidity of quasi-Fuchsian representations*, Ann. Math. **143** (1996), 331–355.

[19] C.-B. Yue, *Webster curvature and Hausdorff dimension of complex hyperbolic Kleinian groups*, preprint.

Graduate School of Mathematics
Nagoya University
Chikusa-ku, Nagoya 464-8602, Japan

E-mail: nayatani@math.nagoya-u.ac.jp

CHAPTER XI

Pseudoconvex Domains in $\mathbb{P}^n$: A Question on the 1-Convex Boundary Points

Takeo Ohsawa

Introduction

Let $\mathbb{P}^n$ be n-dimensional complex projective space, let $\Omega \subsetneq \mathbb{P}^n$ be a pseudoconvex domain, and let $\delta(z)$ be the distance from $z \in \Omega$ to the boundary of Ω with respect to the Fubini-Study metric. According to a fundamental theorem of A. Takeuchi [T], the function $-\log \delta$ is plurisubharmonic and enjoys an estimate

$$\sqrt{-1}\partial\bar{\partial}(-\log \delta) \geq \frac{1}{3}\omega_{\mathrm{FS}},$$

where ω_{FS} denotes the Kähler form of the Fubini-Study metric, and the left hand side of the inequality is defined as a current. From Takeuchi's theorem it follows that Ω admits a strictly plurisubharmonic exhaustion function, so that Ω is a Stein manifold. This means in particular that the boundary of Ω is connected if $n \geq 2$ (cf. [G-R]).

Such an observation was applied by Lins Neto [LN] to prove certain nonseparation properties of complex analytic foliations with singularities in $\mathbb{P}^n$. He showed in particular that if $n \geq 3$, then there exist no real-analytic compact Levi-flat hypersurfaces in $\mathbb{P}^n$. Here a real hypersurface of a complex manifold is said to be Levi-flat if it is the union of (not necessarily closed) complex submanifolds of codimension one. This partially answers a question raised by Cerveau [C] which asks whether or not there exist compact Levi-flat hypersurfaces in $\mathbb{P}^n$ for $n \geq 2$.

The purpose of the present article is to introduce an attempt to solve the following.

Conjecture. *Let $\Omega \subsetneq \mathbb{P}^n$ be a pseudoconvex domain with C^∞-smooth boundary. Then the set of 1-convex boundary points of Ω is nonempty.*

Here we say that a boundary point of Ω, say x, is 1-convex if x becomes a strictly convex boundary point of Ω after a local biholomorphic coordinate change around x.

The corresponding assertion for the bounded domains of $\mathbb{C}^n$ is trivially true, because the farthest boundary point from the origin is strictly convex. On the other hand, the result obviously fails for the products of compact complex manifolds of positive dimensions. Moreover, there are in fact various kinds of pseudoconvex domains with Levi-flat boundaries (cf. [D-O]). For instance, the product of an annulus and the punctured plane is biholomorphically equivalent to a domain in $\mathbb{P}^1 \times \{\mathbb{C}/(\mathbb{Z}+\sqrt{-1}\mathbb{Z})\}$ with Levi-flat boundary (cf. [O-1]). Therefore, to approach the conjecture we need to deduce from Takeuchi's theorem more than the Steinness of the pseudoconvex domains. We shall only consider the case $n \geq 3$. We shall show that there exist nontrivial $\overline{\partial}_b$ cohomology classes of type $(0, n-2)$ with coefficients in sufficiently positive line bundles on $\partial\Omega$. Provided that the representatives of these are of class C^∞, we can show that these representatives extend to $\overline{\partial}$-closed forms on Ω. Since Ω is Stein, this will yield a contradiction.

The organization of the paper is as follows. In section one, we shall give an elementary proof of Takeuchi's inequality based on Royden's theorem on the embedded discs. In section two, we shall prove a variant of the Bocher-Hartogs extension theorem.

1. A variational inequality

We shall present an elementary proof of Takeuchi's inequality based on the existence of a special coordinate around geodesic intervals on Hermitian manifolds. A variational inequality for the distance function will be first derived with respect to this special coordinate. It will then be noted that the inequality thus obtained can be written in a coordinate free way when the metric is Kählerian. Takeuchi's inequality will be deduced as a corollary of the latter.

Let (M, g) be Hermitian manifold with real-analytic metric g, and let $\Gamma \subset M$ be a geodesic interval, i.e., Γ is the image of a C^1-imbedding

$$\gamma\colon [0,1] \to M$$

such that γ is the critical point of the energy functional

$$e_g(\gamma) = \int_0^1 (\gamma^* g) \cdot (ds)^{-1}.$$

Since the Euler-Lagrange equation of e_g is elliptic of second order and with real-analytic coefficients, γ is real-analytic. It is easy to verify that the arc-length from $\gamma(0)$ to $\gamma(s)$ is proportional to s (cf. [J]).

Let β be a holomorphic map into M defined on a neighbourhood of $[0,1]$ in the complex plane such that $\beta|[0,1] = \gamma$, and let $U \supset [0,1]$ be a neighbourhood such that β maps U biholomorphically into M. We may assume by Riemann's mapping theorem that U is biholomorphically equivalent to the unit disc $\Delta = \{z \in \mathbb{C} \mid |z| < 1\}$. It is known from the basic theory of several complex variables that, for any compact subset $K \subset \beta(U)$, one can find a neighbourhood $W \supset K$ in M and a biholomorphic map $\rho\colon \Delta^n \to W$ satisfying $\rho(\{0,\dots,0)\} \times \Delta) = W \cap \beta(U)$. This is known as Royden's theorem whose proof relies essentially on the triviality of holomorphic vector bundles over Δ and the existence of a Stein neighbourhood basis of $\beta(U)$ (cf. [R], [D-G] and [S]).

Therefore we may choose a neighbourhood of Γ in M, say V, a neighbourhood $D \supset [0,1]$ and a biholomorphic map

$$t\colon V \to \Delta^{n-1} \times D$$

such that

$$t(p) = (0,\dots,0,\gamma^{-1}(p))$$

for any $p \in \Gamma$.

We put $t = (t_1,\dots,t_n)$ and $t' = (t_1,\dots,t_{n-1})$, and express the metric as

$$g = \sum_{i,j=1}^{n} g_{i\bar{j}}(t)dt_i d\bar{t}_j.$$

Let

$$\tilde{t}_n = t_n + \sum_{i=1}^{n-1} t_i \int_{t_n}^{1} \frac{\partial g_{n\bar{n}}}{\partial t_i}(0,s)ds$$

and

$$u = (u_1,\dots,u_n) := (t', \tilde{t}_n).$$

Then if we express g in u as

$$g = \sum_{i,j=1}^{n} \tilde{g}_{i\bar{j}}(u)du_i d\bar{u}_j$$

$\tilde{g}_{n\bar{n}}(u', \mathrm{Re} u_n)$ contains no linear terms with respect to u'. Once and for all we shall fix such a coordinate u around Γ. Moreover, multiplying a constant by u_n if necessary, we may assume that

$$(\star)\ \ \tilde{g}_{n\bar{n}}(u) = 1 - \sum_{i,j=1}^{n-1} \lambda_{ij} u_i \bar{u}_j - 4\,\mathrm{Re} \sum_{i=1}^{n-1} \lambda_{in} u_i (\mathrm{Im}\, u_n) - 2\lambda_{nn}(\mathrm{Im}\, u_n)^2 + \mu(u).$$

Here $\lambda_{ij}(=\lambda_{ij}(u',\overline{u}',\operatorname{Re} u_n))$ and the μ are real-analytic and μ is of order at least 3 in $(u',\overline{u}')$.

Let $\delta(u)$ be the distance between u and the set $\{u \mid \tilde{t}_n = 1\}$ with respect to the metric g. Then our first variational formula follows directly from $(\star)$.

Proposition 1.

$$\sum_{i,j=1}^{n} \frac{\partial^2}{\partial u_i \partial \overline{u}_j}\left(\log \frac{1}{\delta(u)}\right) \xi_i \overline{\xi}_j \geq \frac{1}{6}\kappa|\xi|^2 \quad \textit{for any } \xi \in \mathbb{C}^n.$$

Here

$$\kappa = \inf_{\substack{u\in\Gamma \\ \xi\neq 0}} \left(\sum_{i,j=1}^{n} \lambda_{ij}(u)\xi_i\overline{\xi}_j\right)/|\xi|^2.$$

To get a more coordinate free expression of Proposition 1, let S be any complex submanifold of M which intersects with Γ orthogonally at $\gamma(1)$. Then the order of contact between S and $\{u \mid u_n = 1\}$ is at least two at $\gamma(1)$. Hence Proposition 1 implies the following.

Corollary 2. *Under the above situation,*

$$\sum_{i,j=1}^{n} \frac{\partial^2}{\partial u_i \partial \overline{u}_j}\left(\log \frac{1}{\delta_S(u)}\right) \xi_i \overline{\xi}_j \geq \frac{1}{6}\kappa|\xi|^2 \quad \textit{for } \xi \in \mathbb{C}^n.$$

Here $\delta_S(u)$ denotes the distance from u to S with respect to g.

In case g is a Kähler metric, κ can also be replaced by a coordinate free geometric quantity. Namely, in this case we are allowed to assume in advance that

$$\nabla \tilde{g}_{ij} = 0$$

for all i and j along Γ, replacing u_i $(1 \leq i \leq n-1)$ by

$$u_i + \sum_{j,k=1}^{n-1} q_{i,jk}(u_n)u_j u_k$$

for some holomorphic functions $q_{i,jk}$ in u_n. (Clearly such a coordinate change does not influence the second derivatives of $\tilde{g}_{nn}$ along Γ.) Therefore the curvature form

$$\Theta = \left(\sum_{i,j=1}^{n} \Theta^{\mu}_{ij\nu} du_i \wedge d\overline{u}_j\right)_{1\leq\mu,\nu\leq n}$$

of g satisfies

$$\Theta^n_{ijn} = \lambda_{ij}$$

along Γ. In particular if we put

$$\kappa_M = \inf_{\substack{M \\ |\xi|=|\eta|=1}} \sum_{i,j,\mu,\nu,\sigma} \Theta^{\mu}_{ij\nu}\tilde{g}_{\mu\sigma}\xi^i\overline{\xi}^j\eta^\nu\overline{\eta}^\sigma,$$

the inequality

$$\kappa \geq \kappa_M$$

holds whenever g is a Kähler metric. Thus we have reached as a conclusion the following generalization of Takeuchi's inequality.

Theorem 3. (cf. [E], [Su]) *Let (M, g) be a connected Kähler manifold whose holomorphic bisectional curvature is bounded from below by a constant c. Then, for any pseudoconvex domain $\Omega \subsetneq M$,*

$$\sqrt{-1}\partial\overline{\partial}\log\frac{1}{\delta_\Omega} \geq \frac{c}{3}\omega_g$$

holds. Here ω_g denotes the fundamental form of g.

Takeuchi's inequality follows from Theorem 3 because the holomorphic bisectional curvature of $\partial\overline{\partial}\log(1+|z|^2)$ ranges between 1 and 2. Here and in what follows we often identity $\partial\overline{\partial}f$ with the complex Hessian of f.

2. Extension from the boundary

We shall establish an extension theorem for the $\overline{\partial}_b$-closed forms on the boundary of a pseudoconvex domain.

Let M be a connected complex manifold of dimension $n \geq 2$, and let $\Omega \subset M$ be any domain with C^∞-smooth boundary. Let $C^{p,q}(\overline{\Omega})$ be the set of C^∞ (p,q)-forms on the closure $\overline{\Omega}$ of Ω, and let

$$C^{p,q}_0(\overline{\Omega}) = \{u \in C^{p,q}(\overline{\Omega}) \mid u|_{\partial\Omega} = 0\}.$$

Here the restriction $u|_{\partial\Omega}$ is to be understood as a differential form on $\partial\Omega$.

We put

$$C^{p,q}(\partial\Omega) = C^{p,q}(\overline{\Omega})/C^{p,q}_0(\overline{\Omega})$$

and denote by

$$\pi^{p,q}\colon C^{p,q}(\overline{\Omega}) \to C^{p,q}(\partial\Omega)$$
$$\sigma^{p,q}\colon \oplus_{s,t}C^{s,t}(\overline{\Omega}) \to C^{p,q}(\partial\Omega)$$

the natural projections. For simplicity we put

$$\pi^{p,q}(u) = u|_{\partial\Omega}.$$

Let us define the $\overline{\partial}_b$-operator

$$\overline{\partial}_b\colon C^{p,q}(\partial\Omega) \to C^{p,q+1}(\partial\Omega)$$

by

$$\overline{\partial}_b = \sigma^{p,q+1} \circ d \circ (\pi^{p,q})^{-1}.$$

Differentiable functions f on $\partial\Omega$ satisfying $\overline{\partial}_b f = 0$ are called CR functions on $\partial\Omega$. It is clear that f is CR if there exists a differentiable function $\tilde{f}$ on $\overline{\Omega}$ satisfying $\tilde{f}|_{\partial\Omega} = f$ and $\overline{\partial}\tilde{f} = 0$. Let E be a holomorphic vector bundle over M, and let $C^{p,q}(\overline{\Omega}, E)$ be the set of E-valued C^∞ (p,q)-forms on $\overline{\Omega}$. Then the space $C^{p,q}(\partial\Omega, E)$ and the operator

$$\overline{\partial}_b\colon C^{p,q}(\partial\Omega, E) \to C^{p,q+1}(\partial\Omega, E)$$

are defined similarly as above.

Let us denote the kernel and the image of $\overline{\partial}_b$ by $\mathrm{Ker}^{p,q}\overline{\partial}_b$ and $\mathrm{Im}^{p,q+1}\overline{\partial}_b$, respectively, and put

$$H^{p,q}_{\overline{\partial}_b}(\partial\Omega, E) = \mathrm{Ker}^{p,q}\overline{\partial}_b / \mathrm{Im}^{p,q}\overline{\partial}_b.$$

Lemma 4. *For any $\alpha \in C^{p,q}(\partial\Omega, E) \cap \mathrm{Ker}\overline{\partial}_b$, there exists an $\hat{\alpha} \in C^{p,q}(\overline{\Omega}, E)$ such that $\hat{\alpha}|_{\partial\Omega} = \alpha$ and that $\overline{\partial}\hat{\alpha}$ vanishes to the infinite order on $\partial\Omega$.*

Proof. An exercise for the undergraduates.

By virtue of a theory of Kodaira-Andreotti-Vesentini (cf. [K], [A-V-1]), a sufficient condition for the C^k-extendability can be stated as follows.

Theorem 5. *Let (M, g) be a connected Kähler manifold of dimension n, let $\Omega \subset M$ be a relatively compact pseudoconvex domain with C^∞-smooth boundary, and let E be a holomorphic vector bundle over M. Suppose that Ω admits a C^∞ defining function r such that*

$$(*) \qquad \partial\overline{\partial}(-\log(-r)) > c\,(\partial\log(-r)\overline{\partial}\log(-r) + g)$$

holds on Ω for some positive constant c. Then, for any $\alpha \in C^{p,q}(\partial\Omega, E) \cap \mathrm{Ker}\overline{\partial}_b$ with $q < n-1$, and for any nonnegative integer k, there exists a $\overline{\partial}$-closed E-valued (p,q)-form $\tilde{\alpha}_k$ of class C^k on $\overline{\Omega}$ satisfying $\tilde{\alpha}_k|_{\partial\Omega} = \alpha$.

Proof. Let $\hat{\alpha}$ be as in Lemma 4. It suffices to show that for each positive integer k, one can find a solution β to the equation $\overline{\partial}\beta = \overline{\partial}\hat{\alpha}$ in such a way that β is of class C^k on $\overline{\Omega}$, and vanishes on $\partial\Omega$.

As usual we shall regard $\overline{\partial}\hat{\alpha}$ as a $(0, q+1)$-form in the following manner. Let $T_M^{1,0}$ be the holomorphic tangent bundle of M. Then we have a natural identification between $C^{p,q}(\cdot, E)$ and $C^{0,q}(\cdot, \bigwedge^p(T_M^{1,0})^* \otimes E)$ for any p and q. We may regard $\overline{\partial}\hat{\alpha}$ as an element of $C^{0,q+1}(\overline{\Omega}, \bigwedge^p(T_M^{1,0})^* \otimes E)$. To show the existence of the above β, we take advantage of the fact that $\overline{\partial}\hat{\alpha}$ is square integrable on Ω with respect to the Kähler metric $\partial\overline{\partial}(-\log(-r))$ and for any fiber metric of $(\bigwedge^p(T_M^{1,0})^* \otimes E)|_\Omega$ of the form $(-r)^{-N}h$ for $N > 0$. Here h denotes any C^∞ fiber metric of $\bigwedge^p(T_M^{1,0})^* \otimes E$.

From the assumption it follows that there exists a positive integer $N_0 > 0$ such that

$$(\ddagger) \qquad N_0 \mathrm{Id} \otimes \partial\overline{\partial}\log(-r) + \Theta_h < 0$$

on Ω, where Id stands for the identity homomorphism of $\bigwedge^p(T_M^{1,0})^* \otimes E$, Θ_h for the curvature form of h, and the inequality is in the sense of Nakano (cf.[N] or [O]). Since $(*)$ implies that $\partial\overline{\partial}(-\log(-r))$ is a complete Kähler metric, it follows from $(\ddagger)$ that the L^2 à priori estimate for the $\overline{\partial}$-operator holds with respect to the metrics $(\partial\overline{\partial}(-\log(-r)), (-r)^{-N}h)$ at the degree $(0, q+1)$, if $N > N_0$ and $q < n-1$ (cf. [A-V-2] or [O-2]).

Therefore, for any $N > N_0$ one can find a C^∞ E-valued (p,q)-form β_N on Ω such that $\overline{\partial}\beta_N = \overline{\partial}\hat{\alpha}$ and β_N is square integrable and orthogonal to the kernel of $\overline{\partial}$ with respect to $\partial\overline{\partial}(-\log(-r))$ and $(-r)^{-N}h$. One can see that the trivial extension of β_N to M is of class C^k if $N \gg k$. In fact, this is a consequence of the ellipticity of $\overline{\partial}$ as in the standard case up to the following small modification.

Let $\overline{\partial}^*_{(N)}$ and $\overline{\partial}^*_h$ denote respectively the L^2 adjoints of $\overline{\partial}$ with respect to $(\partial\overline{\partial}(-\log(-r)), (-r)^{-N}h)$ and (g, h). Then, as differential operators on Ω,

$$\overline{\partial}^*_{(N)} = \overline{\partial}^*_h + Ne\left(\frac{\overline{\partial}r}{r}\right)^*$$

where $e(\overline{\partial}r/r)$ denotes exterior multiplication by $\overline{\partial}r/r$ from the left hand side, and $e(\overline{\partial}r/r)^*$ denotes the adjoint of $e(\overline{\partial}r/r)$.

Using the assumption on the metric $\partial\overline{\partial}(-\log(-r))$ again, it is easy to see that all the covariant derivatives of $\overline{\partial}r/r$ with respect to $\partial\overline{\partial}(-\log(-r))$ are bounded by const $\cdot (-r)^m$. Here the constants and m depend on the order of differentiation.

Therefore one can deduce from the equations

$$\begin{cases} \overline{\partial}\beta_N = \overline{\partial}\hat{\alpha} \\ \overline{\partial}^*_{(N)}\beta_N = 0 \end{cases}$$

that, for any $N_1 > 0$, there exists N such that the trivial extension of β_N belongs to the Sobolev space of order N_1 for the E-valued (p,q)-forms on M. Hence, by the Sobolev embedding theorem, for any k one can find $N \gg k$ so that the trivial extension of β_N is of class C^k on M.

3. Nontrivial cohomology classes

Let $\Omega \subsetneq \mathbb{P}^n$ be a pseudoconvex domain with C^∞-smooth boundary. Let us assume that there exist no 1-convex boundary points of Ω.

Let $H \subset \mathbb{P}^n$ be a hyperplane such that $\Omega' := H \cap \Omega$ has a C^∞-smooth boundary in H. Let x be any point of Ω' and let U be any neighbourhood of x in H. For the hyperplane section bundle $[H]$ over $\mathbb{P}^n$ and an integer m, we denote by $C^{p,q}(\overline{\Omega'} \setminus \{x\}, [H]^{\otimes m})$ the set of $[H]^{\otimes m}$-valued C^∞ (p,q)-forms on $\overline{\Omega'} \setminus \{x\}$. Let $z = (z_1, \dots, z_{n-1})$ be a local coordinate around x, and fix a C^∞ function ψ on $H \setminus \{x\}$ such that $\psi - \log|z|$ is bounded on a neighbourhood of x.

Then, for any positive integer k, let $C^{p,q,m}_{\langle k\rangle}(\overline{U})$ be the subset of $C^{p,q}(\overline{\Omega'} \setminus \{x\}, [H]^{\otimes m}|_H)$ consisting of those elements u such that $e^{k\psi}|u|^2$ is integrable on U with respect to the Fubini-Study metric.

Then we put

$$H^{p,q,m}_{\langle k\rangle}(\overline{U}) := C^{p,q,m}_{\langle k\rangle}(\overline{U}) \cap \operatorname{Ker}\bar\partial / \{v \in C^{p,q,m}_{\langle k\rangle}(\overline{U}) \mid \exists u \in C^{p,q-1,m}_{\langle k\rangle}(\overline{U}) \text{ s.t. } \bar\partial u = v\}.$$

Since the $\bar\partial$-equation is always solvable at the top degrees on noncompact domains, it is easy to see that the natural restriction map

$$\xi \colon H^{p,n-2,m}_{\langle k\rangle}(\overline{\Omega'}) \to \varinjlim_{U \ni x} H^{p,n-2,m}_{\langle k\rangle}(\overline{U})$$

is surjective. Here U runs through the neighbourhoods of x. Clearly $H^{p,n-2,m}_{\langle k\rangle}(\overline{U}) \neq 0$ if $k \gg 1$ (actually $k > 2n-3$ suffices.) Hence one can fix k so that, for any m there exists a

$$u_k \in C^{0,n-2,m}_{\langle k\rangle}(\overline{\Omega'}) \cap \operatorname{Ker}\bar\partial$$

such that the cohomology class represented by u_k is mapped by ξ to a nonzero element. Since we assumed that there exist no 1-convex boundary points of Ω, one may expect that

$$H^{0,n-1}_{\bar\partial_b}(\partial\Omega, [H]^{\otimes m}) = 0$$

holds for sufficiently large m. (This holds in the L^2 sense, but the regularity question is essential here.)

Given such a vanishing theorem exists, one will have a

$$\hat{u}_k \in C^{0,n-2}\left(\partial\Omega, [H]^{\otimes m}\right) \cap \mathrm{Ker}\bar{\partial}_b \quad (m \gg 1)$$

such that $\hat{u}_k|_{\partial\Omega'} = u_k|_{\partial\Omega'}$. It was proved in [O-S] that Takeuchi's inequality guarantees the hypothesis of Theorem 5 for $M = \mathbb{P}^n$. Therefore, for any positive integer l, one can find a

$$\tilde{u}_{k,l} \in C^{0,n-2}\left(\Omega, [H]^{\otimes m}\right) \cap \mathrm{Ker}\bar{\partial}$$

such that $\tilde{u}_{k,l}$ can be extended to $\overline{\Omega}$ in the C^l sense and satisfies $\tilde{u}_{k,l}|_{\partial\Omega} = \hat{u}_1$. This means that $\tilde{u}_{k,l}|_{\overline{\Omega'}} - u_k$ is of class C^l on $\overline{\Omega'} \setminus \{x\}$ and vanishes on $\partial\Omega'$.

If $n \geq 3$, by choosing the above m sufficiently large in advance, one can find a $w \in C^{0,n-3}\left(\Omega' \setminus \{x\}, [H]^{\otimes m}\right)$ such that $e^{k\psi}|w|^2$ is integrable around x and

$$\bar{\partial} w = \tilde{u}_{k,l}|_{\overline{\Omega'}} - u_k.$$

However this contradicts the choice of u_k.

4. On the vanishing of the $\bar{\partial}_b$-cohomology

Let M be a connected complex manifold of dimension n, and let S be a connected and orientable real hypersurface of class C^∞ in M. Let r be a C^∞ defining function of S. The Levi form of r at a point $x \in S$ is by definition the Hermitian form $\partial\bar{\partial}r$ restricted to the holomorphic tangent space

$$T_x^{1,0}S := T_x^{1,0}M \cap T_x^{\mathbb{C}}S.$$

Here $T_x^{\mathbb{C}}S$ denotes the complexification of T_xS.

We shall say that S is partially Levi flat if the Levi form of r is everywhere degenerate on S. Since S is the boundary of some domain in a neighbourhood of S, we may use the notations as $C^{p,q}(S, E)$, for holomorphic vector bundles E over M. For each point $x \in S$ we denote by $N_x \subset T_x^{1,0}S$ the kernel of the linear map

$$v \mapsto \langle \partial\bar{\partial}r, v\rangle \in (T_x^{0,1}S)^*.$$

Since $\langle \partial r, v\rangle = 0$ for any $v \in T^{\mathbb{C}}S$,

$$\langle \partial\bar{\partial}r, \xi, \bar{\xi}\rangle = [\xi, \bar{\xi}]r$$

for any $T^{0,1}S$ valued vector field ξ of class C^1. Hence, for any open subset $U \subset S$ on which $\dim N_x$ is constant, the distribution $x \mapsto N_x + \overline{N}_x$ is involutive, so that there is a foliation on U whose leaves are complex submanifolds of dimension l having N_x as tangent spaces. Note that one may take $U = S$ if $l = n - 1$.

Let L be a holomorphic line bundle over M and let $\boldsymbol{a}$ be a C^∞ fiber metric of L. We consider the restriction of the curvature form $\Theta_{\boldsymbol{a}}$ of $\boldsymbol{a}$ to the spaces N_x, and its eigenvalues $\gamma_1 \le \gamma_2 \le \cdots \le \gamma_{l'}$ ($l' = \dim N_x$) with respect to some C^∞ Hermitian metric g on M. We shall say that L is q-positive along S if L admits a fiber metric $\boldsymbol{a}$ such that $\gamma_q > 0$.

Lemma 6. *If $(L, \boldsymbol{a})$ is q-positive along S, then there exists a C^∞ Hermitian metric g_0 on M such that $\sum_{i=1}^{q} \gamma_i > 0$ everywhere on S.*

Proof. An exercise how to use the partition of unity.

Proposition 7. *Let S be a compact and partially Levi flat real hypersurface of a complex manifold M, and let $l = \inf_{x \in S} \dim_{\mathbb{C}} N_x$. Let E and L be holomorphic vector bundles over M such that L is of rank one and q-positive along S. Then, for any C^∞ fiber metric h of E there exists a C^∞ Hermitian metric g_0 on M and a C^∞ fiber metric $\boldsymbol{a}$ of L such that, for some $k_0 \in \mathbb{N}$ and $c_0 > 0$,*

$$kc_0||u||^2 \le ||\overline{\partial}_b u||^2 + ||\overline{\partial}_b^* u||^2$$

holds for any $u \in C^{0,q'}(S, E \otimes L^k)$ with $k > k_0$ and $q \ge q'$. Here the L^2 norm $||\ ||$ is defined with respect to g_0 and $h\boldsymbol{a}^k$, and $\overline{\partial}_b^*$ denotes the L^2 adjoint of $\overline{\partial}_b$.

Proof. We fix fiber metrics of E and L so that the conclusion of Lemma 6 is valid. To exhibit the basic computation, we consider first the case where S is a real-analytic hypersurface. In this case, the union of N_x for all $x \in S$ forms a real-analytic fiber space over S, say $W \to S$. Hence one can find a real-analytic subset $A \subset S$ with $\operatorname{codim}_{\mathbb{R}} A \ge 2$, such that W contains a complex vector bundle of rank l over $S \setminus A$, say $X \to S \setminus A$.

Then, from the well known graded commutation relations between $\overline{\partial}$ and $\overline{\partial}^*$, applied to the leaves of foliation tangent to N_x, we obtain an inequality

$$(\diamond) \qquad (kc - d)||u||^2 \le ||\overline{\partial}_b u||^2 + ||\overline{\partial}_b^* u||^2$$

for any $u \in C^{0,q'}(S \setminus A, E \otimes L^k)$ with $\operatorname{supp} u \Subset S \setminus A$ if $q' \ge q$. Here $c := \inf_S \sum_{i=1}^{q} \gamma_i$, and d is a positive number depending on g_0 and S, but not on k. Since A is of real codimension at least 2, ($\diamond$) still holds for any $u \in C^{0,q'}(S, E \otimes L^k)$ with $q' \ge q$. Thus the required estimate is true if S is real-analytic.

In the general case, let $\{S_\mu\}_{\mu=1}^{\infty}$ be a sequence of compact real-analytic hypersurfaces, not necessary partially Levi flat, which approximates S in the C^∞ topology. Then, for each μ, we can find a (not necessarily involutive) real-analytic distribution of l-dimensional subspaces of $T_x^{1,0}S_\mu$ outside a real-analytic subset of codimension 2, say A_μ, in such a way that, as $\mu \to \infty$ the sequence of these subspaces of $T_x^{1,0}S_\mu$ has a subsequence which converges to a subspace of N_x if $x \notin \cup_{\mu=1}^{\infty} A_\mu$.

Then, applying the same method of computation as above, we obtain for any $u \in C^{0,q'}(S_\mu, E \otimes L^k)$ with $q' \geq q$ that

$$(q'c_\mu - d_\mu)||u||^2 \leq ||\overline{\partial}_b u||^2 + ||\overline{\partial}_b^* u||^2 + \epsilon_\mu ||u||_1^2.$$

Here $|| \ ||_1$ denotes the Sobolev 1-norm, and $c_\mu \to c$, $d_\mu \to d$ and $\epsilon_\mu \to 0$ as $\mu \to \infty$. Hence, by taking the limit we obtain the desired estimate. □

It will follow from the assumption of Proposition 7 that

$$H_{\overline{\partial}_b}^{0,q'}(S, E \otimes L^k) = 0$$

for $q' \geq q$ and for sufficiently large k, if we can show that the Sobolev norm estimates

$$(\sharp) \qquad ||u||_m^2 \leq C_m ||\overline{\partial}_b \overline{\partial}_b^* u||_m^2 \quad m \in \mathbb{N}$$

hold for $u \in C^{0,q'}(S, E \otimes L^k)$, $q' \geq q$ and $k \gg 1$. Here C_m is allowed to depend on m, but not on k. However the author does not know how to prove it, although it is very likely to be true.

References

[A-V-1] Andreotti, A. and Vesentini, E., *Sopra un teorema di Kodaira*, Ann. Sci. Norm. Sup. Pisa **15** (1961), 283–309.

[A-V-2] ———, *Carleman estimates for the Laplace-Beltrami equation on complex manifolds*, Publ. Math. IHES **25** (1965), 81–130.

[C] Cerveau, D., *Minimaux des feuilletages algébriques de* $\mathbb{CP}(n)$, Ann. Inst. Fourier. Grenoble **43** (1993), 1535–1543.

[D-O] Diederich, K. and Ohsawa, T., *Harmonic mappings and the disc bundles over compact Kähler manifolds*, Publ. RIMS, Kyoto Univ. **21** (1985), 819–833.

[D-G] Docquier, F. and Grauert, H., *Levisches Problem und Rungescher Satz für Teilgebiete Steinschen Mannigfaltigkeiten*, Math. Ann. **140** (1960), 94–123.

[E] Elencwajg, G., *Pseudoconvexité locale dans les variétés Kählériennes*, Ann. Inst. Fourier **25** (1975), 295–314.

[G-R] Gunning, R. C. and Rossi, H., *Analytic functions of several complex variables*, Prentice-Hall, Inc. Englewood Cliffs, N. J. 1965.

[J] Jost, J., *Riemannian geometry and geometric analysis*, Springer, 1995.

[K] Kodaira, K., *On Kähler varieties of restricted type*, Ann. of Math. **60** (1954), 28–48.

[LN] Lins Neto, A., *A note on projective Levi-flats and minimal sets of algebraic functions*, preprint.

[N] Nakano, S., *On complex analytic vector bundles*, J. M. S. Japan **7** (1955), 1–12.

[O-1] Ohsawa, T., *A Stein domain with smooth boundary which has a product structure*, Publ. RIMS, Kyoto Univ. **18** (1982), 1185–1186.

[O-2] ———, *Cohomology vanishing theorems on weakly 1-complete manifolds*, Publ. RIMS, Kyoto Univ. **19** (1983), 1181–1201.

[O-S] Ohsawa, T. and Sibony, N., *Bounded P. S. H. functions and pseudoconvexity in Kähler manifolds*, Nagoya Math. J. **249**(1998), 1–8.

[R] Royden, H. L., *The extension of regular holomorphic maps*, Proc. Amer. Math. Soc. **43** (1974), 306–310.

[S] Siu, Y.-T., *Every Stein subvariety admits a Stein neighbourhood*, Invent. Math. **38** (1976), 89–100.

[Su] Suzuki, O., *Pseudoconvex domains on a Kähler manifold with positive holomorphic bisectional curvature*, Publ. RIMS, **12** (1976), 191–214.

[T] Takeuchi, A., *Domains pseudoconvexes infinis et la métrique riemannienne dans un espace projectif*, J. Math. Soc. Japan, **16** (1964), 159–181.

Addendum

After submitting this article, the author noticed that the conjectural statement in the last paragraph is false, by finding the following counterexample.

Counterexample. Let R be a compact Riemann surface of genus at least 2, and let Γ be a discrete subgroup of the automorphism group of the unit disc Δ such that $\Delta/\Gamma \cong R$. Let

$$\iota_1 \colon \Delta \to \Delta \times \Delta$$

and

$$\iota_2 \colon \Gamma \to \Gamma \times \Gamma$$

be the diagonal embeddings, and let

$$\begin{aligned} R^{(2)} &:= \Delta \times \Delta / \iota_2(\Gamma) \\ R_0 &:= \iota_1(\Delta)/\iota_2(\Gamma). \end{aligned}$$

Since the action of Γ naturally extends to $\mathbb{P}^1$, $\iota_2(\Gamma)$ acts properly discontinuously on $\mathbb{P}^1 \times \Delta$. We put $M = \mathbb{P}^1 \times \Delta / \iota_2(\Gamma)$.

Clearly M is a compact complex surface and $R^{(2)}$ is a pseudoconvex domain in M with a C^ω-smooth Levi flat boundary.

We put $S = \partial R^{(2)}$. Suppose that the $\overline{\partial}_b$-equations were smoothly solvable for sufficiently positive line bundles. Then, in particular we would be able to produce C^∞ CR sections of $K_M^{\otimes \mu}|S$ for $\mu \gg 1$ in such a way that

$$\dim H^{0,0}_{\overline{\partial}_b}(S,\, K_M^{\otimes \mu}|S) = \infty.$$

On the other hand, we put

$$\varphi(z,w) = -\log\left(1 - \left|\frac{w-z}{\overline{w}z-1}\right|^2\right), \qquad z, w \in \Delta.$$

Then φ is invariant under the action of $\iota(\Gamma)$, so that $\varphi = \psi \circ \pi$ for some $\psi \colon R^{(2)} \to [0,\infty)$, where $\pi \colon \Delta \times \Delta \to R^{(2)}$ denotes the canonical projection.

ψ is plurisubharmonic on $R^{(2)}$. Indeed, we have

$$\begin{aligned} &\partial\overline{\partial}\left(-\log\left(1 - \left|\frac{w-z}{\overline{w}z-1}\right|^2\right)\right) \\ &= \partial\overline{\partial}\left(-\log\left(|\overline{w}z-1|^2 - |w-z|^2\right) + \log|\overline{w}z-1|^2\right) \\ &= \partial\overline{\partial}\left(-\log\left(|w|^2|z|^2 - |w|^2 - |z|^2 + 1\right) + \log|\overline{w}z-1|^2\right) \\ &= \left(1-|z|^2\right)^{-2} dz\, d\overline{z} + \left(1-|w|^2\right)^{-2} dw\, d\overline{w} - (w\overline{z}-1)^2 dw\, d\overline{z} \\ &\quad -(\overline{w}z-1)^{-2} dz\, d\overline{w} \end{aligned}$$

and

$$\begin{aligned}&\left(1-|z|^2\right)^{-2}\left(1-|w|^2\right)^{-2}-|\overline{w}z-1|^{-4}\\&\quad=\left(1-|z|^2\right)^{-2}\left(1-|w|^2\right)^{-2}|\overline{w}z-1|^{-4}|z-w|^4.\end{aligned}$$

Thus ψ is a plurisubharmonic exhaustion function which is of logarithmic growth near $\partial R^{(2)}$ and

$$\partial\overline{\partial}\,\psi > c\,(\partial\psi\,\overline{\partial}\psi + g)$$

holds for some positive constant c, outside a neighbourhood of R_0. Here g denotes a Hermitian metric on M.

Moreover, since the normal bundle of R_0 in M is negative, ψ can be modified near R_0 to a plurisubharmonic function

$$\hat{\psi}\colon R^{(2)}\backslash R_0 \to [-\infty,\infty)$$

such that

$$\partial\overline{\partial}\,\hat{\psi} > c'g \quad \text{on} \quad R^{(2)}\backslash R_0$$

for some $c' > 0$ and $\hat{\psi} = \psi$ near $\partial R^{(2)}$.

Hence, by the same extension technique as in the proof of Theorem 5, we have extensions of the elements of $H^{0,0}_{\overline{\partial}_b}(S,\, K_M^{\otimes\mu}|S)$ to holomorphic sections of $K_M^{\otimes\mu}\otimes[R_0]^{\otimes k}$ which are continuous on $\overline{R^{(2)}}$. Here $[R_0]$ denotes the line bundle associated to the divisor R_0, and k is an integer independent of μ.

It follows in particular that the restriction map

$$H^{0,0}(M,\, K_M^{\otimes\mu}\otimes[R_0]^{\otimes k}) \to H^{0,0}_{\overline{\partial}_b}(S,\, K_M^{\otimes\mu}|S)$$

is surjective.

This contradicts that

$$\dim H^{0,0}(M,\, K_M^{\otimes\mu}\otimes[R_0]^{\otimes k}) < \infty$$

and

$$\dim H^{0,0}_{\overline{\partial}_b}(S,\, K_M^{\otimes\mu}|S) = \infty.$$

Graduate School of Mathematics
Nagoya University
Chikusa-ku, Nagoya 464-8602, Japan

CHAPTER XII

Existence and Applications of Analytic Zariski Decompositions

Hajime Tsuji

Abstract

We construct a kind of Zariski decomposition as a singular Hermitian metric on a big line bundle on a smooth projective variety defined over $\mathbf{C}$ and give applications.

1. Introduction

Let X be a smooth projective variety and let L be a line bundle on X. L is said to be big if

$$\limsup_{m\to\infty} m^{-\dim X} \dim H^0(X, \mathcal{O}_X(mL)) > 0$$

holds. L is said to be nef[2] if

$$L \cdot C \geq 0$$

holds for every irreducible curve C on X.

L (resp. divisor D) is said to be pseudoeffective if $c_1(L)$ (resp. $c_1(D)$) is a limit of the classes of effective $\mathbf{Q}$-divisors on X.

To study a large line bundle we introduce the notion of analytic Zariski decompositions. By using analytic Zariski decompositions, we can handle a big line bundle as if it were nef and big.

Definition 1.1 *Let L be a line bundle on a complex manifold M. A singular Hermitian metric h is given by*

$$h = e^{-\varphi} \cdot h_0,$$

where h_0 is a C^∞- Hermitian metric on L and $\varphi \in L^1_{loc}(M)$ is an arbitrary function on M.

[2]Roughly speaking nef means a semipositivity in the algebra-geometric sense.

The curvature current Θ_h of the singular Hermitian line bundle (L,h) is defined by

$$\Theta_h := \Theta_{h_0} + \sqrt{-1}\partial\bar{\partial}\varphi,$$

where $\partial\bar{\partial}$ is taken in the sense of a current. The L^2-sheaf $\mathcal{L}^2(L,h)$ of the singular Hermitian line bundle (L,h) is defined by

$$\mathcal{L}^2(L,h) := \{\sigma \in \Gamma(U, \mathcal{O}_M(L)) \mid h(\sigma,\sigma) \in L^1_{loc}(U)\},$$

where U runs over open subsets of M. In this case there exists an ideal sheaf $\mathcal{I}(h)$ such that

$$\mathcal{L}^2(L,h) = \mathcal{O}_M(L) \otimes \mathcal{I}(h)$$

holds. We call $\mathcal{I}(h)$ the multiplier ideal sheaf of (L,h) ([16]). If we write h as

$$h = e^{-\varphi} \cdot h_0,$$

where h_0 is a C^∞ Hermitian metric on L and $\varphi \in L^1_{loc}(M)$ is the weight function, we see that

$$\mathcal{I}(h) = \mathcal{L}^2(\mathcal{O}_M, e^{-\varphi})$$

holds. It is known that if Θ_h is positive, $\mathcal{I}(h)$ is a coherent sheaf of $\mathcal{O}_M$-ideals ([16]).

Now we define the notion of analytic Zariski decompositions.

Definition 1.2 *Let M be a compact complex manifold and let L be a line bundle on M. A singular Hermitian metric h on L is said to be an analytic Zariski decomposition (AZD), if the following hold.*

1. *the curvature Θ_h is a closed positive current,*
2. *for every $m \geq 0$, the natural inclusion*

$$H^0(M, \mathcal{O}_M(mL) \otimes \mathcal{I}(h^m)) \rightarrow H^0(M, \mathcal{O}_M(mL))$$

is an isomorphim, where $\mathcal{I}(h^m)$ denotes the multiplier ideal sheaf of h^m.

If a line bundle L admits an AZD h, to study

$$H^0(M, \mathcal{O}_M(mL))$$

it is sufficient to study

$$H^0(M, \mathcal{O}_M(mL) \otimes \mathcal{I}(h^m)).$$

Hence we can consider L as if it were a singular Hermitian line bundle (L, h) whose curvature current Θ_h is a closed positive current. This is very convenient for applying various vanishing theorems (for example Nadel's vanishing theorem [16, p.561]).

The above definition of an AZD is inspired by the Zariski decomposition defined by O. Zariski. The following definition of a Zariski decomposition is the refined one by Y. Kawamata.

Definition 1.3 ([13]) *Let X be a smooth projective variety and let D be a pseudoeffective divisor on X. The expression*

$$D = P + N(P, N \in Div(X) \otimes \mathbf{R})$$

is said to be a Zariski decomposition of D, if the following conditions are satisfied.

1. *P is nef,*
2. *N is effective,*
3. *$H^0(X, \mathcal{O}_X([mP])) \simeq H^0(X, \mathcal{O}_X(mD))$ holds for every integer $m \geq 0$,*

where the integral part $[mP]$ of mP is given by

$$[mP] = [m \sum_i a_i P_i] = \sum_i [ma_i] P_i$$

($[ma_i]$ denotes the Gauss symbol of ma_i).

The existence of a Zariski decomposition was first proved for effective divisors on a smooth projective surfaces by O. Zariski([26]). If a Zariski decomposition exists it is very useful to study the ring

$$R(X, D) = \oplus_{m \geq 0} H^0(X, \mathcal{O}_X(mD)).$$

For example see [13]. But unfortunately this is the only general existence result.

One of the advantage of an AZD is that we have the following general existence theorem.

Theorem 1.1 ([21]) *Let L be a big line bundle on a smooth projective variety M. Then L has an AZD.*

In Section 3, we apply Theorem 1.1 to study pluricanonical systems of varieties of general type. In this paper, most of the proofs are only outlined. The full proofs will be published elsewhere.

2. Analytic Zariski decomposition

2.1. Existence of an AZD.

Here we shall give a sketch of the proof of Theorem 1.1 by using the L^2-estimates for $\bar{\partial}$-operators. Let g be a C^∞ Kähler metric on M. Let h_L be a C^∞ Hermitian metric on L. For every $m \geq 1$, we choose an orthonormal basis $\{\varphi_0^{(m)}, \ldots, \varphi_{N(m)}^{(m)}\}$ of $H^0(M, \mathcal{O}_M(mL))$ with respect to the L^2-inner product with respect to h_L^m and g. We define a section of $mL \otimes m\bar{L}$ by

$$K_m(z, w) := \sum_{j=0}^{N(m)} \varphi_j^{(m)}(z)\bar{\varphi}_j^{(m)}(w).$$

It is easy to see that K_m is independent of the choice of the orthonormal basis. We call $K_m(z, w)$ the Bergman kernel of (mL, h_L^m) with respect to g or m-th Bergman kernel of (L, h_L). For simplicity we denote $K_m(z, z)$ by $K_m(z)$. We consider $K_m(z)$ as a Hermitian metric on $\mathcal{O}_M(-mL)$.

Let ν be a sufficiently large positive integer such that

$$\begin{aligned} |\nu L| &= \mathbf{P}(H^0(M, \mathcal{O}_M(\nu L))) \\ &= H^0(M, \mathcal{O}_M(\nu L)) - \{0\}/\mathbf{C}^* \end{aligned}$$

gives a birational rational map. Let $p_\nu : M_\nu \longrightarrow M$ be a resolution of the base locus

$$\mathrm{Bs}|\nu L| = \cap_{\sigma \in |\nu L|}(\sigma)$$

of $|\nu L|$, where (σ) is the divisor of σ and $\cap$ means the scheme theoretic intersection. Let

$$p_\nu^*|\nu L| = |P_\nu| + F_\nu$$

be the decomposition into the free part $|P_\nu|$ and the fixed part F_ν, i.e., for a section $\sigma \in H^0(X, \mathcal{O}_X(\nu L))$

$$p_\nu^*(\sigma) = (\sigma') + F_\nu$$

where

$$\sigma' \in H^0(M_\nu, \mathcal{O}_{M_\nu}(P_\nu)).$$

By the construction $\mathrm{Bs}|P_\nu|$ is empty and the divisor F_ν is given by

$$F_\nu = \cap_{\sigma \in |\nu L|}(p_\nu^* \sigma).$$

By Kodaira's lemma (cf. [14, Appendix]), there exists an effective **Q**-divisor E_ν on M_ν such that $P_\nu - E_\nu$ is an ample **Q**-divisor. Let a be a positive integer such that aE_ν is a divisor with integer coefficients. Then by Kodaira's embedding theorem, there is a C^∞-Hermitian metric on $aP_\nu - aE_\nu$, say h_A with strictly positive curvature. Let h_{P_ν} be a C^∞-Hermitian metric on $\mathcal{O}_{M_\nu}(P_\nu)$ which is defined by the pullback of the Fubini-Study metric on the hyperplane bundle on the projective space $|P_\nu|$. Then

$$h_{aE_\nu} := h_{P_\nu}^a / h_A$$

is a metric on the line bundle aE_ν. Therefore the Hermitian metric

$$h_A^{1/a} = h_{P_\nu} / (h_{aE_\nu})^{1/a}$$

on $P_\nu - E_\nu$ is of strictly positive curvature. Let σ_{aE_ν} be a holomorphic section of $\mathcal{O}_{M_\nu}(aE_\nu)$ with divisor aE_ν. Therefore we see that by the above fact, the singular Hermitian metric

$$h_1 := h_{P_\nu} / h_{aE_\nu}(\sigma_{aE_\nu}, \sigma_{aE_\nu})^{1/a}$$

has strictly positive curvature. Then we see that for every $0 < \varepsilon < 1$,

$$h_\nu := h_{P_\nu} / h_{aE_\nu}(\sigma_{aE_\nu}, \sigma_{aE_\nu})^{\varepsilon/a}$$

is a singular Hermitian metric on $\mathcal{O}_{M_\nu}(P_\nu)$. with strictly positive curvature (in the sense of current) by the equality

$$\Theta_{h_\nu} = (1 - \varepsilon)\Theta_{h_{P_\nu}} + \varepsilon \Theta_{h_1}.$$

We shall fix ε for a moment. $\sqrt[\nu]{h_\nu}$ defines a metric on L on $M - \mathrm{Bs}|\nu L|$ and has a pole singularity on $\mathrm{Bs}|\nu L|$. Therefore we may consider $\sqrt[\nu]{h_\nu}$ a singular Hermitian metric on L. Multiplying by a suitable positive constant if necessary, we may assume that

$$h_L \leq \sqrt[\nu]{h_\nu}$$

holds on M.

Let us define the Bergman kernel $K_{m,\nu}(z,w)$ by taking an orthonormal basis for

$$H^0(M, \mathcal{O}_M(mL) \otimes \mathcal{I}(h_\nu^{m/\nu}))$$

with repect to g and $h_\nu^{m/\nu}$. We note that

$$K_m(z) = \sup\{|\sigma|^2 | \sigma \in H^0(M, \mathcal{O}_M(mL)), \| \sigma \| = 1\}$$

holds, where $\| \sigma \|$ denotes the L^2-norm of σ with respect to h_0^m and g. Similary

$$K_{m,\nu}(z) = \sup\{|\sigma|^2 | \sigma \in H^0(M, \mathcal{O}_M(mL) \otimes \mathcal{I}(h_\nu^{m/\nu})), \| \sigma \|_{\nu,m} = 1\}$$

holds, where $\| \sigma \|_{m,\nu}$ denotes the L^2-norm of σ with respect to $h_\nu^{m/\nu}$ and g. Then since $\sqrt[\nu]{h_\nu}$ is larger than h_L, we have that

$$K_{m,\nu}(z) \leq K_m(z) \ (z \in M)$$

holds.

Now we quote the following lemma.

Lemma 2.1 (cf. [9, Section 3] and [20])

$$\frac{1}{\sqrt[\nu]{h_\nu}}(z) = \lim_{m\to\infty} K_{m,\nu}(z)^{\frac{1}{m}}$$

holds.

Remark 2.1 In Lemma 2.1, the essential fact used here is that h_ν has strictly positive curvature. In fact this lemma is local and can be stated more generally as an approximation theorem for plurisubharmonic functions (see [9, Section 3]). Also this lemma follows easily from the L^2-extention theorem of Ohsawa-Takegoshi ([17], cf. also [9]). This lemma also follows the earlier work of Tian ([20]), where he used the theory of peak functions which is one of the main tools to study Bergman kernels. In the paper G. Tian considered the convergence of the curvature.

Hence this implies the following corollary.

Corollary 2.1 ([9])

$$\limsup_{m\to\infty} K_m(z)^{\frac{1}{m}} \geq \frac{1}{\sqrt[\nu]{h_\nu(z)}} \ (z \in M)$$

holds.

We shall show that $\limsup_{m\to\infty} K_m(z)^{\frac{1}{m}}$ is finite.

Let Ω be a subdomain of M. Then we define the m-th Bergman kernel $K_{m,\Omega}(z,w)$ of $(L|_\Omega, h_L|_\Omega)$ with respect to the volume form associated with $g|_\Omega$. Then clearly we have that

$$K_m(z) \leq K_{m,\Omega}(z)(= K_{m,\Omega}(z,z))$$

holds for every $z \in \Omega$. Let us take an arbitrary point $z \in X$ and let U be a coordinate neighbourhood of z which is biholomorphic to a polydisk Δ^n. Let h_U be a C^∞ Hermitian metric of strictly positive curvature on the closure of U such that

$$h_U < h_L|_U$$

holds on U. Let $K^*_{m,U}$ be the m-th Bergman kernel of $(L|_U, h_U)$ with respect to $g|_U$ and h_U. Since h_U is smaller than $h_L|_U$,

$$K_m(y) \leq K_{m,U}(y) \leq K^*_{m,U}(y)$$

hold for every $y \in U$. We note that applying Lemma 2.1 for $K^*_{m,U}$ (see also Remark 2.1) we obtain that

$$\lim_{m\to\infty} K^*_{m,U}(y)^{\frac{1}{m}} = h_U^{-1}(y)$$

holds (actually Demailly proved this fact in [9, Section 3]). Hence in particular $K_m(y)^{\frac{1}{m}}$ is uniformly bounded on every compact subset of U. Shirinking U, we obtain the following lemma.

Lemma 2.2

$$\limsup_{m\to\infty} K_m(z)^{1/m} \leq h_L^{-1}$$

holds for every $z \in M$. Also $\sup_{m\geq 1} K_m(z)^{\frac{1}{m}}$ is bounded from above on M.

Now we quote the following theorem.

Theorem 2.1 ([15, p.26, Theorem 5]) *Let $\{\varphi_t\}_{t\in T}$ be a family of plurisubharmonic functions on a domain Ω which is uniformly bounded from above on every compact subset of Ω. Then $\psi = \sup_{t\in T}\varphi_t$ has a minimum uppersemicontinuous majorant ψ^* which is plurisubharmonic.*

Remark 2.2 In the above theorem the equality $\psi = \psi^*$ holds outside of a set of measure 0 (cf.[15, p.29]).

Combining Lemma 2.1 and Lemma 2.2 with Theorem 2.1, we see that

$$h := (\{\limsup_{m\to\infty} K_m^{\frac{1}{m}}\}^*)^{-1}$$

is a singular Hermitian metric, where $\{\ \}^*$ denotes the minimum uppersemicoutinuous majorant as above.

The reason why we take the majorant is to assure that h has positive curvature, in other word $\log\{\limsup_{m\to\infty} K_m^{\frac{1}{m}}\}^*$ is plurisubharmonic. This operation $*$ is called superior regularization.

Now we shall show that h is an AZD of L. By taking $\varepsilon > 0$ sufficiently small in the construction of h_ν, we see that the natural inclusion

$$H^0(M, \mathcal{O}_M(\nu L) \otimes \mathcal{I}(h_\nu)) \to H^0(M, \mathcal{O}_M(\nu L))$$

is an isomorphism. Then by Corollary 2.1, we see that

$$H^0(M, \mathcal{O}_M(\nu L) \otimes \mathcal{I}(h^\nu)) \to H^0(M, \mathcal{O}_M(\nu L))$$

is an isomorphism. Since we can take ν arbitrary large, we see that

$$H^0(M, \mathcal{O}_M(mL) \otimes \mathcal{I}(h^m)) \to H^0(M, \mathcal{O}_M(mL))$$

is an isomorphism for every sufficiently large m. But since

$$R(M, L) := \oplus_{m\geq 0} H^0(M, \mathcal{O}_M(mL))$$

is a ring, by the definition of multiplier ideal sheaves and the Hölder inequality, we see that the above homomorphism is an isomorphism for every $m \geq 0$. This completes the proof of Theorem 1.1.

2.2. A property of AZD.

Definition 2.1 *Let T be a closed positive $(1,1)$-current on the unit open polydisk Δ^n with center O. Then by $\partial\bar{\partial}$-Poincaré lemma there exists a plurisubharmonic function φ on Δ^n such that*

$$T = \frac{\sqrt{-1}}{\pi}\partial\bar{\partial}\varphi.$$

We define the Lelong number $\nu(T, O)$ at O by

$$\nu(T, O) = \liminf_{x\to O} \frac{\varphi(x)}{\log|x|},$$

where $|x| = (\sum |x_i|^2)^{1/2}$. It is easy to see that $\nu(T, O)$ is independent of the choice of φ and local coordinate around O. Let V be a subvariety of Δ^n. Then we define the Lelong number $\nu(T, V)$ by

$$\nu(T, V) = \inf_{x\in V} \nu(T, x).$$

Remark 2.3 More generally the Lelong number is defined for a closed positive (k, k)-current on a complex manifold.

The following theorem is an easy consequence of Kodaira's lemma ([14, Appendix]) and L^2-estimate for $\bar{\partial}$-operator.

Theorem 2.2 *Let L be a big line bundle on a smooth projective variety X and let h be an AZD of L. Then for every $x \in X$, there exists a positive constant $C(x)$ such that*

$$mult_x Bs|mL| - m\nu(\Theta_h, x) \leq C(x)$$

holds for every $m \geq 0$.

This theorem implies that the Lelong number $\nu(\Theta_h, x)$ controls the asymptotic behavior of the base locus $\mathrm{Bs}|mL|$.

3. Pluricanonical systems of varieties of general type

Now let us discuss applications of AZD.

3.1. Effective birationality of pluricanonical maps.

Let X be a smooth projective variety and let K_X be the canonical bundle of X. X is said to be of general type, if K_X is big, i.e.,

$$\limsup_{m \to \infty} m^{-\dim X} \dim H^0(X, \mathcal{O}_X(mK_X)) > 0$$

holds. The following problem is fundamental to study projective varieties of general type.

Problem 1 *Let X be a smooth projective variety of general type. Find a positive integer m_0 such that for every $m \geq m_0$, $|mK_X|$ gives a birational rational map from X into a projective space.*

If $\dim X = 1$, it is well known that $|3K_X|$ gives a projective embedding. In the case of smooth projective surfaces of general type, E. Bombieri showed that $|5K_X|$ gives a birational rational map from X into a projective space ([5]). But for the case of $\dim X \geq 3$, very little is known about Problem 1. One of the main application of AZD is the following theorem.

Theorem 3.1 *There exists a positive integer ν_n which depends only on n such that for every smooth projective n-fold X of general type defined over* **C**, *$|mK_X|$ gives a birational rational map from X into a projective space for every $m \geq \nu_n$.*

For smooth projective varieties with ample canonical bundle, Theorem 3.1 has already been known ([8, 1]). Obviously the main difficulty is the fact that K_X is not ample in general. In the case of projective surfaces of general type, we can overcome this difficulty by taking the minimal models of the surfaces. In higher dimesion, there is also the notion of minimal model, i.e., for a projective variety X, a minimal model X_{min} is a projective variety birationally equivalent to X with only terminal singularities and nef canonical divisor. But at present there is no minimal model theory in dimension ≥ 4. Moreover in this case the canonical divisor of X_{min} is not a Cartier divisor in general.

To overcome this difficulty we use an AZD on K_X. By using the AZD we can handle K_X as if K_X were nef and big. And we can work on smooth varieties. Hence we can handle X as if it were minimal. At this stage the main difference from the case that K_X is ample is that K_X may have very small positivity on some subvarieties, i.e., there may exist a subvariety Y in X such that $\deg \Phi_{|mK_X|}(Y)$ grows very slowly as m tends to infinity, where $\Phi_{|mK_X|}$ denotes the rational map from X into a projective space associated with the linear system $|mK_X|$. We prove Theorem 3.1 using the fact that such subvarieties are birationally bounded.

3.2. Volume of subvarieties.

Let L be a big line bundle on a smooth projective variety X and let h be an AZD of L. To measure the total positivity of L on a subvariety of X. We define the following notion.

Definition 3.1 ([24]) *Let L be a big line bundle on a smooth projective variety X. Let Y be a subvariety of X of dimension r. We define the volume $\mu(Y, L)$ of Y with respect to L by*

$$\mu(Y, L) := r! \limsup_{m\to\infty} m^{-r} \dim H^0(Y, \mathcal{O}_Y(mL) \otimes \mathcal{I}(h^m)).$$

Remark 3.1 If we define $\mu(Y, L)$ by

$$\mu(Y, L) := r! \limsup_{m\to\infty} m^{-r} \dim H^0(Y, \mathcal{O}_Y(mL))$$

then it is totally different unless $Y = X$.

3.3. Stratification of varieties of gneneral type.

Let X be a smooth projective n-fold of general type. Let h be an AZD of K_X. By using h, we can obtain a lower bound of m such that $|mK_X|$ gives a birational rational embbeding of X in terms of the volume of subvarieties which appear in some canonical stratification of X.

Let us denote $\mu(X, K_X)$ by μ_0.

Let U_0 be a Zariski open subset of X such that $\Phi_{|mK_X|}$ is an embedding on U_0 for some m. Let $x, y \in U_0$ be distinct points and let ε be a sufficiently small positive number. As in [22] we construct the following inductively:

1. singular Hermitian metrics

$$h_0, h_1, \dots, h_r$$

on K_X defined by

$$h_i = \frac{1}{|\sigma_i|^{2/m_i}},$$

where m_i is a sufficiently large positive integer and

$$\sigma_i \in H^0(X, \mathcal{O}_X(m_i K_X))$$

such that

$$\sigma_i|_{X_i(x,y)} \in H^0(X_i(x,y), \mathcal{O}_{X_i(x,y)}(m_i K_X) \otimes \mathcal{M}_{x,y}^{[\sqrt[n_i]{\mu_i}(1-\varepsilon)m_i]}),$$

2. invariants

$$\alpha_0(x,y), \alpha_1(x,y), \dots, \alpha_r(x,y),$$

defined by

$$\alpha_i(x,y) := \inf\{\alpha > 0 | x, y \in \mathrm{Spec}(\mathcal{O}_X/\mathcal{I}(h_0^{\alpha_0 - \varepsilon_0} \cdots h_{i-1}^{\alpha_{i-1}-\varepsilon_{i-1}} \cdot h_i^{\alpha}))\},$$

where $\varepsilon_0, \dots, \varepsilon_{i-1}$ are sufficiently small positive numbers,

3. a strictly decreasing sequence of subvarieties

$$X = X_0(x,y) \supset X_1(x,y) \supset \cdots$$
$$\cdots \supset X_r(x,y) \supset X_{r+1}(x,y) = \{x\} \text{ or } \{x,y\},$$

where $X_i(x,y)$ is a branch of

$$\lim_{\delta \downarrow 0} \mathrm{Spec}(\mathcal{O}_X/\mathcal{I}(h^{\alpha_0 - \varepsilon_0} \cdots h_{i-1}^{\alpha_{i-1}-\varepsilon_{i-1}} \cdot h_i^{\alpha_i + \delta})) \ni y$$

containing x[3] (we note that $\mathcal{I}(h^{\alpha_0-\varepsilon_0} \cdots h_{i-1}^{\alpha_{i-1}-\varepsilon_{i-1}} \cdot h_i^{\alpha_i+\delta})$ is increasing as δ goes to 0 and stable for every sufficiently small δ),

[3]If necessary, we exchange x for y. But if $x, y \in U_0$ is very general, we do not need to exchange.

4.

$$\mu_0, \mu_1(x,y), \ldots, \mu_r(x,y)$$

defined by

$$\mu_i = \mu(X_i(x,y), K_X),$$

5.

$$n = n_0 > n_1 > \cdots > n_r,$$

defined by

$$n_i = \dim X_i(x,y).$$

Let x, y be two distinct points on U_0. Then for every

$$m \geq \lceil \sum_{i=0}^{r} \alpha_i(x,y) \rceil + 1,$$

$\Phi_{|mK_X|}$ separates x and y. For simplicity let us denote $\alpha_i(x,y)$ by α_i. Let us define the singular Hermitian metric $h_{x,y}$ of $(m-1)K_X$ defined by

$$h_{x,y} = \left(\prod_{i=0}^{r-1} h_i^{\alpha_i - \varepsilon_i}\right) \cdot h_r^{\alpha_r + \varepsilon_r} h^{(m-1-(\sum_{i=0}^{r-1}(\alpha_i - \varepsilon_i)) - (\alpha_r + \varepsilon_r))},$$

where ε_r is a sufficiently small positive number. Then we see that $\mathcal{I}(h_{x,y})$ defines a subscheme of X with isolated support around x or y by the definition of the invariants $\{\alpha_i\}$'s. Then by Nadel's vanishing theorem ([16, p. 561]) we see that

$$H^1(X, \mathcal{O}_X(mK_X) \otimes \mathcal{I}(h_{x,y})) = 0.$$

This implies that $\Phi_{|mK_X|}$ separates x and y.

We note that for a fixed x, $\sum_{i=0}^{r} \alpha_i(x,y)$ depends on y. We set

$$\alpha(x) = \sup_{y \in U_0} \sum_{i=0}^{r} \alpha_i$$

and let

$$X = X_0 \supset X_1 \supset X_2 \supset \cdots X_r \supset X_{r+1} = \{x\} \text{ or } \{x, y\}$$

be the stratification which attains $\alpha(x)$. In this case we call it the maximal stratification at x. We see that there exists a nonempty open subset U in countable Zariski topology of X such that on U the function $\alpha(x)$ is constant and there exists an irreducible family of stratification which attains $\alpha(x)$ for every $x \in U$.

In fact this can be verified as follows. We note that the cardinarity of

$$\{X_i(x,y) \,|\, x, y \in X, x \neq y (i = 0, 1, \dots)\}$$

is uncountably many, while the cardinarity of the irreducible components of Hilbert scheme of X is countably many. We see that for a fixed i and very general x, $\{X_i(x,y)\}$ should form a family on X. Similary we see that for very general x, we may assume that the maximal stratification $\{X_i(x)\}$ forms a family. This implies the existence of U.

We may also assume that the corresponding invariants $\{\alpha_0, \dots, \alpha_r\}$, $\{\mu_0, \dots, \mu_r\}$, $\{n = n_0 \dots, n_r\}$ are constant on U. Hereafter we denote these invariants by the same notations for simplicity. The following lemma can be proved as in [22, p.12, Lemma 5].

Lemma 3.1

$$\alpha_i \leq \frac{n_i \sqrt[n_i]{2}}{\sqrt[n_i]{\mu_i}} + O(\varepsilon_{i-1})$$

holds for $1 \leq i \leq r$.

By the above lemma we obtain the following estimate.

Proposition 3.1 *For every*

$$m > \lceil \sum_{i=0}^{r} \alpha_i \rceil + 1$$

$|mK_X|$ *gives a birational rational map from* X *into a projective space.*

3.4. Fibration theorem.

Using the stratifiacation in the last subsection, we obtain the following theorems. Roughly speaking we can single out the obstruction to prove Theorem 1.1 as subvarieties of bounded degree.

Theorem 3.2 *For every smooth projective n-fold X of general type, one of the following holds.*

1. *for every* $m \geq n^2 + n + 2$, $|mK_X|$ *gives a birational rational map from* X *into a projective space,*

2. X *is dominated by a family of positive dimensional subvarieties which are birational to subvarieties of degree less than or equal to* $2^n n^{2n}(n+1)^n$ *in a projective space.*

Here we shall show the outline of the proof of Theorem 3.2. As in the last subsection we construct a strictly decreasing sequence of subvarieties (maximal stratification):

$$X = X_0 \supset X_1 \supset \cdots \supset X_r \supset X_{r+1} = \{x\} \text{ or } \{x, y\}.$$

Let $\alpha_0, \ldots, \alpha_r > 0$ be positive numbers as in the last section. If every $\mu_i \geq 1$ for $0 \leq i \leq r$, we see that $|mK_X|$ gives a birational morphism for every $m \geq n^2 + n + 2$. Hence we may assume that

$$\min_i \mu_i < 1.$$

Let $0 \leq k \leq r$ be the number such that

$$\frac{\alpha_k}{\sqrt[n_k]{2n_k}} = \max_i \frac{\alpha_i}{\sqrt[n_i]{2n_i}}.$$

Then we see that $|mK_X|$ gives a birational rational map for every

$$m \geq 2\lceil n^2\alpha_k \rceil + 2.$$

We note that by Lemma 3.1,

$$\alpha_k \leq \frac{n_k \sqrt[n_k]{2}}{\sqrt[n_k]{\mu_k}} + O(\varepsilon_{k-1}).$$

Suppose that $|(n^2 + n + 2)K_X|$ does not give a birational rational map from X into a projective space. Then by Proposition 3.1, we see that

$$\max_i \frac{\alpha_i}{\sqrt[n_i]{2n_i}} \geq 1$$

holds. By Lemma 3.1 we see that

$$\mu_k < 1$$

holds. Now we see that

$$\deg \Phi_{\lceil 2n^2\alpha_k \rceil + 2}(X_k) \leq (\lceil 2n^2\alpha_k \rceil + 2)^{n_k} \mu_k$$

$$\leq 2^{n_k}(n^2\frac{n_k+1}{\sqrt[n_k]{\mu_k}})^{n_k}\mu_k$$

$$\leq 2^{n_k}n^{2n_k}(n_k+1)^{n_k}$$

$$\leq 2^n n^{2n}(n+1)^n$$

hold. This completes the proof of Theorem 3.2.

The following theorem is a refinement of Theorem 3.2

Theorem 3.3 *For every smooth projective n-fold X of general type, one of the following holds.*

1. *for every* $m \geq n^2+n+2$, $|mK_X|$ *gives a birational rational map from* X *into a projective space,*

2. X *is birational to an algebraic fiber space whose fibers are birational to (positive dimensional) subvarieties of degree less than or equal to* $2^n n^{2n}(n+1)^n$ *in a projective space.*

3.5. A nonvanishing theorem.

To study the stable base locus of K_X we use the following theorem.

Theorem 3.4 *Let* X *be a smooth projective variety and let* (L, h_L) *be a singular Hermitian line bundle on* X *such that the curvature current* Θ_L *is positive*[4]. *Let* (A, h_A) *be a singular Hermitian line bundle on* X *with strictly positive curvature current* Θ_A. *Then one of the followings holds.*

1. $H^0(X, \mathcal{O}_X(K_X+A+mL)\otimes\mathcal{I}(h_A h_L^m)) \neq 0$ *holds for every sufficiently large* m,

2. *there exists a rational fibration* $f : X - \cdots \rightarrow Y$ *such that* $\dim Y < \dim X$ *and on the general fiber* F *the restriction* $\Theta_L|_F$ *has 0 absolutely continuous part.*

The proof of Theorem 3.4 is a combination of the stratification method as above and the following lemma.

[4]Here positive means only semipositive or pseudoeffective in the context of algebraic geometry. This teminology may be confusing for algebraic geometers. But this terminology is traditional.

Lemma 3.2 *Suppose that Θ_L has nonzero absolutely continuous part. Then*

$$\lim_{\nu\to\infty} \mu(X, A+\nu L) = \infty$$

holds, where $\mu(X, A+\nu L)$ is defined by

$$\mu(X,A+\nu L) := (\dim X)! \limsup_{m\to\infty} m^{-\dim X} \dim H^0(X, \mathcal{O}_X(A+\nu L)\otimes \mathcal{I}((h_A h_L^{\nu})^m)).$$

3.6. Structure of pluricanonical systems of varieties of general type.

Let X be a smooth projective variety of general type and let h be an AZD of K_X. Let Θ_h be the curvature current of h. We set

$$S := \{x \in X | \nu(\Theta_h, x) > 0\}.$$

Then since S is a union of subvarieties contained in the stable base locus of K_X, S is a countable union of subvarieties in X.

By Nadel's vanishing theorem we obtain the following theorem.

Theorem 3.5 *Let V be an irreducible divisorial component of S. Then*

$$\mu(V, K_X) = 0$$

holds[5].

This theorem means the V is asymptotically contracted by $R(X, K_X)$.

By applying Theorem 3.4 to a divisorial component of S, we obtain the following theorem which singles out the direction of the contraction.

Theorem 3.6 *Let V be an irreducible component of S. Assume that V is a divisor in X. Then there exists a rational fibration $\phi : V - \cdots \to W$ such that*

1. $\dim W < \dim V$,
2. *let E be a general fiber of f, then there exists a positive constant C such that*

$$\deg \Phi_{|mK_X|}(E) \leq C$$

holds for every positive integer m.

---

[5]For the definition of $\mu(V, K_X)$ see Section 3. Please do not confuse $\mu(V, K_X)$ with $\mu(V, K_X|_V)$.

This theorem is fundamental to study $R(X, K_X)$ ([24]).

3.7. Outline of the proof of Theorem 3.1.

Let X be a smooth projective n-fold of general type. Suppose that $|(n^2 + n + 2)K_X|$ does not give a birational rational map. Then by Theorem 3.3, we see that there exists a rational fibration

$$f : X - \cdots \to Y$$

such that $\dim Y < \dim X$ and a general fiber F is birational to a subvariety of a projective space of degree $\leq 2^n n^{2n}(n+1)^n$. To prove Theorem 3.1 we may assume that f is a morphism.

Suppose that there exists a positive constant c_n depending only on n such that

$$\mu(F, K_X) \geq c_n \mu(F, K_F).$$

Then the proof of Theorem 3.1 follows easily from the birational boundedness of F. Hence we may assume that

$$\frac{\mu(F, K_X)}{\mu(F, K_F)}$$

is very small. In this case we see that there exists an irreducible fixed component V such that V is horizontal with respect to f, i.e. $f(V) = Y$. Then by Theorem 3.6, we see that there exists a rational fibration

$$\phi : V - \cdots \to W$$

such that $\dim W < \dim V$ and for every $m \geq 1$ a general fiber E satisfies that

$$\deg \Phi_{|m!K_X|}(E) \leq C.$$

By the refinement of of Theorem 3.6, we may assume that C depends only on n. Let y be a general point on Y. Let $B(y)$ be set of points on Y which can be connected with y by a chain of $\phi(E)$'s. Then $f^{-1}(B(y))$ is birationally bounded by the boundedness of a general fiber of f and the boundedness of E. If $B(y)$ is equal to Y, then X is birationally bounded, hence Theorem 3.1 holds. If $B(y)$ is a proper subvariety of Y, we have a new fibration

$$g : X - \cdots \to Z$$

such that a general fiber is $f^{-1}(B(y))$. We can apply the same argument as before by replacing f by g. Continuing this process we obtain either the birational boundedness of X or the existence of c_n. This completes the proof of Theorem 3.1.

References

[1] U. Anghern-Y.-T. Siu, *Effective freeness and point separation for adjoint bundles,* Invent. Math. **122** (1995), 291–308.

[2] T. Bandeman- G. Detholoff, *Estimates on the number of rational maps to varieties of general type,* to appear.

[3] E. Bombieri, *Algebraic values of meromoprhic maps,* Invent. Math. **10** (1970), 267–287.

[4] E. Bombieri, *Addendum to my paper: Algebraic values of meromorphic maps*, Invent. Math. **11** (1970), 163–166.

[5] E. Bombieri, *Canonical models of surfaces of general type,* Publ. I.H.E.S. **42** (1972), 171–219.

[6] F. Campana, *Une version géometrique généralisée du théoreme du produit de Nadel,* Bull. Soc. Math. France **119** (1991), 479–493.

[7] L. Ein-O. Küchle-R. Lazarsfeld, *Local positivity of ample line bundles,* Jour. Diff. Geom. **42** (1995), 193–219.

[8] J.P. Demailly, *A numerical criterion for very ample line bundles*, J. Diff. Geom. **37** (1993), 323–374.

[9] J.P. Demailly, *Regularization of closed positive currents and intersection theory,* J. of Alg. Geom. **1** (1992), 361–409.

[10] T. Fujita, *On Kähler fiber spaces over curves*, J. Math. Soc. of Japan **30** (1978), 779–794.

[11] T. Fujita, *Approximating Zariski decomposition of big line bundles,* Kodai Math. J. **17** (1994), 1–3.

[12] L. Hörmander, An Introduction to Complex Analysis in Several Variables 3-rd ed., North-Holland, 1990.

[13] Y. Kawamata, *The Zariski decomposition of logcanonical divisors,* Collection of Alg. geom. Bowdowin 1985, Proc. of Sym. Pure Math. **46** Part 1, (1987), 425–433.

[14] S. Kobayashi-T. Ochiai, *Mappings into compact complex manifolds with negative first Chern class*, Jour. Math. Soc. Japan **23** (1971), 137–148.

[15] P. Lelong, Fonctions Plurisousharmoniques et Formes Différentielles Positives, Gordon and Breach (1968).

[16] A.M. Nadel, *Multiplier ideal shaves and existence of Kähler-Einstein metrics of positive scalar curvature*, Ann. of Math. **132** (1990), 549–596.

[17] T. Ohsawa and K. Takegoshi, *L^2-extention of holomorphic functions*, Math. Z. **195** (1987), 197–204.

[18] V.V. Shokurov, *The nonvanishing theorem*, Izv. Nauk USSR **26** (1986), 510–519.

[19] Y.-T. Siu, *Analyticity of sets associated to Lelong numbers and the extension of closed positive currents*, Invent. Math. **27** (1974), 53–156.

[20] G. Tian, *On a set of polarized Kähler metrics on algebraic manifolds*, Jour. Diff. Geom. **32** (1990), 99–130.

[21] H. Tsuji, *Analytic Zariski decomposition*, Proc. of Japan Acad. **61** (1992), 161–163.

[22] H. Tsuji, *Global generation of adjoint bundles*, Nagoya J. of Math. **142**, 5–16 (1996).

[23] H. Tsuji, *Analytic Zariski decomposition*, preprint (1997).

[24] H. Tsuji, *Finite generation of canonical rings*, preprint (1997).

[25] H. Tsuji, *On the structure of pluricanonical systems of projective varieties of general type*, preprint (1997).

[26] O. Zariski, *The theorem of Riemann-Roch for high multiplicities of an effective divisor on an algebraic surface*, Ann. of Math. **76** (1962), 560–615.

Department of Mathematics
Tokyo Institute of Technology
2-12-1 Ohokayama, Megro 152, Japan

E-mail: tsuji@math.titech.ac.jp

CHAPTER XIII

Segre Polar Correspondence and Double Valued Reflection for General Ellipsoids

S. M. Webster†

Introduction

A number of problems concerning an analytic real hypersurface in complex space have been treated by means of its complexification. This manifests itself as a family of complex hypersurfaces, one attached to each point of space. This association of a complex variety to a point is the Segre polar correspondence. Originally B. Segre and E. Cartan used it to attach differential invariants to a nondegenerate real hypersurface. More recently, it has been used in establishing boundary regularity, holomorphic continuation, as well as algebraicity of holomorphic mappings. It also plays a key role in several biholomorphic classification problems. In this paper we shall give a more complete version of the Segre polar correspondence for algebraic real hypersurfaces, in order to treat the phenomenon of double valued reflection.

Stationary curves were first introduced by Lempert [1] in his work on the Kobayashi extremal discs of a smooth, bounded strictly convex domain in $\mathbf{C}^n$. In [7] we established a qualitative procedure to describe the stationary curves and extremal discs for such domains with real algebraic boundary M admitting double valued reflection. We review this here in section one. On the complexification $\mathcal{M}$ of such an M are defined two meromorphic involutions τ_1, τ_2 and their composition $\sigma = \tau_1\tau_2$, which is a reversible map [2]. The main problem [7] is to understand the dynamics of σ, and, in particular, to find invariant curves which will project to stationary curves. Ellipsoids are the primary examples of real hypersurfaces admitting double valued reflection [8].

As in [3] the defining function for a general ellipsoid M can be put into the form

$$r(z, \overline{z}) = z \cdot \overline{z} + Az \cdot z + A\overline{z} \cdot \overline{z} - 1, \tag{0.1}$$

†Partially supported by NSF

where, for some l, $1 \leq l \leq n$,

$$(0.2) \quad Az \cdot z = \sum_{j=1}^{n} A_j z_j^2, \; 1/2 > A_1 \geq \cdots \geq A_l > A_{l+1} = \cdots = A_n = 0.$$

The Segre polar varieties Q_ζ (see (1.3) below) form a (nonlinear, rational) family of complex quadrics. We study this family and the polar mapping $\zeta \mapsto Q_\zeta$ in detail in sections two and three. We also consider the map which assigns to a point z the set of polar varieties containing it.

In [8] we treated the case $l = n$, in which all A_j are nonzero. Then the Segre polar varieties Q_ζ are complex spheres, after a coordinate change. The polar correspondence takes σ into a generalized "null cone billiard map" relative to two nondegenerate complex n-dimensional quadrics in the (n+1)-dimensional Moebius space of spheres. A generalized confocal theory was then developed and applied in [8] to determine more qualitatively the dynamics of σ, when all the coefficients A_j are distinct.

However, if $l < n$, then the construction in [8] breaks down, and a more general framework is needed. We develop this here in sections three and four. In particular, it leads to a pair of singular complex quadrics in dual projective spaces. We define a natural pair of "billiard" involutions, τ_1, τ_2, relative to these quadrics and construct invariant subvarities for them, using results from [8]. These involutions are then identified with those on $\mathcal{M}$ associated to double valued reflection. This yields our main result.

Theorem 0.1 *Suppose that $l < n$, and that $A_1, \ldots, A_l$, are distinct, for the ellipsoidal real hypersurface M. Then, off a proper subvariety, its complexification $\mathcal{M}$ is foliated by an l-parameter family of subvarities of codimension l, which are invariant by the involutions of double valued reflection.*

This differs somewhat from the case $l = n$, where we have an $(n-1)$-parameter family of n-dimensional invariant varieties [8]. Also, it remains to study the dynamics in more detail on the invariant sets. This will be carried out in a future work. We only point out here that for $l < n$ we have, in addition, the unitary symmetry in the last $n - l$ variables to exploit.

We thank the referee for comments which improved the accuracy and clarity of our presentation.

1. Double valued reflection

We consider an algebraic real hypersurface M, that is, one defined by a real polynomial equation, which we arrange in powers of z and $\overline{z}$,

$$(1.1) \qquad M = \{z \in \mathbf{C}^n \mid r(z, \overline{z}) = 0\}.$$

Its complexification is the complex algebraic variety with anti-holomorphic involution ρ, which fixes a copy M_0 of M,

$$(1.2) \quad \begin{aligned} \mathcal{M} &= \{(z, \zeta) \in \mathbf{C}^{2n} \mid r(z, \zeta) = 0\}, \ \rho(z, \zeta) = (\overline{\zeta}, \overline{z}), \ r \circ \rho = \overline{r}, \\ M &\cong M_0 = FP(\rho) = \mathcal{M} \cap \{\zeta = \overline{z}\}. \end{aligned}$$

The Segre polar variety associated to the point $\zeta \in \mathbf{C}^n$ is

$$(1.3) \qquad Q_\zeta = \{z \in \mathbf{C}^n \mid r(z, \zeta) = 0\}, \ \zeta = \overline{w}.$$

By the reality condition on r, we have

$$(1.4) \qquad z \in Q_{\overline{w}} \Leftrightarrow w \in Q_{\overline{z}}; \ \ z \in M \Leftrightarrow z \in Q_{\overline{z}}.$$

This says that the association $z \mapsto Q_{\overline{z}}$ is an anti-holomorphic, involutive correspondence, with fixed-point set M. It is the Segre polar correspondence.

For $n = 1$, $Q_{\overline{z}}$ is generically a finite set of m points, $m = \deg_z r$. For z near a nonsingular point of M, the analytic implicit function theorem gives a unique point $w \in Q_{\overline{z}}$ near z. The locally defined map $z \mapsto w$ is the classical Schwarz-Caratheodory reflection about M. It is globally defined and single valued on $\mathbf{P}_1 \supseteq \mathbf{C}$, if $m = 1$, i. e. for a circle or line. In the next simplest case, $m = 2$, this reflection is globally defined and double valued. This occurs for the ellipse [6] and leads to the explicit formula of H. A. Schwarz for the Riemann map of the interior to the disc.

For $n > 1$, an analytic real hypersurface $M \subset \mathbf{C}^n$ has the wrong dimension and structure to be the fixed-point set of an anti-holomorphic involution. Therefore, one considers [5] the set of its holomorphic tangent spaces,

$$(1.5) \qquad \tilde{M} = \{(z, H_z) \mid z \in M\} \subseteq \mathbf{C}^n \times \mathbf{P}^*_{n-1},$$

contained in the (2n-1)-dimensional complex manifold of holomorphic contact elements. It is totally real precisely when the Levi form of M is nondegenerate [5]. Then we have local single valued reflection about $\tilde{M}$, which will, in the general algebraic case, be multiple valued in the large. It may be described [5] as the correspondence

$$(1.6) \qquad (z, T_z Q_{\overline{w}}) \longleftrightarrow (w, T_w Q_{\overline{z}}),$$

where we note that $H_zM = T_zQ_{\bar{z}}$, for $z \in M$. This is globally single valued on the manifold of all contact elements of $\mathbf{P}_n \supset \mathbf{C}^n$, in the case of the unit sphere, $M = S^{2n-1}$. It is globally defined and double valued in the case of ellipsoids (0.1) [8].

Complexification of the map $z \mapsto (z, H_zM)$, from M to $\tilde{M}$, gives rise to the two rational maps $\pi_i : \mathcal{M} \to \mathbf{C}^n \times \mathbf{P}^*_{n-1}$, $i = 1, 2$,

$$\pi_1(z,\zeta) = (z, T_zQ_\zeta),\ \pi_2(z,\zeta) = (\zeta, T_\zeta Q_z),\ \pi_1 \circ \rho = \overline{\pi}_2. \tag{1.7}$$

The reflection is double valued when these are rational maps of degree two, thus 2-to-1 on a dense open set. We then have the two meromorphic covering involutions τ_i, $i = 1, 2$,

$$\pi_i \circ \tau_i = \pi_i,\ \tau_i^2 = I,\ \tau_2 = \rho\tau_1\rho. \tag{1.8}$$

We further define

$$\rho_1 = \tau_1\rho\tau_1,\ M_1 = FP(\rho_1) = \tau_1 M_0, \tag{1.9}$$

so that

$$\pi_1^{-1}\tilde{M} = M_0 \cup M_1. \tag{1.10}$$

The map

$$\sigma = \tau_1\tau_2 = \rho_1\rho \tag{1.11}$$

is said to be reversible [2], since it is conjugate to its inverse by an involution: $\sigma^{-1} = \tau_2\sigma\tau_2^{-1} = \rho\sigma\rho^{-1}$. Explicit formulae for these maps are given in [8] for the general ellipsoid.

After Lempert [1] we call a stationary curve for M, or for the domain it bounds, any compact irreducible 1-dimensional analytic set with boundary, $L \subset \mathbf{C}^n \times \mathbf{P}^*_{n-1}$, when the boundary curves lie on $\tilde{M}$. The preimage $A = \pi_1^{-1}(L) \subset \mathcal{M}$ is a 1-dimensional analytic set with boundary on $M_0 \cup M_1$. The extension of L by double valued reflection in $\tilde{M}$ [7] is given by the extension of A by double reflection in M_0 and M_1, i. e. by repeated application of the anti-holomorphic involutions ρ and ρ_1. In case A is also irreducible, this leads to a global immersed Riemann surface $\mathcal{A} \subset \mathcal{M}$, which is invariant by τ_1, ρ, and hence the other maps. To find and describe such curves $\mathcal{A}$ requires a detailed understanding of the dynamics of the map σ. In particular, we want to construct σ-invariant subvarieties of $\mathcal{M}$, amenable to explicit geometric description. For generic ellipsoids (all A_j distinct and nonzero) this was shown in [8] using the Segre polar correspondence. We consider a more appropriate version of this in the next section.

2. The Segre polar mappings

Next we consider the Segre polar correspondence from a more global point of view for a general algebraic real hypersurface. For this we introduce homogeneous coordinates $x, y \in \mathbf{C}^{n+1}$,

$$z = x'/x_0, \quad \zeta = y'/y_0, \tag{2.1}$$

and write

$$\begin{aligned} R(x,y) &= (x_0 y_0)^m r(x'/x_0, y'/y_0), \quad m = \deg_z r, \\ \overline{R(x,y)} &= R(\overline{y}, \overline{x}), \text{ or } R \circ \rho = \overline{R}, \ \rho(x,y) = (\overline{y}, \overline{x}), \\ \mathcal{M} &= \{(x,y) \,|\, R(x,y) = 0\}, \\ Q_y &= \{x \,|\, R(x,y) = 0\}. \end{aligned} \tag{2.2}$$

For generic y, Q_y is a complex algebraic hypersurface in $\mathbf{P}_n$ of degree m.

We denote by $\mathbf{P}_N \supseteq \{Q_y\}$ the smallest linear family of such hypersurfaces containg all Segre polar varieties. Let

$$\psi_\alpha(x) = R(x, y_\alpha), \quad 0 \le \alpha \le N, \tag{2.3}$$

for suitable $y_\alpha \in \mathbf{P}_n$, be a basis of homogeneous polynomials for this linear space. We write

$$R(x,y) = \sum_{\alpha=0}^{N} \psi_\alpha(x) g_\alpha(y), \tag{2.4}$$

for certain homogeneous polynomials $g_\alpha(y)$ of degree m. Note that $g_\alpha(y_\beta) = \delta_{\alpha\beta}$. Let $\eta \in \mathbf{C}^{N+1}$ be homogeneous coordinates for $\mathbf{P}_N$ relative to the basis (2.3). We then write

$$Q(\eta) = \{x \in \mathbf{P}_n \,|\, \sum \eta_\alpha \psi_\alpha(x) = 0\}. \tag{2.5}$$

The correspondence $y \mapsto Q_y$ defines a rational mapping, $G : \mathbf{P}_n \to \mathbf{P}_N$,

$$G: \ \eta_\alpha = g_\alpha(y), \ 0 \le \alpha \le N, \ \mathcal{Q} = G(\mathbf{P}_n) = \{Q_y\}, \tag{2.6}$$

$\mathcal{Q}$ being (the closure of) the image of G.

If we apply the reality condition (2.2) on R to the equation (2.4), and set $y = y_\beta$, we get

$$\psi_\alpha(\overline{y}) = \sum h_{\alpha\overline{\beta}} \overline{g_\beta(y)}, \tag{2.7}$$

where $(h_{\alpha\overline{\beta}})$ is a nondegenerate Hermitian matrix. Thus, on $\mathbf{P}_N$ we have the nondegenerate Hermitian form and real hyperquadric

$$h(\xi, \eta) = \sum h_{\alpha\overline{\beta}} \xi_\alpha \overline{\eta}_\beta, \ \mathcal{H} = \{\eta \in \mathbf{P}_N \,|\, h(\eta,\eta) = 0\}, \tag{2.8}$$

and the relations

$$R(x,\overline{y}) = H(G(x),G(y)),\ \ G(M)\subset \mathcal{Q}\cap\mathcal{H}. \tag{2.9}$$

Thus far, we have adapted the scheme of [4].

The requirement that $x \in Q(\eta)$ places a linear condition on η, and so defines a hyperplane $V_x \subset \mathbf{P}_N$. This gives us a second rational map, $F: x \mapsto V_x$. Let $\hat{\xi} \in \mathbf{C}^{N+1}$ denote homogeneous coordinates for the dual space, relative to the same basis (2.3). Then the map $F: \mathbf{P}_n \to \mathbf{P}_N^*$ is given by

$$F:\ \hat{\xi}_\alpha = \psi_\alpha(x),\ \ 0\le\alpha\le N,\ \ \hat{\mathcal{S}} = F(\mathbf{P}_n) = \{V_x\}. \tag{2.10}$$

We denote the dual pairing by

$$\langle\hat{\xi},\eta\rangle = \sum_{\alpha=0}^{N}\hat{\xi}_\alpha\eta_\alpha, \tag{2.11}$$

and

$$\hat{\xi}^\perp = \{\eta \,|\, \langle\hat{\xi},\eta\rangle = 0\},\ \ \eta^\perp = \{\hat{\xi}\,|\,\langle\hat{\xi},\eta\rangle = 0\}. \tag{2.12}$$

Then we also have

$$R(x,y) = \langle F(x),G(y)\rangle. \tag{2.13}$$

By combining F and G we get the representation

$$(F\times G)(\mathcal{M}) = (\hat{\mathcal{S}}\times\mathcal{Q})_0 \equiv \{(\hat{\xi},\eta)\in\hat{\mathcal{S}}\times\mathcal{Q}) \,|\, \langle\hat{\xi},\eta\rangle = 0\}. \tag{2.14}$$

The Hermitian form h induces the usual anti-linear isomorphism $H: \mathbf{P}_N \to \mathbf{P}_N^*$,

$$H:\ \hat{\xi}_\alpha = \sum h_{\alpha\overline{\beta}}\overline{\eta}_\beta. \tag{2.15}$$

It follows from (2.7) that

$$F(\overline{y}) = H(G(y)),\ \ \hat{\mathcal{S}} = H(\mathcal{Q}). \tag{2.16}$$

The map ρ (2.2) on $\mathbf{P}_n\times\mathbf{P}_n$ is readily seen to correspond under $F\times G$ to the restriction to $\hat{\mathcal{S}}\times\mathcal{Q}$ of the following map (also denoted ρ),

$$\rho(\hat{\xi},\eta) = (H(\eta), H^{-1}(\hat{\xi})). \tag{2.17}$$

3. The general ellipsoid

Now we specialize to the defining functions (0.1) and carry out the theory of the last section. There are many important additional special features in this case. In particular, $\hat{S}$ and $\mathcal{Q}$ are themselves quadrics, and we shall find their duals.

a) The linear family of quadrics. Introducing the homogeneous coordinates (2.1) into (0.1) gives

$$R(x,y) = y_0^2 Ax' \cdot x' + y_0 y' \cdot x' x_0 + (Ay' \cdot y' - y_0^2) x_0^2. \tag{3.1}$$

Since the coefficients are real, we have the additional symmetry,

$$R \circ \kappa = R, \quad \rho = \overline{\kappa}, \quad \kappa(x,y) = (y,x). \tag{3.2}$$

For the basis (2.3) we shall take

$$\psi_0(x) = Ax' \cdot x', \quad \psi_j(x) = x_0 x_j, \ 1 \le j \le n, \quad \psi_*(x) = x_0^2; \tag{3.3}$$

and the corresponding homogeneous coordinates,

$$\eta = (\eta_0, \eta', \eta_*), \ \eta' = (\eta_1, \ldots, \eta_n). \tag{3.4}$$

Thus $N = n+1$, and we identify $\eta \in \mathbf{P}_{n+1}$ with the quadric

$$Q(\eta) = \{x \in \mathbf{P}_n \,|\, \eta_0 Ax' \cdot x' + \eta' \cdot x' x_0 + \eta_* x_0^2 = 0\}. \tag{3.5}$$

We further define

$$\begin{aligned} H_\infty &= \{x \in \mathbf{P}_n \,|\, x_0 = 0\}, \\ Q_\infty &= \{x \in H_\infty \,|\, Ax' \cdot x' = 0\}, \\ N_\infty &= \{x \in \mathbf{P}_n \,|\, x_0 = x_1 = \cdots = x_l = 0\}. \end{aligned} \tag{3.6}$$

Note that $N_\infty \subset Q_\infty$ is the vertex, i. e. singular set, of Q_∞, and is empty if $l = n$. $Q(\eta)$ consists of H_∞ and another hyperplane if $\eta_0 = 0$, and 2 copies of H_∞ if $\eta_0 = 0$ and $\eta' = 0'$. Otherwise $Q(\eta)$ meets H_∞ in the points of Q_∞, being tangent to H_∞ along N_∞. In fact, we may characterize $\mathbf{P}_{n+1} = \{Q(\eta)\}$ as the family of quadrics in $\mathbf{P}_n$ passing through Q_∞.

We further denote

$$\begin{aligned} L &= \{\eta \in \mathbf{P}_{n+1} \,|\, \eta_{l+1} = \cdots = \eta_n = 0\}, \\ \mathcal{S} &= \{\eta \in L \,|\, -4\eta_0\eta_* + \sum_{j=1}^{l} A_j^{-1} \eta_j^2 = 0\}, \\ p_\infty &= (0, 0', 1) \in L, \end{aligned} \tag{3.7}$$

and observe that $\dim L = l+1$ and $\dim S = l$. We readily see, as in [3], that

$$(3.8)\quad \begin{aligned} rk(\eta) &= 1, && \text{for } \eta = p_\infty, \\ rk(\eta) &= 2, && \text{for } \eta_0 = 0,\ \eta \neq p_\infty, \\ rk(\eta) &= l, && \text{for } \eta \in S,\ \eta_0 \neq 0, \\ rk(\eta) &= l+1, && \text{for } \eta \notin S,\ \eta_0 \neq 0, \end{aligned}$$

where $rk(\eta) = \operatorname{rank} Q(\eta)$ is the rank of the quadratic form (3.5) in x.

b) The polar images. For the polynomials g_α in (2.4) we have

$$(3.9)\quad g_0(y) = y_0^2,\ g_j(y) = y_0 y_j,\ 1 \le j \le n,\ g_*(y) = Ay' \cdot y' - y_0^2;$$

so that the map $G : \mathbf{P}_n \to \mathbf{P}_{n+1}$ is given by

$$(3.10)\quad G:\ \eta_0 = y_0^2,\ \eta' = y_0 y',\ \eta_* = Ay' \cdot y' - y_0^2.$$

If we write the coordinates on $\mathbf{P}^*_{n+1}$ dual to η as

$$(3.11)\quad \hat{\xi} = (\hat{\xi}_0, \hat{\xi}', \hat{\xi}_*),\ \ \hat{\xi}' = (\hat{\xi}_1, \ldots, \hat{\xi}_n),$$

then $F : \mathbf{P}_n \to \mathbf{P}^*_{n+1}$ is given by

$$(3.12)\quad F:\ \hat{\xi}_0 = Ax' \cdot x',\ \hat{\xi}' = x_0 x',\ \hat{\xi}_* = x_0^2.$$

The maps F and G are related by the linear map $K : \mathbf{P}_{n+1} \to \mathbf{P}^*_{n+1}$,

$$(3.13)\quad F = KG,\ \ \hat{\xi} = K\eta,\ \ K = \begin{bmatrix} 1 & 0 & 1 \\ 0 & I & 0 \\ 1 & 0 & 0 \end{bmatrix},$$

where I is the $n \times n$ identity matrix. The map G is one-to-one, except that $H_\infty - Q_\infty$ is collapsed to the point p_∞, and each point of Q_∞ is blown up; and similarly for F. We readily find that κ in (3.2) corresponds to

$$(3.14)\quad \kappa(\hat{\xi}, \eta) = (K\eta, K^{-1}\hat{\xi}).$$

Clearly, (3.9) gives

$$(3.15)\quad g_0(x) = \psi_*(x),\ g_j(x) = \psi_j(x),\ g_*(x) = \psi_0(x) - \psi_*(x),$$

so that the Hermitian form (2.8) is

$$(3.16)\quad h(\eta, \eta) = \eta_0 \overline{\eta}_* + \eta' \cdot \overline{\eta}' + (\eta_0 + \eta_*)\overline{\eta}_0.$$

As noted in [4], h has signature $(n+1,1)$, so that $\mathcal{H} \cong S^{2n+1}$ is strongly pseudoconvex.

To get the image of G, we eliminate y from (3.10). This gives the quadric

$$\begin{aligned} \mathcal{Q} &= \{\eta \in \mathbf{P}_{n+1} \mid q(\eta,\eta) = 0\}, \\ q(\xi,\eta) &= -\xi_0\eta_0 - (1/2)(\xi_0\eta_* + \xi_*\eta_0) + A\xi' \cdot \eta', \end{aligned} \tag{3.17}$$

which has rank $l+2$. Its singular locus, or vertex, is the $(n-l-1)$-plane

$$N = \{\eta \in \mathbf{P}_{n+1} \mid \eta_0 = \eta_* = \eta_1 = \ldots = \eta_l = 0\}, \tag{3.18}$$

which is complementary to the plane L (3.7). Thus, $\mathcal{Q}$ is the cone

$$\mathcal{Q} = N * \mathcal{Q}^{(0)}, \quad \mathcal{Q}^{(0)} = \mathcal{Q} \cap L. \tag{3.19}$$

Notice that retracting $\mathcal{Q} - N$ onto $\mathcal{Q}^{(0)}$ simply amounts to translating the ζ of Q_ζ until $\zeta_{l+1} = \ldots = \zeta_n = 0$. We also remark that $\{\eta_0 = 0\} = T_{p_\infty}\mathcal{Q}$ is the space of rank-two quadrics $Q(\eta)$ (3.5).

Similarly, elimination of x from (3.12) shows that the image of F is the quadric

$$\begin{aligned} \hat{\mathcal{S}} &= \{\hat{\xi} \in \mathbf{P}^*_{n+1} \mid \hat{s}(\hat{\xi},\hat{\xi}) = 0\}, \\ \hat{s}(\hat{\xi},\hat{\eta}) &= -(1/2)(\hat{\xi}_0\hat{\eta}_* + \hat{\xi}_*\hat{\eta}_0) + A\hat{\xi}' \cdot \hat{\eta}', \end{aligned} \tag{3.20}$$

which also has rank $l+2$. Its singular locus is the $(n-l-1)$-plane

$$\hat{N} = \{\hat{\xi} \in \mathbf{P}^*_{n+1} \mid \hat{\xi}_0 = \hat{\xi}_* = \hat{\xi}_1 = \ldots = \hat{\xi}_l = 0\}, \tag{3.21}$$

which has the complement

$$\hat{L} = \{\hat{\xi} \in \mathbf{P}^*_{n+1} \mid \hat{\xi}_{l+1} = \ldots = \hat{\xi}_n = 0\}. \tag{3.22}$$

Again we have a cone

$$\hat{\mathcal{S}} = \hat{N} * \hat{\mathcal{S}}^{(0)}, \quad \hat{\mathcal{S}}^{(0)} = \hat{\mathcal{S}} \cap \hat{L}. \tag{3.23}$$

Furthermore,

$$\hat{s}(K\xi, K\eta) = q(\xi,\eta), \quad \hat{\mathcal{S}} = K\mathcal{Q}. \tag{3.24}$$

c) The dual quadrics. The quadratic form q associates to each point $\xi \in \mathbf{P}_{n+1}$ the set $q(\cdot,\xi) = 0$, which is a hyperplane containing N, whenever $\xi \notin N$. Thus, we have a well defined map into $\mathbf{P}^*_{n+1}$,

$$\begin{aligned} \tilde{A} : \mathbf{P}_{n+1} - N \to N^\perp &\equiv \{\hat{\xi} \mid \langle \hat{\xi}, N \rangle = 0\} = \hat{L}, \\ \xi \mapsto \hat{\xi}, \qquad & q(\cdot,\xi) = \langle \hat{\xi}, \cdot \rangle, \end{aligned} \tag{3.25}$$

or more explicitly,

$$\hat{\xi} = \tilde{A}\xi, \quad \tilde{A} = \begin{bmatrix} -1 & 0 & -1/2 \\ 0 & A & 0 \\ -1/2 & 0 & 0 \end{bmatrix}. \tag{3.26}$$

The dual quadric $\hat{\mathcal{Q}}$ of $\mathcal{Q}$, the set of hyperplanes tangent to $\mathcal{Q}$, is the image of $\mathcal{Q} - N$ under this map. Eliminating ξ from (3.26) and $q(\xi, \xi) = 0$ gives

$$\begin{aligned} \hat{\mathcal{Q}} &= \{\hat{\xi} \in \hat{L} \,|\, \hat{q}(\hat{\xi}, \hat{\xi}) = 0\}, \\ \hat{q}(\hat{\xi}, \hat{\eta}) &= -2(\hat{\xi}_0\hat{\eta}_* + \hat{\xi}_*\hat{\eta}_0) + 4\hat{\xi}_*\hat{\eta}_* + \sum_{j=1}^{l} A_j^{-1}\hat{\xi}_j\hat{\eta}_j. \end{aligned} \tag{3.27}$$

Notice that $\hat{\mathcal{Q}}$ is an l-dimensional, nondegenerate quadric in the $(l+1)$-dimensional space $\hat{L}$.

Proceeding similarly with $\hat{\mathcal{S}}$, we get the map into $\mathbf{P}_{n+1}$,

$$\begin{aligned} \tilde{B} : \mathbf{P}^*_{n+1} - \hat{N} \to \hat{N}^{\perp} &\equiv \{\eta \,|\, \langle \hat{N}, \eta \rangle = 0\} = L, \\ \hat{\xi} \mapsto \xi, \qquad & \hat{s}(\cdot, \hat{\xi}) = \langle \cdot, \xi \rangle, \end{aligned} \tag{3.28}$$

or more explicitly,

$$\xi = \tilde{B}\hat{\xi}, \quad \tilde{B} = \begin{bmatrix} 0 & 0 & -1/2 \\ 0 & A & 0 \\ -1/2 & 0 & 0 \end{bmatrix}. \tag{3.29}$$

Eliminating $\hat{\xi}$ from (3.29) and $\hat{s}(\hat{\xi}, \hat{\xi}) = 0$ gives the dual quadric, which happens to agree with (3.7),

$$\begin{aligned} \mathcal{S} &= \{\xi \in L \,|\, s(\xi, \xi) = 0\}, \\ s(\xi, \eta) &= -2(\xi_0\eta_* + \xi_*\eta_0) + \sum_{j=1}^{l} A_j^{-1}\xi_j\eta_j. \end{aligned} \tag{3.30}$$

It is l-dimensional and nondegenerate in the space L.

For later use we observe that

$$\tilde{A} = K\tilde{B}K. \tag{3.31}$$

Finally, we remark that (3.28) gives an intrinsic definition of the subspace L.

4. Billiards

In this section we define a natural pair of rational involutions τ_1, τ_2 relative to the two quadrics $\hat{\mathcal{S}}$, $\mathcal{Q}$ of the last section. We refer to them as "billiard" involutions because of the close analogy of $\sigma = \tau_1\tau_2$ to the billiard map associated to an ellipsoid in real Euclidean space. We show that they correspond, under the mapping $F \times G$ (2.14), to the involutions on $\mathcal{M}$ arising from double valued reflection. Finally, we construct invariant subvarieties for these involutions, which leads to the proof of Theorem 0.1.

a) Billiard involutions. Let $(\hat{\xi}, \eta) \in (\hat{\mathcal{S}} \times \mathcal{Q})_0$, i. e. $\hat{\xi} \in \hat{\mathcal{S}}$ and $\eta \in \mathcal{Q} \cap \hat{\xi}^\perp$. By duality $\hat{\xi}^\perp \cap L$ is tangent to $\mathcal{S} \subset L$ at $\xi = \tilde{B}\hat{\xi}$. Consider the (complex projective) line $[\eta\xi] \subset \mathbf{P}_{n+1}$. In general, it will meet the quadric $\mathcal{Q}$ in a second point $\eta^\sim$. We define $\tau_1(\hat{\xi}, \eta) = (\hat{\xi}, \eta^\sim)$, or more explicitly,

$$
\begin{aligned}
\tau_1(\hat{\xi}, \eta) &= (\hat{\xi}, \eta + b_1\xi), \ \xi = \tilde{B}\hat{\xi}, \\
b_1(\hat{\xi}, \eta) &= -2q(\xi, \eta)/q(\xi, \xi).
\end{aligned} \tag{4.1}
$$

Alternatively, we may consider $\eta \in \mathcal{Q}$ and $\hat{\xi} \in \hat{\mathcal{S}} \cap \eta^\perp$. By duality $\eta^\perp \cap \hat{L}$ is tangent to $\hat{\mathcal{Q}} \subset \hat{L}$ at $\hat{\eta} = \tilde{A}\eta$. The line $[\hat{\xi}\hat{\eta}] \subset \mathbf{P}^*_{n+1}$ meets $\hat{\mathcal{S}}$ in a second point $\hat{\xi}^\sim$, in general. We define $\tau_2(\hat{\xi}, \eta) = (\hat{\xi}^\sim, \eta)$, or

$$
\begin{aligned}
\tau_2(\hat{\xi}, \eta) &= (\hat{\xi} + b_2\hat{\eta}, \eta), \ \hat{\eta} = \tilde{A}\eta, \\
b_2(\hat{\xi}, \eta) &= -2\hat{s}(\hat{\eta}, \hat{\xi})/\hat{s}(\hat{\eta}, \hat{\eta}).
\end{aligned} \tag{4.2}
$$

We claim that these two involutions are related by

$$\tau_2 = \kappa\tau_1\kappa, \tag{4.3}$$

where κ is given by (3.14). To see this note that

$$\kappa\tau_1\kappa(\hat{\xi}, \eta) = (K(K^{-1}\hat{\xi} + b_1\tilde{B}K\eta), K^{-1}K\eta) = (\hat{\xi} + b_1\tilde{A}\eta, \eta)$$

by (3.31), where

$$b_1 = b_1(K\eta, K^{-1}\hat{\xi}) = b_2(\hat{\xi}, \eta).$$

This proves (4.3).

b) Correspondence of involutions. We shall prove the following result.

Proposition 4.1 *Let M be a general ellipsoid with involutions τ_1, τ_2 on its complexification $\mathcal{M}$ resulting from the double valued reflection. Then τ_1, τ_2 correspond to the above billiard involutions under the map $F \times G$.*

For the proof, we note that we have already shown the correspondence of the two maps κ, (3.2) and (3.14). Since both pairs satisfy (4.3) (see formula (3.9) of [8]), it suffices to show that the two τ_1's correspond. By formula (3.7) of [8], we have on $\mathcal{M}$,

$$\tau_1(z,\zeta) = (z,\tilde{\zeta}),\ \tilde{\zeta} = \zeta + a_1(\zeta + 2Az),$$
$$a_1(z,\zeta) = -\frac{(z+2A\zeta)\cdot(\zeta+2Az)}{A(\zeta+2Az)\cdot(\zeta+2Az)}. \tag{4.4}$$

We must show that the line $[\eta\eta^{\sim}] \subset \mathbf{P}_{n+1}$ determined by the two points

$$\eta = G(\zeta) = (1,\zeta, A\zeta\cdot\zeta - 1),$$
$$\eta^{\sim} = G(\tilde{\zeta}) = (1,\tilde{\zeta}, A\tilde{\zeta}\cdot\tilde{\zeta} - 1) \tag{4.5}$$

also contains the point $\xi = \tilde{B}\hat{\xi}$, where $\hat{\xi} = F(z)$. But $\eta^{\sim} + t\eta \in L$ happens if and only if $t = -1 - a_1$, so the point in question is

$$(-a_1, 2a_1Az, c),\ \ c = A\tilde{\zeta}\cdot\tilde{\zeta} - 1 - (1+a_1)(A\zeta\cdot\zeta - 1). \tag{4.6}$$

Substuting from (4.4) and simplifying gives

$$c = a_1(-z\cdot\zeta - 2Az\cdot z - A\zeta\cdot\zeta + 1) = -a_1Az\cdot z, \tag{4.7}$$

where we have used $r(z,\zeta) = 0$. On the other hand,

$$\xi = \tilde{B}F(z) = (1/2)(-1, 2Az, -Az\cdot z)^t. \tag{4.8}$$

Hence, the two points agree, and the proof is complete.

c) Invariant subvarieties. In studying the maps τ_1, τ_2 and their composition σ on the variety $(\hat{\mathcal{S}}\times\mathcal{Q})_0$ (2.14), it is natural to look for invariant subvarieties. The classical way to specify such subvarieties of $(\hat{\mathcal{S}}\times\mathcal{Q})_0$ is to require that the line $[\eta\xi]$ remain tangent to a fixed hypersurface in $\mathbf{P}_{n+1}$. Such a condition is clearly invariant under τ_1, which does not change the line. Thus, one has only to consider τ_2.

The two quadrics $\mathcal{S}$ and $\mathcal{Q}^{(0)}$ are l-dimensional and nondegenerate in the $(l+1)$-dimensional space L. Hence, we may form their confocal family of quadrics $\mathcal{Q}^{(0)}_\lambda \subset L$, $\lambda \in \mathbf{C}$, as in [8] (see also below). Over them we form the family of cones from N,

$$\mathcal{Q}_\lambda = N * \mathcal{Q}^{(0)}_\lambda, \tag{4.9}$$

which are quadratic hypersurfaces in $\mathbf{P}_{n+1}$. The main result is that tangency to these hypersurfaces is invariant.

Proposition 4.2 *Let $\tau_2(\hat{\xi}, \eta) = (\hat{\xi}^{\sim}, \eta)$ and $\xi = \tilde{B}\hat{\xi}$, $\xi^{\sim} = \tilde{B}\hat{\xi}^{\sim}$. Then for any $\lambda \neq 0$, the line $[\eta\xi]$ is tangent to $\mathcal{Q}_\lambda$ if and only if $[\eta\xi^{\sim}]$ is tangent to $\mathcal{Q}_\lambda$.*

The proof occupies the rest of this subsection.

Let π denote projection from N onto L,

$$\pi : \mathbf{P}_{n+1} - N \to L; \quad \eta^{(0)} = \pi(\eta) \tag{4.10}$$

is gotten by setting $\eta_{l+1} = \cdots = \eta_n = 0$. Then $[\eta\xi]$ is tangent to $\mathcal{Q}_\lambda$ if and only if $\pi[\eta\xi] = [\eta^{(0)}\xi]$ is tangent to $\mathcal{Q}_\lambda^{(0)}$. We shall show that the projected point-line pair $(\eta^{(0)}, [\eta^{(0)}\xi])$ undergoes a billiard transformation which preserves tangency of the line to the confocal quadric $\mathcal{Q}_\lambda^{(0)}$, for each fixed $\lambda \neq 0$. Since

$$\tilde{A}\eta = \tilde{A}\eta^{(0)}, \quad \xi^{\sim} = \xi + b_2\tilde{B}\tilde{A}\eta^{(0)}, \quad \xi, \xi^{\sim} \in \mathcal{S}, \tag{4.11}$$

we must show that the point $\tilde{B}\tilde{A}\eta^{(0)}$ is the pole relative to $\mathcal{S}$ of the tangent space $T_{\eta^{(0)}}\mathcal{Q}^{(0)}$.

For the rest of the proof we work in the space L, setting $\eta = \eta^{(0)}$, and suppressing the components $\eta_{l+1}, \ldots, \eta_n$. We have

$$s(\xi, \eta) = -2(\xi_0\eta_* + \xi_*\eta_0) + \sum_{j=1}^{l} A_j^{-1}\xi_j\eta_j, \tag{4.12}$$

and

$$q(\xi, \eta) = -\xi_0\eta_0 - (1/2)(\xi_0\eta_* + \xi_*\eta_0) + \sum_{j=1}^{l} A_j\xi_j\eta_j. \tag{4.13}$$

As in [8] we express q via an s-symmetric operator B,

$$q(\xi, \eta) = s(B^{-1}\xi, \eta) = s(\xi, B^{-1}\eta), \quad B = \begin{bmatrix} 4 & 0 & 0 \\ 0 & A^{-2} & 0 \\ -8 & 0 & 4 \end{bmatrix}. \tag{4.14}$$

The confocal quadrics are given by

$$\begin{aligned} \mathcal{Q}_\lambda^{(0)} &= \{\eta \in L \,|\, q_\lambda(\eta, \eta) = 0\}, \\ q_\lambda(\xi, \eta) &= s((\lambda - B)^{-1}\xi, \eta). \end{aligned} \tag{4.15}$$

Let $\eta \in \mathcal{Q}^{(0)}$ and $v \in L$ be the pole of $T_\eta\mathcal{Q}^{(0)}$ relative to $\mathcal{S}$. This means that $v^{\perp_s} = \eta^{\perp_q}$, or

$$v = B^{-1}\eta. \tag{4.16}$$

But we readily check that

$$\tilde{B}\tilde{A} = B^{-1}, \tag{4.17}$$

which is what was needed above. Thus, under the action of τ_2, the projected points $\eta \in \mathcal{Q}^{(0)}$ and $\xi \in \mathcal{S}$ undergo the involution

$$\begin{aligned} \tau(\xi,\eta) &= (\xi^{\sim},\eta),\ \xi^{\sim} = \xi + av, \\ a(\xi,\eta) &= -2s(\xi,v)/s(v,v). \end{aligned} \tag{4.18}$$

Now we have the situation of section 5 of [8], except that the line $[\eta\xi]$ need *not* be tangent to $\mathcal{S}$ at ξ. We must show that the arguments of [8] carry over to our more general situation. Tangency of $[\eta\xi]$ to $\mathcal{Q}_\lambda^{(0)}$ is given by the condition

$$\Delta_\lambda(\xi,\eta) \equiv q_\lambda(\xi,\xi)q_\lambda(\eta,\eta) - q_\lambda(\xi,\eta)^2 = 0. \tag{4.19}$$

Therefore, the following lemma implies that tangency of the projected line to the confocal quadric is preserved by τ, and hence, that the proposition holds.

Lemma 4.3 $\Delta_\lambda(\xi^{\sim},\eta) = \Delta_\lambda(\xi,\eta)$, *if* $\lambda \neq 0$.

To prove this, we substute (4.18) into (4.19) getting

$$\begin{aligned} \Delta_\lambda(\xi^{\sim},\eta) = {} & \Delta_\lambda(\xi,\eta) \\ & +2a\{q_\lambda(\xi,v)q_\lambda(\eta,\eta) - q_\lambda(\xi,\eta)q_\lambda(v,\eta)\} \\ & +a^2\{q_\lambda(v,v)q_\lambda(\eta,\eta) - q_\lambda(v,\eta)^2\}. \end{aligned} \tag{4.20}$$

By definition

$$\begin{aligned} q_\lambda(\xi,v) &= s(B^{-1}(\lambda - B)^{-1}\xi,\eta), \\ q_\lambda(v,\eta) &= s(B^{-1}(\lambda - B)^{-1}\eta,\eta), \\ q_\lambda(v,v) &= s(B^{-1}(\lambda - B)^{-1}\eta, B^{-1}\eta). \end{aligned} \tag{4.21}$$

We use the formula

$$B^{-1}(\lambda - B)^{-1} = \lambda^{-1}(B^{-1} + (\lambda - B)^{-1}), \tag{4.22}$$

which is valid for $\lambda \neq 0$. Applying it once in the first two and twice in the third equation, and using $s(B^{-1}\eta,\eta) = q(\eta,\eta) = 0$ gives

$$\begin{aligned} q_\lambda(\xi,v) &= \lambda^{-1}[q_\lambda(\xi,\eta) + q(\xi,\eta)], \\ q_\lambda(v,\eta) &= \lambda^{-1}q_\lambda(\eta,\eta), \\ q_\lambda(v,v) &= \lambda^{-2}q_\lambda(\eta,\eta) + \lambda^{-1}s(v,v). \end{aligned} \tag{4.23}$$

Substuting these into (4.20) and simplifying with (4.18) gives the lemma, and hence the proposition.

d) Proof of Theorem 0.1. We fix a generic line $\gamma \subset L$. By the results of [8] γ is tangent to precisely l of the confocal quadrics, $\mathcal{Q}^{(0)}_{\lambda_1}, \ldots, \mathcal{Q}^{(0)}_{\lambda_l}$, $\lambda_i \in \mathbf{C}$. Furthermore, the λ_i are locally single valued, independent functions of the line, and the set of all lines in L tangent to l generic confocal quadrics forms an l-dimensional variety.

We note parenthetically for the case $l = n$ treated in [8], that $L = \mathbf{P}_{n+1}$, and there is no projection (4.10). Moreover, we must restrict to lines γ which are tangent to $\mathcal{S} = \mathcal{Q}_\infty$. Hence $\lambda_l \equiv \infty$, and we have an $(n-1)$-parameter family of n-dimensional invariant subvarieties in the result in [8] corresponding to Theorem 0.1.

We now assume $l < n$. Given γ as above, we choose points

$$\eta^{(0)} \in \gamma \cap \mathcal{Q}^{(0)}, \quad \xi \in \gamma \cap \mathcal{S}. \tag{4.24}$$

The set of points $\eta \in \mathbf{P}_{n+1}$ such that $\pi(\eta) = \eta^{(0)}$ is the $(n-l)$-dimensional linear space $N * \{\eta^{(0)}\} \subset \mathcal{Q}$. The set of all $\hat{\xi} \in \mathbf{P}^*_{n+1}$ such that $\xi = \tilde{B}\hat{\xi}$, that is, the set of all hyperplanes in $\mathbf{P}_{n+1}$ containing the l-dimensional tangent plane $T_\xi S$, is an $(n-l)$-dimensional linear space contained in $\hat{\mathcal{S}}$. Since $(\hat{\mathcal{S}} \times \mathcal{Q})_0$ is defined by the single further condition $\langle \hat{\xi}, \eta \rangle = 0$, we get a $2(n-l)-1$ dimensional subset over γ. Since γ moves in an l-dimensional variety, we get a $2n-1-l$ dimensional invariant subvariety corresponding to fixed generic values of $\lambda_1, \ldots, \lambda_l$. The images in $\mathcal{M}$ of these varieties by the birational map $(F \times G)^{-1}$ give the invariant foliation of Theorem 0.1, the proof of which is now complete.

References

[1] L. Lempert, *La métrique de Kobayashi et la représentation des domaines sur la boule*, Bull. Soc. Math. de France **109** (1981), 427–474.

[2] J. K. Moser and S. M. Webster, *Normal forms for real surfaces in* $\mathbf{C}^2$ *near complex tangents and hyperbolic surface transformations*, Acta Math. **150** (1983), 255–296.

[3] S. M. Webster, *On the mapping problem for algebraic real hypersurfaces*, Invent. Math. **43** (1977), 53–68.

[4] ———, *Some birational invariants for algebraic real hypersurfaces*, Duke Math. Jour. **45** (1978), 39–46.

[5] ———, *On the reflection principle in several complex variables*, Proc. AMS **71** (1978), 26–28.

[6] ———, *Double valued reflection in the complex plane*, l'Enseign. Math. **42** (1996), 25–48.

[7] ———, *A note on extremal discs and double valued reflection*, AMS Contemp. Math. **205** (1997), 271–276.

[8] ———, *Real ellipsoids and double valued reflection in complex space*, Amer. Jour. Math. **120** (1998), 757–809.

UNIVERSITY OF CHICAGO
CHICAGO, ILLINOIS 60637

E-mail: webster@math.uchicago.edu

CHAPTER XIV

G_2-Geometry of Overdetermined Systems of Second Order

Keizo Yamaguchi

Introduction

The main theme of this paper is "Contact Geometry of Second Order". This topic has its origin in the following paper of E. Cartan.

[C1] *Les systèmes de Pfaff à cinq variables et les équations aux derivées partielles du second ordre*, Ann. Ec. Normale, 27 (1910), 109–192.

In this paper, following the tradition of geometric theory of partial differential equations of the 19th century, E. Cartan dealt with the equivalence problem of two classes of second order partial differential equations in two independent variables under "contact transformations". One class consists of overdetermined systems, which are involutive, and the other class consists of single equations of Goursat type, i.e., single equations of parabolic type whose Monge characteristic systems are completely integrable. Especially in the course of the investigation, he found out the following: the symmetry algebras (i.e., the Lie algebra of infinitesimal contact transformations) of the following overdetermined system (involutive system) (A) and the single Goursat type equation (B) are both isomorphic to the 14-dimensional exceptional simple Lie algebra G_2.

$$\frac{\partial^2 z}{\partial x^2} = \frac{1}{3}\left(\frac{\partial^2 z}{\partial y^2}\right)^3, \quad \frac{\partial^2 z}{\partial x \partial y} = \frac{1}{2}\left(\frac{\partial^2 z}{\partial y^2}\right)^2. \tag{A}$$

$$9r^2 + 12t^2(rt - s^2) + 32s^3 - 36rst = 0, \tag{B}$$

where

$$r = \frac{\partial^2 z}{\partial x^2}, \qquad s = \frac{\partial^2 z}{\partial x \partial y}, \qquad t = \frac{\partial^2 z}{\partial y^2}$$

are the classical terminology.

Our aim here is to clarify the contents of "Contact Geometry of Second Order" in the course of showing how to recognize the above facts.

1. Second Order Contact Manifolds

We will here recall the basic facts about the geometry of second order Jet spaces ([Y1], [Y3]).

1.1. Space of Contact Elements (Grassmannian Bundles). The notion of contact manifolds originates from the following space $J(M,n)$ of contact elements: Let M be a C^∞- manifold of dimension $m+n$. We put

$$J(M,n) = \bigcup_{x\in M} J_x, \qquad J_x = \mathrm{Gr}(T_x(M),n),$$

where $\mathrm{Gr}(T_x(M),n)$ denotes the Grassmann manifold consisting of n-dimensional subspaces in $T_x(M)$ (i.e., n-dimensional contact elements to M at x). $J(M,n)$ is endowed with the canonical subbundle C of $T(J(M,n))$ as follows: Let π be the projection of $J(M,n)$ onto M. Each element $u \in J(M,n)$ is a linear subspace of $T_x(M)$ of codimension m, where $x = \pi(u)$. Hence we have a subspace $C(u)$ of codimension m in $T_u(J(M,n))$ by putting

$$C(u) = \pi_*^{-1}(u) \subset T_u(J(M,n)).$$

C is called the canonical system on $J(M,n)$. We have an inhomogeneous Grassmann coordinate system of $J(M,n)$ as follows: Let us fix $u_o \in J(M,n)$ and take a coordinate system $U' : (x_1,\dots,x_n,z^1,\dots,z^m)$ of M around $x_o = \pi(u_o)$ such that $dx_1\wedge\cdots\wedge dx_n\,|_{u_o} \neq 0$. Then we have the coordinate system $(x_1,\cdots,x_n,z^1,\cdots,z^m,p_1^1,\cdots,p_n^m)$ on the neighborhood $U = \{u \in \pi^{-1}(U') \mid \pi(u) = x \in U' \text{ and } dx_1\wedge\cdots\wedge dx_n\,|_u \neq 0\}$ of u_o by

$$dz^\alpha\,|_u = \sum_{i=1}^n p_i^\alpha(u)\,dx_i\,|_u \qquad (\alpha = 1,\dots,m).$$

Clearly the canonical system C is given in this coordinate system by

$$C = \{\varpi^1 = \cdots = \varpi^m = 0\},$$

where $\varpi^\alpha = dz^\alpha - \sum_{i=1}^n p_i^\alpha\,dx_i$ $(\alpha = 1,\cdots,m)$.

$(J(M,n),C)$ is the (geometric) 1-jet space and especially, in case $m = 1$, is the so-called contact manifold. Let M, $\hat{M}$ be manifolds (of dimension $m+n$) and $\varphi : M \to \hat{M}$ be a diffeomorphism between them. Then φ induces the isomorphism $\varphi_* : (J(M,n),C) \to (J(\hat{M},n),\hat{C})$, i.e., the differential map $\varphi_* : J(M,n) \to J(\hat{M},n)$ is a diffeomorphism sending C onto $\hat{C}$. The reason that the case $m = 1$ is special is explained by the following theorem of Bäcklund (cf. Theorem 1.4 [Y3]).

Theorem 1.1 (Bäcklund) *Let M and $\hat{M}$ be manifolds of dimension $m+n$. Assume $m \geqq 2$. Then, for an isomorphism $\Phi : (J(M,n),C) \to (J(\hat{M},n),\hat{C})$, there exists a diffeomorphism $\varphi : M \to \hat{M}$ such that $\Phi = \varphi_*$.*

In case $m = 1$, it is a well known fact that the group of isomorphisms of $(J(M,n),C)$, i.e., the group of contact transformations, is really larger than the group of diffeomorphisms of M. Therefore, when we consider the geometric 2-jet spaces, the situation differs according to whether the number m of unknown functions is 1 or greater. In case $m = 1$, we should start from a contact manifold (J,C) of dimension $2n+1$, which can be regarded locally as a space of 1-jets for one unknown function by Darboux's theorem. Then we can construct the geometric second order jet space $(L(J),E)$ as follows: We consider the Lagrange-Grassmann bundle $L(J)$ over J consisting of all n-dimensional integral elements of (J,C);

$$L(J) = \bigcup_{u \in J} L_u,$$

where L_u is the Grassmann manifolds of all Lagrangian (or Legendrian) subspaces of the symplectic vector space $(C(u), d\varpi)$. Here ϖ is a local contact form on J. Let π be the projection of $L(J)$ onto J. Then the canonical system E on $L(J)$ is defined by

$$E(v) = \pi_*^{-1}(v) \subset T_v(L(J)) \qquad \text{at} \quad v \in L(J).$$

Starting from a canonical coordinate system $(x_1, \dots, x_n, z, p_1, \dots, p_n)$ of the contact manifold (J,C), we can introduce a coordinate system (x_i, z, p_i, p_{ij}) $(1 \leqq i \leqq j \leqq n)$ of $L(J)$ such that $p_{ij} = p_{ji}$ and E is defined by

$$E = \{\varpi = \varpi_1 = \cdots = \varpi_n = 0\},$$

where $\varpi = dz - \sum_{i=1}^n p_i\, dx_i$, $\varpi_i = dp_i - \sum_{j=1}^n p_{ij}\, dx_j$ $(i = 1, \cdots, n)$. Let (J,C), $(\hat{J},\hat{C})$ be contact manifolds of dimension $2n+1$ and $\varphi : (J,C) \to (\hat{J},\hat{C})$ be a contact diffeomorphism between them. Then φ induces an isomorphism $\varphi_* : (L(J),E) \to (L(\hat{J}),\hat{E})$. Conversely we have the following (cf. Theorem 3.2 [Y1]).

Theorem 1.2 *Let (J,C) and $(\hat{J},\hat{C})$ be contact manifolds of dimension $2n+1$. Then, for an isomorphism $\Phi : (L(J),E) \to (L(\hat{J}),\hat{E})$,there exists a contact diffeomorphism $\varphi : (J,C) \to (\hat{J},\hat{C})$ such that $\Phi = \varphi_*$.*

Our first aim is to formulate the submanifold theory for $(L(J),E)$, which will be given in §4.

1.2. Realization Lemma. We here recall the following Realization Lemma for the Grassmannian construction, which plays the basic role in the discussions of §4 and §5.

Lemma 1.3 (Realization Lemma) *Let R and M be manifolds. Assume that the quadruple (R, D, p, M) satisfies the following conditions :*

(1) p is a map of R into M of constant rank.

(2) D is a differential system on R such that $F = Ker\ p_*$ is a subbundle of D of codimension r.

Then there exists a unique map ψ of R into $J(M, r)$ satisfying $p = \pi \cdot \psi$ and $D = \psi_*^{-1}(C)$, where C is the canonical differential system on $J(M, r)$ and $\pi : J(M, r) \to M$ is the projection. Furthermore, let v be any point of R. Then ψ is in fact defined by

$$\psi(v) = p_*(D(v)) \qquad \text{as a point of } Gr\ (T_{p(v)}(M)),$$

and satisfies

$$Ker\ (\psi_*)_v = F(v) \cap Ch(D)(v).$$

where $Ch(D)$ is the Cauchy characteristic system of D (see §2.1 below).

For the proof, see Lemma 1.5 [Y1].

2. Geometry of Linear Differential Systems (Tanaka Theory)

We will recall here the Tanaka theory for linear differential systems following [T1] and [T2].

2.1. Derived Systems and Characteristic Systems. By a differential system (M, D), we mean a subbundle D of the tangent bundle $T(M)$ of a manifold M of dimension d. Locally D is defined by 1-forms $\omega_1, \ldots, \omega_{d-r}$ such that $\omega_1 \wedge \cdots \wedge \omega_{d-r} \neq 0$ at each point, where r is the rank of D;

$$D = \{\, \omega_1 = \cdots = \omega_{d-r} = 0 \,\}.$$

For two differential systems (M, D) and $(\hat{M}, \hat{D})$, a diffeomorphism φ of M onto $\hat{M}$ is called an isomorphism of (M, D) onto $(\hat{M}, \hat{D})$ if the differential map φ_* of φ sends D onto $\hat{D}$.

By the Frobenius theorem, we know that D is completely integrable if and only if

$$d\omega_i \equiv 0 \pmod{\omega_1, \ldots, \omega_s} \qquad \text{for } i = 1, \ldots, s,$$

or equivalently, if and only if

$$[\mathcal{D}, \mathcal{D}] \subset \mathcal{D},$$

where $s = d - r$ and $\mathcal{D} = \Gamma(D)$ denotes the space of sections of D.

Thus, for a nonintegrable differential system D, we are led to consider the *derived system* ∂D of D, which is defined, in terms of sections, by

$$\partial \mathcal{D} = \mathcal{D} + [\mathcal{D}, \mathcal{D}].$$

Furthermore the *Cauchy characteristic system* $\mathrm{Ch}\,(D)$ of (M, D) is defined at each point $x \in M$ by

$$\mathrm{Ch}\,(D)(x) = \{X \in D(x) \mid \\ X \rfloor d\omega_i \equiv 0 \pmod{\omega_1, \ldots, \omega_s} \quad \text{for } i = 1, \ldots, s\},$$

When $\mathrm{Ch}\,(D)$ is a differential system (i.e., has constant rank), it is always completely integrable (cf. [Y1]). Moreover higher derived systems $\partial^k D$ are usually defined successively (cf. [BCG_3]) by

$$\partial^k D = \partial(\partial^{k-1} D),$$

where we put $\partial^0 D = D$ for convention.

On the other hand we define the k-th weak derived system $\partial^{(k)} D$ of D inductively by

$$\partial^{(k)} \mathcal{D} = \partial^{(k-1)} \mathcal{D} + [\mathcal{D}, \partial^{(k-1)} \mathcal{D}],$$

where $\partial^{(0)} D = D$ and $\partial^{(k)} \mathcal{D}$ denotes the space of sections of $\partial^{(k)} D$. This notion is one of the key points in the Tanaka theory ([T1]).

A differential system (M, D) is called regular if $D^{-(k+1)} = \partial^{(k)} D$ are subbundles of $T(M)$ for every integer $k \geqq 1$. For a regular differential system (M, D), we have ([T2], Proposition 1.1)

($S1$) *There exists a unique integer* $\mu > 0$ *such that, for all* $k \geqq \mu$,

$$D^{-k} = \cdots = D^{-\mu} \supsetneqq D^{-\mu+1} \supsetneqq \cdots \supsetneqq D^{-2} \supsetneqq D^{-1} = D,$$

($S2$) $[\mathcal{D}^p, \mathcal{D}^q] \subset \mathcal{D}^{p+q}$ *for all* $p, q < 0$

where $\mathcal{D}^p$ denotes the space of sections of D^p. ($S2$) can be checked easily by induction on q. Thus $D^{-\mu}$ is the smallest completely integrable differential system which contains $D = D^{-1}$.

2.2. Symbol Algebras. Let (M, D) be a regular differential system such that $T(M) = D^{-\mu}$. As a first invariant for nonintegrable differential systems, we now define the *graded algebra* $\mathfrak{m}(x)$ *associated with a differential system* (M, D) at $x \in M$, which was introduced by N. Tanaka [T2].

We put $\mathfrak{g}_{-1}(x) = D^{-1}(x)$, $\mathfrak{g}_p(x) = D^p(x)/D^{p+1}(x)$ $(p < -1)$ and

$$\mathfrak{m}(x) = \bigoplus_{p=-1}^{-\mu} \mathfrak{g}_p(x).$$

Let ϖ_p be the projection of $D^p(x)$ onto $\mathfrak{g}_p(x)$. Then, for $X \in \mathfrak{g}_p(x)$ and $Y \in \mathfrak{g}_q(x)$, the bracket product $[X, Y] \in \mathfrak{g}_{p+q}(x)$ is defined by

$$[X, Y] = \varpi_{p+q}([\tilde{X}, \tilde{Y}]_x),$$

where $\tilde{X}$ and $\tilde{Y}$ are any element of $\mathcal{D}^p$ and $\mathcal{D}^q$ respectively such that $\varpi_p(\tilde{X}_x) = X$ and $\varpi_q(\tilde{Y}_x) = Y$.

Endowed with this bracket operation, by $(S2)$ above, $\mathfrak{m}(x)$ becomes a nilpotent graded Lie algebra such that $\dim \mathfrak{m}(x) = \dim M$ and satisfies

$$\mathfrak{g}_p(x) = [\mathfrak{g}_{p+1}(x), \mathfrak{g}_{-1}(x)] \qquad \text{for } p < -1.$$

We call $\mathfrak{m}(x)$ the *symbol algebra of* (M, D) at $x \in M$ for short.

Furthermore, let $\mathfrak{m}$ be a FGLA (fundamental graded Lie algebra) of μ-th kind, that is,

$$\mathfrak{m} = \bigoplus_{p=-1}^{-\mu} \mathfrak{g}_p$$

is a nilpotent graded Lie algebra such that

$$\mathfrak{g}_p = [\mathfrak{g}_{p+1}, \mathfrak{g}_{-1}] \qquad \text{for } p < -1.$$

Then (M, D) is called of type $\mathfrak{m}$ if the symbol algebra $\mathfrak{m}(x)$ is isomorphic with $\mathfrak{m}$ at each $x \in M$.

Conversely, given a FGLA $\mathfrak{m} = \bigoplus_{p=-1}^{-\mu} \mathfrak{g}_p$, we can construct a model differential system of type $\mathfrak{m}$ as follows: Let $M(\mathfrak{m})$ be the simply connected Lie group with Lie algebra $\mathfrak{m}$. Identifying $\mathfrak{m}$ with the Lie algebra of left invariant vector fields on $M(\mathfrak{m})$, $\mathfrak{g}_{-1}$ defines a left invariant subbundle $D_{\mathfrak{m}}$ of $T(M(\mathfrak{m}))$. By definition of symbol algebras, it is easy to see that $(M(\mathfrak{m}), D_{\mathfrak{m}})$ is a regular differential system of type $\mathfrak{m}$. $(M(\mathfrak{m}), D_{\mathfrak{m}})$ is called the standard differential system of type $\mathfrak{m}$. The Lie algebra $\mathfrak{g}(\mathfrak{m})$ of all infinitesimal automorphisms of $(M(\mathfrak{m}), D_{\mathfrak{m}})$ can be calculated algebraically as the prolongation of $\mathfrak{m}$ ([T1], cf. [Y5]). We will discuss in §3 the question of when $\mathfrak{g}(\mathfrak{m})$ becomes finite dimensional and simple.

As an example to calculate symbol algebras, let us show that $(L(J), E)$ is a regular differential system of type $\mathfrak{c}^2(n)$:

$$\mathfrak{c}^2(n) = \mathfrak{c}_{-3} \oplus \mathfrak{c}_{-2} \oplus \mathfrak{c}_{-1},$$

where $\mathfrak{c}_{-3} = \mathbb{R}$, $\mathfrak{c}_{-2} = V^*$ and $\mathfrak{c}_{-1} = V \oplus S^2(V^*)$. Here V is a vector space of dimension n and the bracket product of $\mathfrak{c}^2(n)$ is defined accordingly through the pairing between V and V^* such that V and $S^2(V^*)$ are both abelian subspaces of $\mathfrak{c}_{-1}$. This fact can be checked as follows: Let us take a canonical coordinate system $U; (x_i, z, p_i, p_{ij})$ $(1 \leqq i \leqq j \leqq n)$ of $(L(J), E)$. Then we have a coframe $\{\varpi, \varpi_i, dx_i, dp_{ij}\}$ $(1 \leqq i \leqq j \leqq n)$ at each point in U, where $\varpi = dz - \sum_{i=1}^n p_i\, dx_i$, $\varpi_i = dp_i - \sum_{j=1}^n p_{ij}\, dx_j$ $(i = 1, \dots, n)$. Now take the dual frame $\{\frac{\partial}{\partial z}, \frac{\partial}{\partial p_i}, \frac{d}{dx_i}, \frac{\partial}{\partial p_{ij}}\}$, of this coframe, where

$$\frac{d}{dx_i} = \frac{\partial}{\partial x_i} + p_i \frac{\partial}{\partial z} + \sum_{j=1}^n p_{ij} \frac{\partial}{\partial p_j}$$

is the classical notation. Notice that $\{\frac{d}{dx_i}, \frac{\partial}{\partial p_{ij}}\}$ $(i = 1, \dots, n)$ forms a free basis of $\Gamma(E)$. Then an easy calculation shows the above fact. Moreover we see that the derived system ∂E of E satisfies the following :

$$\partial E = \{\varpi = 0\} = \pi_*^{-1} C, \qquad \mathrm{Ch}\,(\partial E) = \mathrm{Ker}\,\pi_*.$$

These are the key facts to Theorem 1.2 (cf. Theorem 3.2 [Y1]).

Similarly we see that $(J(M, n), C)$ is a regular differential system of type $\mathfrak{c}^1(n, m)$:

$$\mathfrak{c}^1(n, m) = \mathfrak{c}_{-2} \oplus \mathfrak{c}_{-1},$$

where $\mathfrak{c}_{-2} = W$ and $\mathfrak{c}_{-1} = V \oplus W \otimes V^*$ for vector spaces V and W of dimension n and m respectively, and the bracket product of $\mathfrak{c}^1(n, m)$ is defined accordingly through the pairing between V and V^* such that V and $W \otimes V^*$ are both abelian subspaces of $\mathfrak{c}_{-1}$.

2.3. Classification of Symbol Algebras of Lower Dimensions. In this section, following a short passage from Cartan's paper [C1], let us classify FGLAs $\mathfrak{m} = \bigoplus_{p=-1}^{-\mu} \mathfrak{g}_p$ such that $\dim \mathfrak{m} \leqq 5$, which gives us the first invariants towards the classification of regular differential systems (M, D) such that $\dim M \leqq 5$.

In the case $\dim \mathfrak{m} = 1$ or 2, $\mathfrak{m} = \mathfrak{g}_{-1}$ should be abelian. To discuss the case $\dim \mathfrak{m} \geqq 3$, we further assume that $\mathfrak{g}_{-1}$ is nondegenerate, i.e., $[X, \mathfrak{g}_{-1}] = 0$ implies $X = 0$ for $X \in \mathfrak{g}_{-1}$. This condition is equivalent to saying that $\mathrm{Ch}\,(D) = \{0\}$ for a regular differential system (M, D) of type $\mathfrak{m}$. When $\mathfrak{g}_{-1}$ is degenerate, $\mathrm{Ch}\,(D)$ is nontrivial, hence at least locally, (M, D) induces a regular differential system (X, D^*) on the lower dimensional space X, where $X = M/\mathrm{Ch}\,(D)$ is the leaf space of the foliation on M defined by $\mathrm{Ch}\,(D)$ and D^* is the differential system on X such that $D = p_*^{-1}(D^*)$. Here $p : M \to X = M/\mathrm{Ch}\,(D)$ is the projection. Moreover, for the following discussion, we first observe that the dimension of $\mathfrak{g}_{-2}$ does not exceed $\binom{m}{2}$, where $m = \dim \mathfrak{g}_{-1}$.

In the case $\dim \mathfrak{m} = 3$, we have $\mu \leqq 2$. When $\mu = 2$, $\mathfrak{m} = \mathfrak{g}_{-2} \oplus \mathfrak{g}_{-1}$ is the contact gradation, i.e., $\dim \mathfrak{g}_{-2} = 1$ and $\mathfrak{g}_{-1}$ is nondegenerate. In the case $\dim \mathfrak{m} = 4$, we see that $\mathfrak{g}_{-1}$ is degenerate when $\mu \leqq 2$. When $\mu = 3$, we have $\dim \mathfrak{g}_{-3} = \dim \mathfrak{g}_{-2} = 1$ and $\dim \mathfrak{g}_{-1} = 2$. Moreover it follows that $\mathfrak{m}$ is isomorphic to $\mathfrak{c}^2(1)$ in this case. In the case $\dim \mathfrak{m} = 5$, we have $\dim \mathfrak{g}_{-1} = 4$, 3 or 2. When $\dim \mathfrak{g}_{-1} = 4$, $\mathfrak{m} = \mathfrak{g}_{-2} \oplus \mathfrak{g}_{-1}$ is the contact gradation. When $\dim \mathfrak{g}_{-1} = 3$, $\mathfrak{g}_{-1}$ is degenerate if $\dim \mathfrak{g}_{-2} = 1$, which implies that $\mu = 2$ and $\dim \mathfrak{g}_{-2} = 2$ in this case. Moreover, when $\mu = 2$, it follows that $\mathfrak{m}$ is isomorphic to $\mathfrak{c}^1(1,2)$. When $\dim \mathfrak{g}_{-1} = 2$, we have $\dim \mathfrak{g}_{-2} = 1$ and $\mu = 3$ or 4. Moreover, when $\mu = 4$, it follows that $\mathfrak{m}$ is isomorphic to $\mathfrak{c}^3(1)$, where $\mathfrak{c}^3(1)$ is the symbol algebra of the canonical system on the third order jet spaces for 1 unknown function (cf. §3 [Y1]).

Summarizing the above discussion, we obtain the following classification of the FGLAs $\mathfrak{m} = \bigoplus_{p=-1}^{-\mu} \mathfrak{g}_p$ such that $\dim \mathfrak{m} \leqq 5$ and $\mathfrak{g}_{-1}$ is nondegenerate.

(1) $\dim \mathfrak{m} = 3 \Longrightarrow \mu = 2$

$$\mathfrak{m} = \mathfrak{g}_{-2} \oplus \mathfrak{g}_{-1} \cong \mathfrak{c}^1(1) : \text{contact gradation}$$

(2) $\dim \mathfrak{m} = 4 \Longrightarrow \mu = 3$

$$\mathfrak{m} = \mathfrak{g}_{-3} \oplus \mathfrak{g}_{-2} \oplus \mathfrak{g}_{-1} \cong \mathfrak{c}^2(1)$$

(3) $\dim \mathfrak{m} = 5$, then $\mu \leqq 4$

(*a*) $\mu = 4$ $\quad \mathfrak{m} = \mathfrak{g}_{-4} \oplus \mathfrak{g}_{-3} \oplus \mathfrak{g}_{-2} \oplus \mathfrak{g}_{-1} \cong \mathfrak{c}^3(1)$

(*b*) $\mu = 3$ $\quad \mathfrak{m} = \mathfrak{g}_{-3} \oplus \mathfrak{g}_{-2} \oplus \mathfrak{g}_{-1}$
such that $\dim \mathfrak{g}_{-3} = \dim \mathfrak{g}_{-1} = 2$ and $\dim \mathfrak{g}_{-2} = 1$

(*c*) $\mu = 2$ $\quad \mathfrak{m} = \mathfrak{g}_{-2} \oplus \mathfrak{g}_{-1} \cong \mathfrak{c}^1(1,2)$

(*d*) $\mu = 2$ $\quad \mathfrak{m} = \mathfrak{g}_{-2} \oplus \mathfrak{g}_{-1} \cong \mathfrak{c}^1(2)$: contact gradation

A notable and rather misleading fact is that, once the dimensions of $\mathfrak{g}_p$ are fixed, the Lie algebra structure of $\mathfrak{m} = \bigoplus_{p=-1}^{-\mu} \mathfrak{g}_p$ is unique in the above classification list. Moreover, except for the cases (*b*) and (*c*), every regular differential system (M, D) of type $\mathfrak{m}$ in the above list is isomorphic with the standard differential system $(M(\mathfrak{m}), D_{\mathfrak{m}})$ of type $\mathfrak{m}$ by Darboux's theorem (cf. Corollary 6.6 [Y1]). The first nontrivial situation that cannot be analyzed on the basis of Darboux's theorem occurs in the cases (*b*) and (*c*) (see [C1], [St]). Regular differential systems of type (*b*) and (*c*) are closely related to each other (cf. §6.3 and [C1]). We shall encounter the type (*b*) fundamental graded Lie algebra in §6.2 in connection with the root space decomposition of the exceptional simple Lie algebra G_2.

3. Differential systems associated with SGLAs

We will classify here the standard differential systems $(M(\mathfrak{m}), D_{\mathfrak{m}})$ for which the prolongation $\mathfrak{g}(\mathfrak{m})$ becomes finite dimensional and simple ([Y5]). In this section we will solely consider Lie algebras over $\mathbb{C}$ for the sake of simplicity.

3.1. Classification of Gradation of Simple Lie Algebras by Root Systems. Let $\mathfrak{g}$ be a finite dimensional simple Lie algebra over $\mathbb{C}$. Let us fix a Cartan subalgebra $\mathfrak{h}$ of $\mathfrak{g}$ and choose a simple root system $\Delta = \{\alpha_1, \dots, \alpha_\ell\}$ of the root system Φ of $\mathfrak{g}$ relative to $\mathfrak{h}$. Then every $\alpha \in \Phi$ is an (all nonnegative or all nonpositive) integer coefficient linear combination of elements of Δ and we have the root space decomposition of $\mathfrak{g}$;

$$\mathfrak{g} = \bigoplus_{\alpha\in\Phi^+} \mathfrak{g}_\alpha \oplus \mathfrak{h} \oplus \bigoplus_{\alpha\in\Phi^+} \mathfrak{g}_{-\alpha},$$

where $\mathfrak{g}_\alpha = \{X \in \mathfrak{g} \mid [h, X] = \alpha(h)X \quad \text{for } h \in \mathfrak{h}\}$ is a (1-dimensional) root space (corresponding to $\alpha \in \Phi$) and Φ^+ denotes the set of positive roots.

Now let us take a nonempty subset Δ_1 of Δ. Then Δ_1 defines the partition of Φ^+ as in the following and induces the gradation of $\mathfrak{g} = \bigoplus_{p\in\mathbb{Z}} \mathfrak{g}_p$ as follows:

$$\Phi^+ = \cup_{p\geqq 0}\Phi_p^+, \qquad \Phi_p^+ = \{\alpha = \sum_{i=1}^{\ell} n_i\alpha_i \mid \sum_{\alpha_i\in\Delta_1} n_i = p\},$$

$$\mathfrak{g}_p = \bigoplus_{\alpha\in\Phi_p^+} \mathfrak{g}_\alpha, \quad \mathfrak{g}_0 = \bigoplus_{\alpha\in\Phi_0^+} \mathfrak{g}_\alpha \oplus \mathfrak{h} \oplus \bigoplus_{\alpha\in\Phi_0^+} \mathfrak{g}_{-\alpha}, \quad \mathfrak{g}_{-p} = \bigoplus_{\alpha\in\Phi_p^+} \mathfrak{g}_{-\alpha},$$

$$[\mathfrak{g}_p, \mathfrak{g}_q] \subset \mathfrak{g}_{p+q} \qquad \text{for} \quad p, q \in \mathbb{Z}.$$

Moreover the negative part $\mathfrak{m} = \bigoplus_{p<0} \mathfrak{g}_p$ satisfies the following generating condition :

$$\mathfrak{g}_p = [\mathfrak{g}_{p+1}, \mathfrak{g}_{-1}] \quad \text{for} \quad p < -1.$$

We denote the SGLA (simple graded Lie algebra) $\mathfrak{g} = \bigoplus_{p=-\mu}^{\mu} \mathfrak{g}_p$ obtained from Δ_1 in this manner by (X_ℓ, Δ_1), when $\mathfrak{g}$ is a simple Lie algebra of type X_ℓ. Here X_ℓ stands for the Dynkin diagram of $\mathfrak{g}$ representing Δ and Δ_1 is a subset of vertices of X_ℓ. Moreover we have

$$\mu = \sum_{\alpha_i\in\Delta_1} n_i(\theta),$$

where $\theta = \sum_{i=1}^{\ell} n_i(\theta)\,\alpha_i$ is the highest root of Φ^+.

Conversely we have (Theorem 3.12 [Y5])

Theorem 3.1 *Let $\mathfrak{g} = \bigoplus_{p\in\mathbb{Z}} \mathfrak{g}_p$ be a simple graded Lie algebra over $\mathbb{C}$ satisfying the generating condition. Let X_ℓ be the Dynkin diagram of $\mathfrak{g}$. Then $\mathfrak{g} = \bigoplus_{p\in\mathbb{Z}} \mathfrak{g}_p$ is isomorphic with a graded Lie algebra (X_ℓ, Δ_1) for some $\Delta_1 \subset \Delta$. Moreover (X_ℓ, Δ_1) and (X_ℓ, Δ_1') are isomorphic if and only if there exists a diagram automorphism ϕ of X_ℓ such that $\phi(\Delta_1) = \Delta_1'$.*

In the real case, we can utilize the Satake diagram of $\mathfrak{g}$ to describe gradations of $\mathfrak{g}$ (Theorem 3.12 [Y5]).

3.2. Differential systems associated with SGLAs. By Theorem 3.1, the classification of gradations $\mathfrak{g} = \bigoplus_{p\in\mathbb{Z}} \mathfrak{g}_p$ of simple Lie algebras $\mathfrak{g}$ satisfying the generating condition coincides with that of parabolic subalgebras $\mathfrak{g}' = \bigoplus_{p\geqq 0} \mathfrak{g}_p$ of $\mathfrak{g}$. Accordingly, to each SGLA (X_ℓ, Δ_1), there corresponds a unique R-space $M_\mathfrak{g} = G/G'$ (compact simply connected homogeneous complex manifold). Furthermore, when $\mu \geqq 2$, there exists the G-invariant differential system $D_\mathfrak{g}$ on $M_\mathfrak{g}$, which is induced from $\mathfrak{g}_{-1}$, and $(M(\mathfrak{m}), D_\mathfrak{m})$ (standard differential system of type $\mathfrak{m}$) becomes an open submanifold of $(M_\mathfrak{g}, D_\mathfrak{g})$. For the Lie algebras of all infinitesimal automorphisms of $(M_\mathfrak{g}, D_\mathfrak{g})$, hence of $(M(\mathfrak{m}), D_\mathfrak{m})$, we have the following theorem (Theorem 5.2 [Y5]).

Theorem 3.2 *Let $\mathfrak{g} = \bigoplus_{p\in\mathbb{Z}} \mathfrak{g}_p$ be a simple graded Lie algebra over $\mathbb{C}$ satisfying the generating condition. Then $\mathfrak{g} = \bigoplus_{p\in\mathbb{Z}} \mathfrak{g}_p$ is the prolongation of $\mathfrak{m} = \bigoplus_{p<0} \mathfrak{g}_p$ except for the following three cases.*

(1) $\mathfrak{g} = \mathfrak{g}_{-1} \oplus \mathfrak{g}_0 \oplus \mathfrak{g}_1$ *is of depth* 1 (*i.e.*, $\mu = 1$).

(2) $\mathfrak{g} = \bigoplus_{p=-2}^{2} \mathfrak{g}_p$ *is a* (*complex*) *contact gradation.*

(3) $\mathfrak{g} = \bigoplus_{p\in\mathbb{Z}} \mathfrak{g}_p$ *is isomorphic with* $(A_\ell, \{\alpha_1, \alpha_i\})$ $(1 < i < \ell)$ *or* $(C_\ell, \{\alpha_1, \alpha_\ell\})$.

Here R-spaces corresponding to the above exceptions (1), (2) and (3) are as follows: (1) corresponds to compact irreducible Hermitian symmetric spaces. (2) corresponds to contact manifolds of Boothby type (standard contact manifolds), which exist uniquely for each simple Lie algebra other than $\mathfrak{sl}(2, \mathbb{C})$ (see §5.1 below). In case of (3), $(J(\mathbb{P}^\ell, i), C)$ corresponds to $(A_\ell, \{\alpha_1, \alpha_i\})$ and $(L(\mathbb{P}^{2\ell-1}), E)$ corresponds to $(C_\ell, \{\alpha_1, \alpha_\ell\})$ $(1 < i < \ell)$, where $\mathbb{P}^\ell$ denotes the ℓ-dimensional complex projective space and $\mathbb{P}^{2\ell-1}$ is the standard contact manifold of type C_ℓ. Here we note that R-spaces corresponding to (2) and (3) are all Jet spaces of the first or second order.

For the real version of this theorem, we refer the reader to Theorem 5.3 [Y5].

4. Geometry of PD-manifolds

We will here formulate the submanifold theory for $(L(J), E)$ as the geometry of PD-manifolds ([Y1]).

4.1. ***PD*-manifolds.** Let R be a submanifold of $L(J)$ satisfying the following condition:

$(R.0)$ $p : R \to J$; submersion,

where $p = \pi \mid_R$ and $\pi : L(J) \to J$ is the projection. There are two differential systems $C^1 = \partial E$ and $C^2 = E$ on $L(J)$. We denote by D^1 and D^2 those differential systems on R obtained by restricting these differential systems to R. Moreover we denote by the same symbols those 1-forms obtained by restricting the defining 1-forms $\{\varpi, \varpi_1, \dots, \varpi_n\}$ of the canonical system E to R. Then it follows from $(R.0)$ that these 1-forms are independent at each point on R and that

$$D^1 = \{\varpi = 0\}, \qquad D^2 = \{\varpi = \varpi_1 = \cdots = \varpi_n = 0\}.$$

In fact $(R; D^1, D^2)$ further satisfies the following conditions:

$(R.1)$ *D^1 and D^2 are differential systems of codimension* 1 *and* $n+1$ *respectively.*

$(R.2)$ $\partial D^2 \subset D^1$.

$(R.3)$ $\mathrm{Ch}\,(D^1)$ *is a subbundle of D^2 of codimension n.*

$(R.4)$ $\mathrm{Ch}\,(D^1)(v) \cap \mathrm{Ch}\,(D^2)(v) = \{0\}$ *at each* $v \in R$.

Conversely these four conditions characterize submanifolds in $L(J)$ satisfying $(R.0)$. In fact we call the triplet $(R; D^1, D^2)$ of a manifold and two differential systems on it a **PD-manifold** if these satisfy the above four conditions $(R.1)$ to $(R.4)$. We have the (local) Realization Theorem for PD-manifolds as follows: From conditions $(R.1)$ and $(R.3)$, it follows that the codimension of the foliation defined by the completely integrable system $\mathrm{Ch}\,(D^1)$ is $2n+1$. Assume that R is regular with respect to $\mathrm{Ch}\,(D^1)$, i.e., the space $J = R/\mathrm{Ch}\,(D^1)$ of leaves of this foliation is a manifold of dimension $2n+1$. Then D^1 drops down to J. Namely there exists a differential system C on J of codimension 1 such that $D^1 = p_*^{-1}(C)$, where $p : R \to J = R/\mathrm{Ch}\,(D^1)$ is the projection. Obviously (J, C) becomes a contact manifold of dimension $2n+1$. Conditions $(R.1)$ and $(R.2)$ guarantee that the image of the following map ι is a Legendrian subspace of (J, C):

$$\iota(v) = p_*(D^2(v)) \subset C(u), \qquad u = p(v).$$

Finally the condition $(R.4)$ shows that $\iota : R \to L(J)$ is an immersion by the Realization Lemma for (R, D^2, p, J) (see §1.2). Furthermore we have (Corollary 5.4 [Y1]) the following.

Theorem 4.1 *Let $(R; D^1, D^2)$ and $(\hat{R}; \hat{D}^1, \hat{D}^2)$ be PD-manifolds. Assume that R and $\hat{R}$ are regular with respect to $Ch(D^1)$ and $Ch(\hat{D}^1)$ respectively. Let (J, C) and $(\hat{J}, \hat{C})$ be the associated contact manifolds. Then an isomorphism $\Phi : (R; D^1, D^2) \to (\hat{R}; \hat{D}^1, \hat{D}^2)$ induces a contact diffeomorphism $\varphi : (J, C) \to (\hat{J}, \hat{C})$ such that the following commutes:*

$$\begin{array}{ccc} R & \xrightarrow{\iota} & L(J) \\ {\scriptstyle \Phi}\downarrow & & \downarrow{\scriptstyle \varphi_*} \\ \hat{R} & \xrightarrow{\iota} & L(\hat{J}). \end{array}$$

By this theorem, the submanifold theory for $(L(J), E)$ is reformulated as the geometry of PD-manifolds.

When $D^1 = \partial D^2$ holds for a PD-manifold $(R; D^1, D^2)$, the geometry of $(R; D^1, D^2)$ reduces to that of (R, D^2) and the Tanaka theory is directly applicable to this case. Concerning this situation, the following theorem is known under the compatibility condition (C) below:

(C) $\quad p^{(1)} : R^{(1)} \to R$ *is onto*

where $R^{(1)}$ is the first prolongation of $(R; D^1, D^2)$ (cf. Proposition 5.11 [Y1]).

Theorem 4.2 *Let $(R; D^1, D^2)$ be a PD-manifold satisfying the condition (C) above. Then the following equality holds at each point v of R:*

$$\dim D^1(v) - \dim \partial D^2(v) = \dim Ch\,(D^2)(v).$$

In particular $D^1 = \partial D^2$ holds if and only if $Ch(D^2) = \{0\}$.

4.2. First Reduction Theorem. When a PD-manifold $(R; D^1, D^2)$ admits nontrivial Cauchy characteristics, i.e., when rank $\mathrm{Ch}\,(D^2) > 0$, the geometry of $(R; D^1, D^2)$ is further reducible to the geometry of single differential systems. Here we will be concerned with the local equivalence of $(R; D^1, D^2)$, hence we may assume that R is regular with respect to $\mathrm{Ch}\,(D^2)$, i.e., the leaf space $X = R/\mathrm{Ch}\,(D^2)$ is a manifold such that the projection $\rho : R \to X$ is a submersion and there exists a differential system D on X satisfying $D^2 = \rho_*^{-1}(D)$. Then the local equivalence of $(R; D^1, D^2)$ is further reducible to that of (X, D) as in the following : We assume that $(R; D^1, D^2)$ satisfies the condition (C) above and $\mathrm{Ch}\,(D^2)$ is a subbundle of rank r $(0 < r < n)$. Then, by Theorem 4.2, ∂D^2 is a subbundle of D^1 of codimension r. From (X, D), at least locally, we can construct a PD-manifold $(R(X); D^1_X, D^2_X)$ as follows. $R(X)$ is the collection of hyperplanes v in each tangent space $T_x(X)$ at $x \in X$ which contains the fibre $\partial D(x)$ of the derived system ∂D of D.

$$R(X) = \bigcup_{x \in X} R_x \subset J(X, m-1),$$
$$R_x = \{v \in \mathrm{Gr}(T_x(X), m-1) \mid v \supset \partial D(x)\},$$

where $m = \dim X$. Moreover D^1_X is the canonical system obtained by the Grassmaniann construction and D^2_X is the lift of D. Precisely, D^1_X and D^2_X are given by

$$D^1_X(v) = \nu_*^{-1}(v) \supset D^2_X(v) = \nu_*^{-1}(D(x)),$$

for each $v \in R(X)$ and $x = \nu(v)$, where $\nu : R(X) \to X$ is the projection. Then we have a map κ of R into $R(X)$ given by

$$\kappa(v) = \rho_*(D^1(v)) \subset T_x(X),$$

for each $v \in R$ and $x = \rho(v)$. By the Realization Lemma for (R, D^1, ρ, X), κ is a map of constant rank such that

$$\mathrm{Ker}\ \kappa_* = \ \mathrm{Ch}\,(D^1) \cap \mathrm{Ker}\,\rho_* = \ \mathrm{Ch}\,(D^1) \cap \mathrm{Ch}\,(D^2) = \{0\}.$$

Thus κ is an immersion and, by a dimension count, in fact, a local diffeomorphism of R into $R(X)$ such that

$$\kappa_*(D^1) = D^1_X \qquad \text{and} \qquad \kappa_*(D^2) = D^2_X.$$

Namely $\kappa : (R, D^1, D^2) \to (R(X), D^1_X, D^2_X)$ is a local isomorphism of PD-manifolds. (Precisely, in general, $(R(X), D^1_X, D^2_X)$ becomes a PD-manifold on an open subset.)

Summarizing the above consideration, we obtain the following Reduction Theorem for PD-manifolds admitting nontrivial Cauchy characteristics.

Theorem 4.3 *Let (R, D^1, D^2) and $(\hat{R}; \hat{D}^1, \hat{D}^2)$ be PD-manifolds satisfying the condition (C) such that $Ch(D^2)$ and $Ch(\hat{D}^2)$ are subbundles of rank r $(0 < r < n)$. Assume that R and $\hat{R}$ are regular with respect to $Ch(D^2)$ and $Ch(\hat{D}^2)$ respectively. Let (X, D) and $(\hat{X}, \hat{D})$ be the leaf spaces, where $X = R/Ch(D^2)$ and $\hat{X} = \hat{R}/Ch(\hat{D}^2)$. Let us fix points $v_o \in R$ and $\hat{v}_o \in \hat{R}$ and put $x_o = \rho(v_o)$ and $\hat{x}_o = \hat{\rho}(\hat{v}_o)$. Then a local isomorphism $\psi : (R; D^1, D^2) \to (\hat{R}; \hat{D}^1, \hat{D}^2)$ such that $\psi(v_o) = \hat{v}_o$ induces a local isomorphism $\varphi : (X, D) \to (\hat{X}, \hat{D})$ such that $\varphi(x_o) = \hat{x}_o$ and $\varphi_*(\kappa(x_o)) = \hat{\kappa}(\hat{x}_o)$, and vice versa.*

The involutive system (A) in the introduction is the example of this situation and we have $\dim X = 5$ and $\operatorname{rank} D = 2$.

5. Contact Geometry of Single Equations of Goursat Type

In order to discuss the generalization of the equation (B) in the introduction, we will define single equations of Goursat type and formulate the Reduction Theorems for the contact equivalence of this type of equations.

5.1. Single Equations of Goursat Type. By a single equation (of second order), we mean a hypersurface R of $L(J)$ satisfying the condition $(R.0)$ in §4. Then, by the Cauchy-Kowalevsky theorem, we see that R also satisfies the compatibility condition (C) and the symbol algebra $\mathfrak{s}(v)$ of (R, D^2) at $v \in R$ is a subalgebra of $\mathfrak{c}^2(n)$ such that

$$\mathfrak{s}(v) = \mathfrak{s}_{-3}(v) \oplus \mathfrak{s}_{-2}(v) \oplus \mathfrak{s}_{-1}(v)$$

where $\mathfrak{s}_{-3}(v) = \mathbb{R}$, $\mathfrak{s}_{-2}(v) = V^*$, $\mathfrak{s}_{-1}(v) = V \oplus \mathfrak{f}(v)$ and $\mathfrak{f}(v)$ is a subspace of $S^2(V^*)$ of codimension 1. Let $(\mathfrak{f}(v))^\perp$ be the annihilator of $\mathfrak{f}(v)$ in $S^2(V)$ under the pairing between $S^2(V)$ and $S^2(V^*)$. Then $\dim (\mathfrak{f}(v))^\perp = 1$.

We say that R is of (weak) parabolic type at v if $(\mathfrak{f}(v))^\perp$ is generated by a symmetric two form of rank 1. When R is defined in a canonical coordinate (x_i, z, p_i, p_{ij}) $(1 \leqq i \leqq j \leqq n)$ by

$$F(x_i, z, p_i, p_{ij}) = 0,$$

then the above condition is equivalent to say that the symmetric matrix $(\frac{\partial F}{\partial p_{ij}}(v))$ has rank 1 (cf. §3.3 [Y1]).

When R is of (weak) parabolic type at each point, (R, D^2) is a regular differential system of type $\mathfrak{s}$:

$$\mathfrak{s} = \mathfrak{s}_{-3} \oplus \mathfrak{s}_{-2} \oplus \mathfrak{s}_{-1},$$

where $\mathfrak{s}_{-3} = \mathbb{R}$, $\mathfrak{s}_{-2} = V^*$, $\mathfrak{s}_{-1} = V \oplus \mathfrak{f}$ and $\mathfrak{f} \subset S^2(V^*)$ is given by $(\mathfrak{f})^\perp = \langle e^2 \rangle \subset S^2(V)$ for a nonzero vector $e \in V$.

Let $A(\mathfrak{s})$ be the group of graded Lie algebra automorphisms of $\mathfrak{s}$ and E be the 1-dimensional subspace of V spanned by e. Then the annihilator subspace $E^\perp$ of E is an $A(\mathfrak{s})$-invariant subspace of $V^* = \mathfrak{s}_{-2}$. Starting from the 1-dimensional subspace $E = \langle e \rangle$ of V, we can construct the first order covariant system $N(E)$ and the Monge characteristic system $M(E)$ as in the following (For details, see §7.3 [Y1]): Let v be any point of R and let $\mathfrak{s}(v)$ be the symbol algebra at v. Take a graded Lie algebra isomorphism ϕ of $\mathfrak{s}(v)$ onto $\mathfrak{s}$. Let $\mathfrak{n}(E)(v)$ denote the linear subspace of $\mathfrak{s}_{-2}(v)$ defined by

$$\mathfrak{n}(E)(v) = \phi^{-1}(E^\perp).$$

Then, since $E^{\perp}$ is $A(\mathfrak{s})$-invariant, it follows that $\mathfrak{n}(E)(v)$ is well-defined. Let κ_{-2} be the projection of $D^1(v)$ onto $\mathfrak{s}_{-2}(v) = D^1(v)/D^2(v)$. We define the linear subspace $N(E)(v)$ of $D^1(v)$ by setting

$$N(E)(v) = (\kappa_{-2})^{-1}(\mathfrak{n}(E)(v)).$$

Then it follows that the assignment $v \mapsto N(E)(v)$ defines a subbundle $N(E)$ of D^1.

Let $\mathfrak{m}(E)$ denote the linear subspace of $\mathfrak{s}_{-1}$ spanned by linear subspaces $\phi(E)$, $\phi \in A(\mathfrak{s})$, i.e.,

$$\mathfrak{m}(E) = \langle\{\phi(E) \subset \mathfrak{s}_{-1} \mid \phi \in A(\mathfrak{s})\}\rangle.$$

$\mathfrak{m}(E)$ is an $A(\mathfrak{s})$-invariant subspace of $\mathfrak{s}_{-1}$ by construction. Taking a graded Lie algebra isomorphism ϕ of $\mathfrak{s}(v)$ onto $\mathfrak{s}$, let $M(E)(v)$ denote the linear subspace of $\mathfrak{s}_{-1}(v) = D^2(v)$ defined by

$$M(E)(v) = \phi^{-1}(\mathfrak{m}(E)).$$

It follows that the assignment $v \mapsto M(E)(v)$ defines a subbundle $M(E)$ of D^2. $M(E)(v)$ is the linear subspace of $D^2(v)$ spanned by the Monge characteristic elements corresponding to E.

We say that R is a (single) equation of Goursat type when R is of (weak) parabolic type and its Monge characteristic system $M(E)$ is completely integrable.

Now let us describe the covariant systems $N = N(E)$ and $M = M(E)$ of (R, D^2) in terms of adapted coframes (cf. [Y4]). Let R be a single equation of (weak) parabolic type, i.e., let (R, D^2) be a regular differential system of type $\mathfrak{s}$. Let v be any point of R. A coframe, i.e., a base of 1-forms $\{\varpi, \varpi_a, \omega_a, \varpi_{1\alpha}, \varpi_{\alpha\beta}\}$ $(1 \leqq a \leqq n, 2 \leqq \alpha \leqq \beta \leqq n)$ on a neighborhood U of v in R is called an adapted coframe if it satisfies the following conditions (5.1) and (5.2) :

$$D^2 = \{\varpi = \varpi_1 = \cdots = \varpi_n = 0\}, \tag{5.1}$$

$$\begin{cases} d\varpi \equiv \omega_1 \wedge \varpi_1 + \cdots\cdots + \omega_n \wedge \varpi_n & (\operatorname{mod} \varpi), \\ d\varpi_1 \equiv \qquad \omega_2 \wedge \varpi_{12} + \cdots + \omega_n \wedge \varpi_{1n} & (\operatorname{mod} \varpi, \varpi_1, \cdots, \varpi_n), \\ d\varpi_\alpha \equiv \omega_1 \wedge \varpi_{\alpha 1} + \cdots\cdots + \omega_n \wedge \varpi_{\alpha n} & (\operatorname{mod} \varpi, \varpi_1, \cdots, \varpi_n), \end{cases} \tag{5.2}$$

where we understand that $\varpi_{\alpha\beta} = \varpi_{\beta\alpha}$ and $\varpi_{1\alpha} = \varpi_{\alpha 1}$ for $2 \leqq \alpha, \beta \leqq n$. The equalities (5.2) are the structure equations of (R, D^2) in the sense of E. Cartan ([C1], [C2]) and describe the structure of the symbol algebra

$\mathfrak{s} = \mathfrak{s}_{-3} \oplus \mathfrak{s}_{-2} \oplus \mathfrak{s}_{-1}$ of (R, D^2). In terms of an adapted coframe, covariant systems N and M are given by (cf. §3 [Y4])

$$N = \{\varpi = \varpi_1 = 0\},$$
$$M = \{\varpi = \varpi_1 = \cdots = \varpi_n = \omega_\alpha = \varpi_{1\alpha} = 0 \quad (2 \leqq \alpha \leqq n)\}.$$

Then, for the structure of N, we obtain the following by Cartan's method (cf. §2, §3 [Y4], [Ts]).

Proposition 5.1 *Let R be a single equation of Goursat type and let v be any point of R. Then there exists an adapted coframe on a neighborhood of v such that the following equality holds :*

$$d\varpi_1 \equiv \omega_2 \wedge \varpi_{12} + \cdots + \omega_n \wedge \varpi_{1n} \pmod{\varpi, \varpi_1}.$$

Especially, $Ch(N) = M$ on R.

5.2. Reduction Theorems. We now describe the two step reduction procedure for the (contact) equivalence problem of single equations of Goursat type, which explains the link between the exceptional simple Lie algebra G_2 and the equation (B) of Goursat type mentioned in the introduction.

Let $R \subset L(J)$ be a single equation of Goursat type. We consider the (involutive) Grassmann bundle $I(J,1)$ of codimension 1 over the contact manifold (J, C) :

$$I(J,1) = \bigcup_{u \in J} I_u, \qquad I_u = \mathrm{Gr}\,(C(u), 2n-1),$$

where $C(u) \subset T_u(J)$ is the fibre of the contact distribution. Here we note that each hyperplane in $C(u)$ is an involutive subspace of the symplectic vector space $(C(u), d\varpi)$. In this sense, I_u is the collection of involutive subspaces of codimension 1 in $(C(u), d\varpi)$.

On $I(J,1)$, we have two differential systems C^* and N^*, where N^* is the canonical system obtained by the Grassmaniann construction and C^* is the lift of C. More precisely, C^* and N^* are given by

$$C^*(w) = \pi_*^{-1}(C(u)) \supset N^*(w) = \pi_*^{-1}(w),$$

for each $w \in I(J,1)$ and $u = \pi(w) \in J$, where $\pi : I(J,1) \to J$ is the bundle projection.

The first order covariant system N of (R, D^2) induces a map φ of R into $I(J,1)$ by

$$\varphi(v) = p_*(N(v)) \subset C(u),$$

for each $v \in R$ and $u = p(v)$, where $p : R \to J$ is the projection. By the Realization Lemma for (R, N, p, J), φ is a map of constant rank such that

$$\text{Ker } \varphi_* = \text{Ch}\,(N) \cap \text{Ker } p_* = \text{Ch}\,(N) \cap \text{Ch}\,(D^1).$$

By Proposition 5.1, we have

$$\text{rank Ker } \varphi_* = \frac{1}{2}n(n-1) = \dim S^2(E^\perp).$$

In the rest of this section, we will be concerned with the local equivalence problem for single equations (R, D^2) of Goursat type. Hence we may assume that the image $W = \text{Im } \varphi$ is a submanifold of $I(J,1)$. Thus φ is a submersion of R onto W such that $p = q \cdot \varphi$, where q is the restriction of the projection $\pi : I(J,1) \to J$ to W. Here we note that $\dim W = 3n$. Moreover we have two differential systems C_W and N_W on W, which are the restrictions to W of C^* and N^* on $I(J,1)$. Then we have

$$\varphi_*^{-1}(N_W) = N, \qquad \text{and} \qquad \varphi_*^{-1}(C_W) = D^1.$$

We call $(W; C_W, N_W)$ the associated involutive bundle of the single equation R of Goursat type.

Now the local equivalence of (R, D^2) is first reducible to that of the involutive bundle $(W; C_W, N_W)$ as in the following: Locally W is the leaf space of the foliation on R defined by $\text{Ch}\,(N) \cap \text{Ch}\,(D^1)$. Conversely, from $(W; C_W, N_W)$, we can construct a PD-manifold $(R(W); D^1_W, D^2_W)$ as follows. First, by Grassmannian construction, we define

$$R(W) = \bigcup_{w \in W} R_w,$$

$$R_w = \{v \in \text{Gr}(N_W(w), 2n-1) \mid \\ v \supset \text{Ch}\,(C_W)(w) \quad \text{and} \quad q_*(v) \in L_u, u = q(w)\},$$

where L_u is the fibre of $L(J)$ at $u \in J$. D^2_W is the canonical system obtained by the Grassmannian construction and D^1_W is the lift of C_W.

Precisely, D^1_W and D^2_W are given by

$$D^1_W(v) = (\varphi_W)_*^{-1}(C_W(w)) \supset D^2_W(v) = (\varphi_W)_*^{-1}(v),$$

for each $v \in R(W)$ and $w = \varphi_W(v)$, where $\varphi_W : R(W) \to W$ is the projection. By definition, we have a map ι_W of $R(W)$ into $L(J)$ given by

$$\iota_W(v) = q_*(v) \in L_u,$$

for each $v \in R(W)$ and $u = q(v) \in J$. Then we note that the image $R^*(W) = \operatorname{Im} \iota_W$ has the following description :

$$R^*(W) = \bigcup_{w \in W} R^*_w, \qquad R^*_w = \{v \in L_u \mid v \subset w \subset C(u), \quad u = q(w)\}.$$

Namely $R^*(W)$ is the collection of Legendrian subspaces of (J, C) contained in involutive subspaces of codimension 1 belonging to $W \subset I(J, 1)$.

Now we have a map κ_1 of R into $R(W)$ given by

$$\kappa_1(v) = \varphi_*(D^2(v)) \subset N_W(w),$$

for each $v \in R$ and $w = \varphi(v)$. By the Realization Lemma for (R, D^2, φ, W), κ_1 is a map of constant rank such that

$$\operatorname{Ker} \kappa_1 = \operatorname{Ch}(D^2) \cap \operatorname{Ker} \varphi_* = \operatorname{Ch}(N) \cap \operatorname{Ch}(D^1) \cap \operatorname{Ch}(D^2) = \{0\}.$$

Thus κ_1 is an immersion and, by a dimension count, in fact, a local diffeomorphism of R into $R(W)$ such that

$$(\kappa_1)_*(D^1) = D^1_W \qquad \text{and} \qquad (\kappa_1)_*(D^2) = D^2_W.$$

Namely $\kappa_1 : (R, D^1, D^2) \to (R(W), D^1_W, D^2_W)$ is a local isomorphism of PD-manifolds. (Precisely $(R(W), D^1_W, D^2_W)$ becomes a PD-manifold on an open subset.)

Summarizing, we obtain the following first Reduction Theorem for contact equivalence of single equations of Goursat type.

Theorem 5.2 *Let R and $\hat{R}$ be single equations of Goursat type. Let $(W; C_W, N_W)$ and $(\hat{W}; C_{\hat{W}}, N_{\hat{W}})$ be the associated involutive bundles of R and $\hat{R}$ respectively. Let κ_1 and $\hat{\kappa}_1$ be defined as above. Let us fix points $v_o \in R$ and $\hat{v}_o \in \hat{R}$ and put $w_o = q(v_o)$ and $\hat{w}_o = \hat{q}(\hat{v}_o)$. Then a local isomorphism $\psi : (R, D^2) \to (\hat{R}, \hat{D}^2)$ such that $\psi(v_o) = \hat{v}_o$ induces a local isomorphism $\varphi : (W; C_W, N_W) \to (\hat{W}; C_{\hat{W}}, N_{\hat{W}})$ such that $\varphi(w_o) = \hat{w}_o$ and $\varphi_*(\kappa_1(w_o)) = \hat{\kappa}_1(\hat{w}_o)$, and vice versa.*

By Proposition 5.1, it follows that rank $\operatorname{Ch}(N_W) = 1$. Then, as in §4.2, the geometry of $(W; C_W, N_W)$ is further reducible to the geometry of regular differential system of type $\mathfrak{c}^1(n-1,2)$ as follows. We may assume that W is regular with respect to $\mathrm{Ch}\,(N_W)$ so that the leaf space $Y = W/\mathrm{Ch}\,(N_W)$ is a manifold such that the projection $\beta : W \to Y$ is a submersion and there exists a differential system D_N on Y satisfying $N_W = \beta_*^{-1}(D_N)$. Moreover, by Proposition 5.1, (Y, D_N) is a regular differential system of type $\mathfrak{c}^1(n-1,2)$ (cf. Theorem 1.6 [Y3]). From (Y, D_N), we can construct $(W(Y); C_Y, N_Y)$ as follows. $W(Y)$ is the collection of hyperplanes w in each tangent space $T_y(Y)$ at $y \in Y$ which contains the fibre $D_N(y)$ of D_N:

$$W(Y) = \bigcup_{y \in Y} W_y \subset J(Y, 3n-2),$$

$$W_y = \{w \in \mathrm{Gr}(T_y(Y), 3n-2) \mid w \supset D_N(y)\}.$$

C_Y is the canonical system obtained by the Grassmannian construction and N_Y is the lift of D_N. Precisely C_Y and N_Y are defined by

$$C_Y(w) = \mu_*^{-1}(w) \supset N_Y(w) = \mu_*^{-1}(D_N(y)),$$

for each $w \in W(Y)$ and $y = \mu(w)$, where $\mu : W(Y) \to Y$ is the projection. Then we have a map κ_2 of W into $W(Y)$ given by

$$\kappa_2(w) = \beta_*(C_W(w)) \subset T_y(Y),$$

for each $w \in W$ and $y = \beta(w)$. By the Realization Lemma for (W, C_W, β, Y), κ_2 is a map of constant rank such that

$$\mathrm{Ker}\ \kappa_2 = \mathrm{Ch}\,(C_W) \cap \mathrm{Ker}\ \beta_* = \mathrm{Ch}\,(C_W) \cap \mathrm{Ch}\,(N_W) = \{0\}.$$

Thus κ_2 is an immersion and, by a dimension count, in fact, a local diffeomorphism of W into $W(Y)$ such that

$$(\kappa_2)_*(C_W) = C_Y \qquad \text{and} \qquad (\kappa_2)_*(N_W) = N_Y.$$

Namely $\kappa_2 : (W; C_W, N_W) \to (W(Y); C_Y, N_Y)$ is a local isomorphism.

Summarizing, we obtain the second Reduction Theorem for contact equivalence of single equations of Goursat type.

Theorem 5.3 *Let R and $\hat{R}$ be single equations of Goursat type. Let $(W; C_W, N_W)$ and $(\hat{W}; C_{\hat{W}}, N_{\hat{W}})$ be the associated involutive bundles of R and $\hat{R}$ respectively. Assume that W and $\hat{W}$ are regular with respect to $Ch\,(N_W)$ and $Ch\,(N_{\hat{W}})$ respectively. Let (Y, D_N) and $(\hat{Y}, \hat{D}_N)$ be the leaf spaces, where $Y = W/Ch\,(N_W)$ and $\hat{Y} = \hat{W}/Ch\,(N_{\hat{W}})$. Let us fix points*

$w_o \in W$ and $\hat{w}_o \in \hat{W}$ and put $y_o = \beta(w_o)$ and $\hat{y}_o = \hat{\beta}(\hat{w}_o)$. Then a local isomorphism $\psi : (W; C_W, N_W) \to (\hat{W}; C_{\hat{W}}, N_{\hat{W}})$ such that $\psi(w_o) = \hat{w}_o$ induces a local isomorphism $\varphi : (Y, D_N) \to (\hat{Y}, \hat{D}_N)$ such that $\varphi(y_o) = \hat{y}_o$ and $\varphi_*(\kappa_2(y_o)) = \hat{\kappa}_2(\hat{y}_o)$, and vice versa.

Thus, finally, the local contact equivalence problem of single equations R of Goursat type reduces to the equivalence of (Y, D_N), which are regular differential systems of type $\mathfrak{c}^1(n-1, 2)$ (cf. [Ts], §3 [Y4]).

6. G_2-geometry

In view of discussions in §3, §4 and §5, we will here consider the generalization of (A) and (B) to other simple Lie algebras.

6.1. Standard Contact Manifolds. Each simple Lie algebra $\mathfrak{g}$ over $\mathbb{C}$ has highest root θ. Let Δ_θ denote the subset of Δ consisting of all vertices which are connected to $-\theta$ in the Extended Dynkin diagram of X_ℓ ($\ell \geqq 2$). This subset Δ_θ of Δ, by the construction in §4, defines a gradation (or a partition of Φ^+), which distinguishes the highest root θ. Then, this gradation (X_ℓ, Δ_θ) turns out to be a contact gradation, which is unique up to conjugacy.

Moreover we have the adjoint (or equivalently coadjoint) representation, which has θ as the highest weight. The R-space $J_\mathfrak{g}$ corresponding to (X_ℓ, Δ_θ) can be obtained as the projectivization of the (co-)adjoint orbit of G passing through the root vector of θ. By this construction, $J_\mathfrak{g}$ has the natural contact structure $C_\mathfrak{g}$ induced from the symplectic structure as the coadjoint orbit, which corresponds to the contact gradation (X_ℓ, Δ_θ) (cf. [Y5, §4]). Standard contact manifolds $(J_\mathfrak{g}, C_\mathfrak{g})$ were first found by Boothby ([Bo]) as compact simply connected homogeneous complex contact manifolds.

6.2. Gradation of G_2. The Dynkin diagram of G_2 is given as follows:

$$\underset{\alpha_1}{\circledcirc} \Lleftarrow \underset{\alpha_2}{\circledcirc}, \qquad \theta = 3\alpha_1 + 2\alpha_2.$$

In this case, from $\Delta = \{\alpha_1, \alpha_2\}$, we have three choices for Δ_1:

$(G1)$ $\Delta_1 = \{\alpha_1\}$. In this case, we have $\mu = 3$, $\dim \mathfrak{g}_{-3} = \dim \mathfrak{g}_{-1} = 2$ and $\dim \mathfrak{g}_{-2} = 1$. Moreover $(M_\mathfrak{g}, D_\mathfrak{g})$ coincides with (X, D) in case of (A).

$(G2)$ $\Delta_1 = \{\alpha_2\}$. In this case, we have the standard contact gradation.

$(G3)$ $\Delta_1 = \{\alpha_1, \alpha_2\}$. In this case, we have $\mu = 5$, $\dim \mathfrak{g}_{-1} = 2$ and $\dim \mathfrak{g}_p = 1$ for others.

Let $(J_{\mathfrak{g}}, C_{\mathfrak{g}})$ be the standard contact manifold of type G_2.If we lift the action of the exceptional group G_2 to $L(J_{\mathfrak{g}})$, then we have the following orbit decomposition:

$$L(J_{\mathfrak{g}}) = O \cup R_1 \cup R_2,$$

where O is the open orbit and R_i is the orbit of codimension i. Here R_1 and R_2 can be considered as the global model of (B) and (A) respectively. Moreover R_2 is compact and is a R-space corresponding to $(G_2, \{\alpha_1, \alpha_2\})$. From this fact, it becomes possible to describe the PD-manifold $(R; D^1, D^2)$ corresponding to (A) in terms of the R-space corresponding to $(G_2, \{\alpha_1, \alpha_2\})$.

Extended Dynkin Diagrams with the coefficient of Highest Root (cf. [Bu])

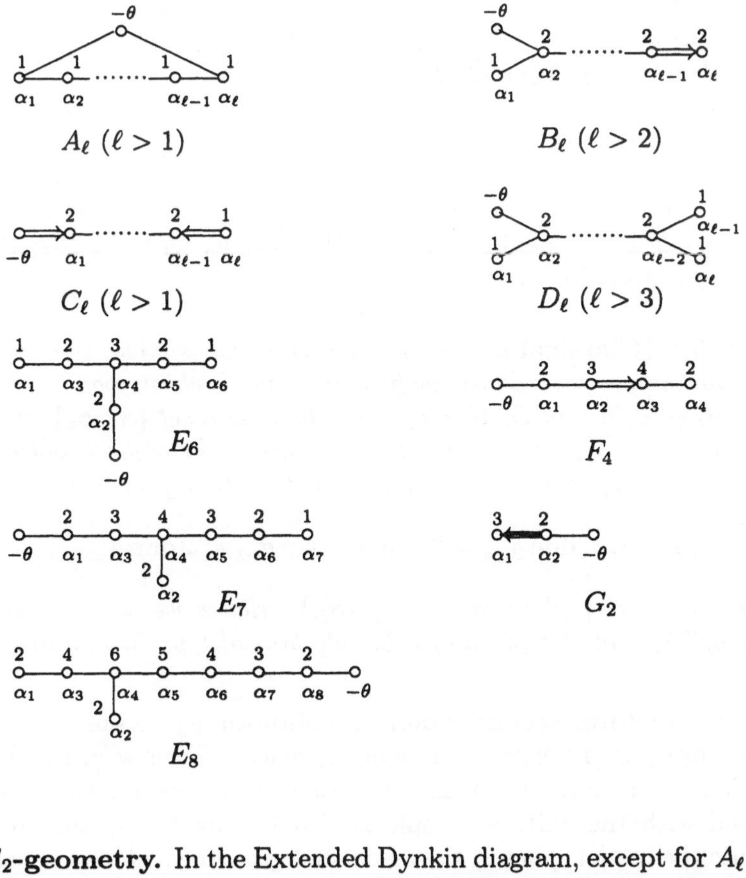

6.3. G_2-geometry. In the Extended Dynkin diagram, except for A_ℓ type, Δ_θ consists of one simple root α_θ. The coefficient of α_θ in the highest root $\theta = \sum_{i=1}^{\ell} n_i(\theta)\, \alpha_i$ is of course 2. Furthermore, for the exceptional simple Lie algebras, there exists, without exception, a unique simple root

α_G next to α_θ such that the coefficient of α_G in the highest root is 3. For $X_\ell \cong E_6, E_7, E_8, G_2, F_4$, the gradation $(X_\ell, \{\alpha_G\})$ has the following property;

$$\mu = 3, \quad \dim \mathfrak{g}_{-3} = 2 \quad \text{and} \quad \dim \mathfrak{g}_{-1} = 2 \dim \mathfrak{g}_{-2}.$$

Moreover, ignoring the bracket product in $\mathfrak{g}_{-1}$, the bracket product of other part can be expressed in terms of pairing by

$$\mathfrak{g}_{-3} = W, \quad \mathfrak{g}_{-2} = V \quad \text{and} \quad \mathfrak{g}_{-1} = W \otimes V^*.$$

Namely the derived system $(M_\mathfrak{g}, \partial D_\mathfrak{g})$ is a regular differential system of type $\mathfrak{c}^1(r, 2)$ for suitable r, where $(M_\mathfrak{g}, D_\mathfrak{g})$ is the standard differential system of type $(X_\ell, \{\alpha_G\})$. This fact assures us to construct the single equation of Goursat type from the differential system $(Y, D_N) \doteq (M_\mathfrak{g}, \partial D_\mathfrak{g})$, which is the generalization of (B).

Obviously the R-space R_G corresponding to $(X_\ell, \{\alpha_\theta, \alpha_G\})$ is a fibre space over the standard contact manifold $(J_\mathfrak{g}, C_\mathfrak{g})$ corresponding to $(X_\ell, \{\alpha_\theta\})$. In fact this R_G can be realized as a compact orbit in $L(J_\mathfrak{g})$, which gives the generalization of (A). Moreover, in this case, we have rank $Ch(D^2) = 1$ as a PD-manifold and (X, D) coincides with the R-space corresponding to $(X_\ell, \{\alpha_G\})$.

Remark 6.1 (Classical cases) *In the classical simple Lie algebras, there is no simple root whose coefficient in the highest root is* 3. *However, in B_ℓ ($\ell \geqq 3$) and D_ℓ ($\ell \geqq 5$) types, there is a set $\{\alpha_1, \alpha_3\}$ of simple roots next to $\alpha_\theta = \alpha_2$ whose sum of coefficients in the highest root is* 3. *In fact, $(B_\ell, \{\alpha_1, \alpha_3\})$ and $(D_\ell, \{\alpha_1, \alpha_3\})$ have the above property;*

$$\mu = 3, \quad \dim \mathfrak{g}_{-3} = 2 \quad \textit{and} \quad \dim \mathfrak{g}_{-1} = 2 \dim \mathfrak{g}_{-2},$$

and the set $\{\alpha_1, \alpha_3\}$ plays the role of $\{\alpha_G\}$. Hence we have the generalizations of (A) and (B) for simple Lie algebras of type B_ℓ and type D_ℓ.

6.4. Nonvanishing second Spencer cohomology. So far in this section, we have just discussed the model R-spaces. In view of the Tanaka theory [T4] of the normal Cartan connections for the geometric structures associated with the SGLAs (simple graded Lie algebras), each R-space represents the model space of the associated geometry. Moreover the fundamental system of invariants (essential part of the curvature) of the normal connection of this geometry takes its values in the second Spencer cohomology associated with the SGLA. Thus it is quite important to have knowledge of this cohomology for each geometry associated with the SGLA (§2.5 [T4], cf. §5.3 [Y5]).

In our previous paper: *Differential Systems Associated with Simple Graded Lie Algebras*, Adv. Studies in Pure Math. **22** (1993), 413–494, the list of the Nonvanishing second Spencer cohomology (Proposition 5.5) contains some misprints and omissions. Here we would like to take this opportunity to correct the following points:

(I) A_ℓ-type.
^{o}(11) is missing, and (2) and (4) lack information for the case $\ell = 3$.

(II) B_ℓ-type.
(7) contains a misprint ($\mu = 3$ shuld be replaced by $\mu = 4$).

(III) C_ℓ-type.
(3) contains a misprint ($p_{21} = 2$ ($\ell = 2$) should be deleted).

(IV) D_ℓ-type.
(3), (5) and (7) lack information for the case $\ell = 4$.

Thus the corrected version of Proposition 5.5 [Y5] should be stated as follows:

Proposition 6.2 *Let (X_ℓ, Δ_1) be a simple graded Lie algebra over $\mathbb{C}$ described in §3.1 (§3.4[Y5]). Then the following are the list of (X_ℓ, Δ_1) and p_{ij} such that $p_{ij} \geqq 0$ holds for the irreducible component $\mathcal{H}^{\sigma_{ij}} \subset C^{p_{ij},2}(\mathfrak{m}, \mathfrak{g})$ of the harmonic space $\mathcal{H}^2 \cong H^2(\mathfrak{m}, \mathfrak{g})$ corresponding to $\sigma_{ij} \in W^0(2)$ in Kostant's theorem (see [Ko]).*

(I) *A_ℓ-type* ($\ell \geqq 2$).

(1) $\{\alpha_1\}$ $p_{12} = 2 \quad (\ell = 2)$,
$p_{12} = 1 \quad (\ell \geqq 3)$.

(2) $\{\alpha_2\}$ $p_{21} = p_{23} = 1 \quad (\ell = 3)$,
$p_{21} = 1, \quad p_{23} = 0 \quad (\ell \geqq 4)$.

(3) $\{\alpha_i\}$ $p_{i\,i-1} = p_{i\,i+1} = 0 \quad (2 < i \leqq [\frac{\ell+1}{2}])$.

(4) $\{\alpha_1, \alpha_2\}$ $p_{12} = p_{21} = 3 \quad (\ell = 2)$,
$p_{12} = 1, \quad p_{21} = 2, \quad p_{23} = 0 \quad (\ell = 3)$,
$p_{12} = 1, \quad p_{21} = 2 \quad (\ell \geqq 4)$.

(5) $\{\alpha_1, \alpha_i\}$ $p_{12} = p_{1i} = 0 \quad (2 < i < \ell - 1)$.

(6) $\{\alpha_1, \alpha_{\ell-1}\}$ $p_{12} = p_{1\,\ell-1} = p_{\ell-1\,\ell} = 0 \quad (\ell \geqq 4)$.

(7) $\{\alpha_1, \alpha_\ell\}$ $p_{12} = p_{\ell\,\ell-1} = 0, \quad p_{1\ell} = 1 \quad (\ell \geqq 3)$.

(8) $\{\alpha_2, \alpha_3\}$ $p_{21} = p_{23} = p_{32} = p_{34} = 0 \quad (\ell = 4)$,
$p_{21} = p_{23} = p_{32} = 0 \quad (\ell \geqq 5)$.

(9) $\{\alpha_2, \alpha_i\}$ $p_{21} = 0 \quad (3 < i < \ell - 1)$.

(10) $\{\alpha_2, \alpha_{\ell-1}\}$ $p_{21} = p_{\ell-1\,\ell} = 0 \quad (\ell \geqq 5)$.

(11) $\{\alpha_i, \alpha_{i+1}\}$ $\quad p_{i\,i+1} = p_{i+1\,i} = 0 \quad (2 < i \leqq [\frac{\ell}{2}])$.
^{o}(11) $\{\alpha_1, \alpha_2, \alpha_i\}$ $\quad p_{12} = 0, \quad p_{21} = 1 \quad (2 < i < \ell)$.
(12) $\{\alpha_1, \alpha_2, \alpha_\ell\}$ $\quad p_{13} = p_{12} = p_{32} = 0, \quad p_{21} = p_{23} = 1 \quad (\ell = 3)$,
$p_{1\ell} = p_{12} = 0, \quad p_{21} = 1 \quad (\ell \geqq 4)$.
(13) $\{\alpha_1, \alpha_i, \alpha_\ell\}$ $\quad p_{1\ell} = 0 \quad (2 < i \leqq [\frac{\ell+1}{2}])$.
(14) $\{\alpha_1, \alpha_2, \alpha_i, \alpha_j\}$ $\quad p_{21} = 0 \quad (2 < i < j \leqq \ell)$.
(15) $\{\alpha_1, \alpha_2, \alpha_{\ell-1}, \alpha_\ell\}$ $\quad p_{21} = p_{\ell-1\,\ell} = 0$.

(II) *B_ℓ-type* $(\ell \geqq 3)$.

(1) $\{\alpha_1\}$ $\quad \mu = 1 \quad p_{12} = 1$.
(2) $\{\alpha_2\}$ $\quad \mu = 2 \quad p_{21} = p_{23} = 0$.
(3) $\{\alpha_3\}$ $\quad \mu = 2 \quad p_{32} = 2 \quad (\ell = 3)$,
$p_{32} = 0 \quad (\ell \geqq 4)$.
(4) $\{\alpha_\ell\}$ $\quad \mu = 2 \quad p_{\ell\,\ell-1} = 0 \quad (\ell \geqq 4)$.
(5) $\{\alpha_1, \alpha_2\}$ $\quad \mu = 3 \quad p_{21} = 0, \quad p_{12} = 1$.
(6) $\{\alpha_1, \alpha_3\}$ $\quad \mu = 3 \quad p_{32} = 1 \quad (\ell = 3)$.
(7) $\{\alpha_2, \alpha_3\}$ $\quad \mu = 4 \quad p_{32} = 2 \quad (\ell = 3)$,
$p_{32} = 0 \quad (\ell \geqq 4)$.
(8) $\{\alpha_1, \alpha_2, \alpha_3\}$ $\quad \mu = 5 \quad p_{32} = 1 \quad (\ell = 3)$.

(III) *C_ℓ-type* $(\ell \geqq 2)$.

(1) $\{\alpha_\ell\}$ $\quad \mu = 1 \quad p_{21} = 2 \quad (\ell = 2), \quad p_{\ell\,\ell-1} = 0 \quad (\ell \geqq 3)$.
(2) $\{\alpha_1\}$ $\quad \mu = 2 \quad p_{12} = 2 \quad (\ell = 2), \quad p_{12} = 1 \quad (\ell \geqq 3)$.
(3) $\{\alpha_2\}$ $\quad \mu = 2 \quad p_{21} = 1, \quad p_{23} = 0 \quad (\ell = 3)$,
$p_{21} = 1 \quad (\ell \geqq 4)$.
(4) $\{\alpha_{\ell-1}\}$ $\quad \mu = 2 \quad p_{\ell-1\,\ell} = 0 \quad (\ell \geqq 4)$.
(5) $\{\alpha_1, \alpha_\ell\}$ $\quad \mu = 3 \quad p_{12} = 2, \quad p_{21} = 3 \quad (\ell = 2)$,
$p_{1\ell} = p_{12} = 0 \quad (\ell \geqq 3)$.
(6) $\{\alpha_2, \alpha_\ell\}$ $\quad \mu = 3 \quad p_{21} = p_{23} = 0 \quad (\ell = 3)$,
$p_{21} = 0 \quad (\ell \geqq 4)$.
(7) $\{\alpha_{\ell-1}, \alpha_\ell\}$ $\quad \mu = 3 \quad p_{\ell-1\,\ell} = 0 \quad (\ell \geqq 4)$.
(8) $\{\alpha_1, \alpha_2\}$ $\quad \mu = 4 \quad p_{12} = 0, \quad p_{21} = 2 \quad (\ell \geqq 3)$.
(9) $\{\alpha_1, \alpha_2, \alpha_\ell\}$ $\quad \mu = 5 \quad p_{21} = 1$.
(10) $\{\alpha_1, \alpha_2, \alpha_i\}$ $\quad \mu = 6 \quad p_{21} = 0 \quad (2 < i < \ell)$.

(IV) *D_ℓ-type* $(\ell \geqq 4)$.

(1) $\{\alpha_1\}$ $\quad \mu = 1 \quad p_{12} = 1$.
(2) $\{\alpha_\ell\}$ $\quad \mu = 1 \quad p_{\ell\,\ell-2} = 0 \quad (\ell \geqq 5)$.
(3) $\{\alpha_2\}$ $\quad \mu = 2 \quad p_{21} = p_{23} = p_{24} = 0 \quad (\ell = 4)$,
$= p_{21} = p_{23} = 0 \quad (\ell \geqq 5)$.
(4) $\{\alpha_3\}$ $\quad \mu = 2 \quad p_{32} = 0 \quad (\ell \geqq 5)$.
(5) $\{\alpha_1, \alpha_\ell\}$ $\quad \mu = 2 \quad p_{12} = p_{42} = 0 \quad (\ell = 4)$,
$p_{12} = 0 \quad (\ell \geqq 5)$.

(6) $\{\alpha_1, \alpha_2\}$ $\mu = 3$ $p_{12} = 1, \quad p_{21} = 0.$
(7) $\{\alpha_1, \alpha_2, \alpha_\ell\}$ $\mu = 4$ $p_{12} = p_{42} = 0 \quad (\ell = 4),$
$p_{12} = 0 \quad (\ell \geqq 5).$
(8) $\{\alpha_2, \alpha_3\}$ $\mu = 4$ $p_{32} = 0 \quad (\ell \geqq 5).$

(V) *Exceptional types.*

(1) $(E_6, \{\alpha_1\}), (E_7, \{\alpha_7\})$ $\mu = 1$ $p_{ij} = 0,$ *where* $\{\alpha_i\} = \Delta_1$ *and* $\langle \alpha_i, \alpha_j \rangle \neq 0.$
(2) $(E_6, \{\alpha_2\}), (E_7, \{\alpha_1\}), (E_8, \{\alpha_8\}), (F_4, \{\alpha_1\})$ *and* $(G_2, \{\alpha_2\}).$
Contact gradations: $\mu = 2$ $p_{ij} = 0,$ *where* $\{\alpha_i\} = \Delta_\theta$ *and* $\langle \alpha_i, \alpha_j \rangle \neq 0.$
(3) $(G_2, \{\alpha_1\})$ $\mu = 3$ $p_{12} = 3.$
(4) $(G_2, \{\alpha_1, \alpha_2\})$ $\mu = 5$ $p_{12} = 3.$

We would like to thank Professor Hajime Sato at Nagoya University for pointing out these omissions and also for providing us the .exe files for computing nonvanishing second Spencer cohomology.

In our G_2-Geometry, we notice from the above list that, except for $(G_2, \{\alpha_1\})$ and $(B_3, \{\alpha_1, \alpha_3\})$, Darboux type theorems hold for the regular differential systems of type $(X_\ell, \{\alpha_G\})$ (cf. §5 [Y5]).

References

[Bo] W. M. Boothby, *Homogeneous complex contact manifolds*, Proc. Symp. Pure Math., Amer. Math. Soc. **3** (1961), 144–154.

[Bu] N.Bourbaki, *Groupes et algèbres de Lie, Chapitre 4,5 et 6*, Hermann, Paris (1968).

[BCG$_3$] R. Bryant, S. S. Chern, R. B. Gardner, H. Goldschmidt, and P. Griffiths, *Exterior differential systems*, Springer-Verlag, New-York (1986).

[C1] E.Cartan, *Les systèmes de Pfaff à cinq variables et les équations aux dérivées partielles du second ordre*, Ann. Ec. Normale **27** (1910), 109–192.

[C2] ———, *Sur les systèmes en involution d'équations aux dérivées partielles du second ordre à une fonction inconnue de trois variables indépendantes*, Bull. Soc. Math. France **39** (1911), 352–443.

[Hu] J. E. Humphreys, *Introduction to Lie Algebras and Representation Theory*, Springer-Verlag, New York (1972).

[Ko] B. Kostant, *Lie algebra cohomology and generalized Borel-Weil theorem*, Ann. Math. **74** (1961), 329–387.

[Ku] M. Kuranishi, *Lectures on involutive systems of partial differential equations*, Pub.Soc.Mat., São Paulo (1967).

[St] S. Sternberg, *Lectures on Differential Geometry*, Prentice-Hall, New Jersey (1964).

[T1] N. Tanaka, *On generalized graded Lie algebras and geometric structures I*, J. Math. Soc. Japan **19** (1967), 215–254.

[T2] ———, *On differential systems, graded Lie algebras and pseudo-groups*, J. Math. Kyoto Univ. **10** (1970), 1–82.

[T3] ———, *On non-degenerate real hypersurfaces, graded Lie algebras and Cartan connections*, Japan. J. Math. **2** (1976), 131–190.

[T4] ———, *On the equivalence problems associated with simple graded Lie algebras*, Hokkaido Math. J. **8** (1979), 23–84.

[T5] ———, *On affine symmetric spaces and the automorphism groups of product manifolds*, Hokkaido Math. J. **14** (1985), 277–351.

[Ts] A.Tsuchiya, *Geometric theory of partial differential equations of second order*, Lecture Note at Nagoya University (in Japanese) (1981).

[Y1] K.Yamaguchi, *Contact geometry of higher order*, Japanese J. of Math. **8** (1982), 109–176.

[Y2] ———, *On involutive systems of second order of codimension 2*, Proc. of Japan Acad. **58**, Ser A, No.7 (1982), 302–305.

[Y3] ———, *Geometrization of Jet bundles*, Hokkaido Math. J. **12** (1983), 27–40.

[Y4] ———, *Typical classes in involutive systems of second order*, Japanese J. Math. **11** (1985), 265–291.

[Y5] ———, *Differential systems associated with simple graded Lie algebras*, Adv. Studies in Pure Math. **22** (1993), 413–494.

Department of Mathematics, Faculty of Science,
Hokkaido University, Sapporo 060, Japan
E-mail:yamaguch@math.sci.hokudai.ac.jp

Trends in Mathematics

Series Titles

Anderson/Lapidus (eds). Progress in Inverse Spectral Geometry

Jain/Rizvi (eds). Advances in Ring Theory

Martino/Kantor/Lunardon/Pasini/Tamburini (eds). Groups and Geometries

Ross/Anderson/Litvinov/Singh/Sunder/Wildberger (eds). Harmonic Analysis and Hypergroups

Karatzas/Rajput/Taqqu (eds). Stochastic Processes and Related Topics: In Memory of Stamatis Cambanis 1943-1995

Komatsu/Kuranishi (eds). Analysis and Geometry in Several Complex Variables; Proceedings of the 40thTaniguchi Symposium